AF685034

Future Civil Engineering: Low-Carbon, High Performance and Strong Durability

Future Civil Engineering: Low-Carbon, High Performance and Strong Durability

Guest Editors

Zhongya Zhang
Minqiang Meng
Xiujiang Shen
Abedulgader Baktheer

MDPI

Basel • Beijing • Wuhan • Barcelona • Belgrade • Novi Sad • Cluj • Manchester

Guest Editors

Zhongya Zhang
School of Civil Engineering
Chongqing Jiaotong University
Chongqing
China

Minqiang Meng
College of Water Conservancy and Architectural Engineering
Northwest A&F University
Xianyang
China

Xiujiang Shen
Sustainable Materials
Flemish Institute for Technological Research
Mol
Belgium

Abedulgader Baktheer
Institute of Structural Concrete
Numerical modelling group
Faculty of Civil Engineering
RWTH Aachen University
Aachen
Germany

Editorial Office
MDPI AG
Grosspeteranlage 5
4052 Basel, Switzerland

This is a reprint of the Special Issue, published open access by the journal *Buildings* (ISSN 2075-5309), freely accessible at: https://www.mdpi.com/journal/buildings/special_issues/JE42072TB8.

For citation purposes, cite each article independently as indicated on the article page online and as indicated below:

Lastname, A.A.; Lastname, B.B. Article Title. *Journal Name* **Year**, *Volume Number*, Page Range.

ISBN 978-3-7258-4485-2 (Hbk)
ISBN 978-3-7258-4486-9 (PDF)
https://doi.org/10.3390/books978-3-7258-4486-9

© 2025 by the authors. Articles in this book are Open Access and distributed under the Creative Commons Attribution (CC BY) license. The book as a whole is distributed by MDPI under the terms and conditions of the Creative Commons Attribution-NonCommercial-NoDerivs (CC BY-NC-ND) license (https://creativecommons.org/licenses/by-nc-nd/4.0/).

Contents

About the Editors

Zhongya Zhang

Zhongya Zhang is an associate professor and doctoral supervisor at Chongqing Jiaotong University. He graduated from Chongqing University in 2019 with a Ph.D. degree in Civil Engineering. His main research focuses on the durability of UHPC and concrete, including UHPC materials and interfacial behavior, novel structures based on UHPC, and UHPC durability in complex environments. He has published 70 papers in journals such as *Cement and Concrete Research, Cement and Cement Concrete Composites, Engineering Structures,* and more, with an h-index of 20 (h-index, 2025). The results of his research were awarded the First Prize of Chongqing Municipal Scientific and Technological Progress.

Minqiang Meng

Minqiang Meng is an associate professor at Northwest A&F University, China. He graduated from Chongqing University with a Ph.D. degree in civil engineering. He focuses on the mechanical behaviors of coarse-grains and the reinforcement and improvement technology of special soils. He published about 30 papers in the *Geotechnical Journal* (including *Acta Geotechnica, International Journal of Geomechanics-ASCE, Rock and Soil Mechanics* (in Chinese), and more) and applied four Chinese patents.

Xiujiang Shen

Xiujiang Shen obtained his PhD degree in Engineering Science at EPFL, Switzerland (2020), focusing on the static and fatigue behavior of UHPFRC material. Since 2022, he has been a postdoctoral researcher at the Flemish Institute for Technological Research (VITO) in Belgium. He has authored 29 publications recognized on Google Scholar, holds two patents in structural engineering, and has been the principal investigator for five research projects.

Abedulgader Baktheer

Abedulgader Baktheer is currently a group leader at the Institute of Mechanics and Computational Mechanics at Leibniz Universität Hannover. His research focuses on modeling degradation processes in concrete, emphasizing advanced physics-based and machine learning-enhanced models for accurate lifetime prediction of critical infrastructure. His work bridges computational modeling and structural engineering, contributing to next-generation digital twins for concrete structures.

Preface

The global transition towards a green and low-carbon economy has catalyzed an unprecedented transformation in civil engineering. Against a backdrop of increasing resource constraints and environmental concerns, the stability and durability of infrastructure will be seriously affected. Innovations and modifications in engineering materials and structures have garnered substantial research attention in the recent years. To address these challenges and advance sustainable development in civil engineering, this Special Issue included 11 papers, which covered several key areas of the topic. Those areas included non-destructive testing, smart building technologies, structural performance optimization, and post-disaster repair. Collectively, these studies elucidate integrated approaches that reconcile engineering efficiency, environmental stewardship, and long-term durability. Ultimately, this research is what will drive the civil engineering industry towards a green and low-carbon perspective, improving intelligence and operational efficiency

Zhongya Zhang, Minqiang Meng, Xiujiang Shen, and Abedulgader Baktheer
Guest Editors

buildings

MDPI

Editorial

Editorial on Future Civil Engineering: Low Carbon, High Performance, and Strong Durability

Zhongya Zhang [1,2,*], Minqiang Meng [3], Xiujiang Shen [4] and Abedulgader Baktheer [5]

1 State Key Laboratory of Mountain Bridge and Tunnel Engineering, Chongqing Jiaotong University, Chongqing 400074, China
2 School of Civil Engineering, Chongqing Jiaotong University, Chongqing 400074, China
3 College of Water Conservancy and Architectural Engineering, Northwest A&F University, Yangling, Xianyang 712100, China
4 Sustainable Materials, VITO, Boeretang 200, 2400 Mol, Belgium
5 Institute of Structural Concrete, RWTH Aachen University, 52062 Aachen, Germany
* Correspondence: zhangzhongya@cqjtu.edu.cn

1. Introduction

The global transition towards a green and low-carbon economy has catalyzed an unprecedented transformation in civil engineering [1–4]. Against a backdrop of increasing resource constraints and environmental concerns, the stability [5,6] and durability [7,8] of infrastructure will be seriously affected. Innovations and modifications in engineering materials and structures have garnered substantial research attention in recent years to address these challenges and advance sustainable development in civil engineering [9–11]. The 11 papers introduced in this paper covered several key areas, including non-destructive testing, smart building technologies, structural performance optimization, and post-disaster repair. Collectively, these studies clarify integrated approaches that reconcile engineering efficiency, environmental stewardship, and long-term durability. Ultimately, these efforts will drive the civil engineering industry towards a green and low-carbon perspective, improving intelligence and operational efficiency.

2. Overview of Contributions

Cheng et al. (contribution 1) investigated the effect of surface conditions on thermal contrast (ΔT) to ensure the effectiveness of the developed thermal contrast nondestructive testing (NDT) technique for the quantification of delamination in concrete structures. Three typical scenarios were presented based on the experimental results to establish the association of δT with delamination detection. Finally, a new approach to quantify the delamination of concrete structures at different time windows was proposed.

Based on Citespace software (version 6.2R4), Dong et al. (contribution 2) performed visualization and analysis of 2947 documents related to construction robots sourced from the CNKI Chinese database and WOS Core database. The summary results emphasized the critical role and importance of construction robots as an essential tool for intelligently upgrading the construction industry. Additionally, it was pointed out that future trends in the field would likely favor human–robot collaboration, intelligent construction, and robotic vision technologies.

A detailed numerical model of steel–UHPC composite slabs was developed by Li et al. (contribution 3), based on previous flexural behavior research, to clarify their flexural failure mechanism. Combined with the results of numerical simulation analyses, the formula, with high accuracy for calculating the flexural load capacity of steel–UHPC composite

https://doi.org/10.3390/buildings15111957

slabs under positive and negative bending moments, was proposed. The results provide theoretical guidance for the design of the flexural performance of steel–UHPC composite slabs.

The accuracy of detection for steel tube concrete de-bonding phenomena via infrared thermography based on heat excitation was ensured by Cai et al. (contribution 4), who first introduced spray cooling as an excitation method during the exothermic hydration stage. A systematic analysis of 39 groups of indoor concrete-filled steel tube (CFST) models was carried out by considering variables such as the atomization level, excitation distance, excitation duration, and water temperature in the tank. Finally, an efficient and stable cooling excitation method applicable to engineering detection was proposed.

Zhao et al. (contribution 5) systematically studied the influence mechanism of vibration loads on the flexural performance of ultra-high-performance concrete (UHPC) wet joints by using four-point flexural experiments combined with theoretical analyses. The results showed that low-amplitude or low-frequency vibrations would increase the flexural strength of UHPC wet joints. Meanwhile, a model for the calculation of flexural strength considering vibration parameters was developed. The flexural strength obtained using the proposed model agreed very well with the experimental observations, demonstrating the validity of the model.

To address the challenge of the insufficient load-carrying capacity of existing structures, Zhu et al. (contribution 6) investigated the strengthening technique of high-strength steel wire mesh and UHPC through axial tensile tests, constitutive modeling, and finite element analysis. The effects of the steel fiber volume ratio and high-tensile wire mesh strengthening ratio on the axial tensile performance of reinforced concrete (RC) beams were clarified. In addition, the influence of key parameters on the strengthening effect of RC beams was analyzed using finite elements, and an optimized combination of strengthening parameters was proposed.

Du et al. (contribution 7) investigated the flexural performance of retard-bonded prestressed ultra-high-performance concrete (RBPU) beams under various reinforcement ratios through experimental studies and theoretical analyses. The results showed that the cracking moment, ultimate moment, and ductility coefficient of RBPU beams increased significantly as the reinforcement ratio increased. This study also established a calculation method for the cracking bending moment and ultimate bending moment of RBPU beams and introduced a correction factor to improve the calculation accuracy.

To address the effect of ground fissure dislocations on the bottom void range of underground pipe gallery structures, Yu et al. (contribution 8) proposed an innovative analytical method and revealed critical mechanical regulations through a combination of theoretical modeling and experiments. The hyperbolic tangent function was used to quantitatively describe the nonlinear relationship between the end displacement of the hanging wall pipe corridor structure (Δ_1) and the dislocation of the ground fissure (Δ_2). In addition, the prediction of the bottom void range was achieved by solving the static equilibrium equations in MATLAB R2022a programming.

Ran et al. (contribution 9) proposed an innovative active underpinning technology that integrates a '井'-shaped cap system, graded preloading of the foundation, and synchronous beam body correction to repair the concrete box girder bridge piers in mountainous highways that were damaged by falling rocks. The safety and effectiveness of the repaired bridge structure were verified through numerical simulation, dynamic monitoring, and load testing. This technology provides an effective solution for the fast and safe repair of damaged bridge piers, which is essential for improving the disaster resistance and emergency response speed of bridge projects.

A dynamic prediction model based on the gray system theory was proposed by Ma et al. (contribution 10). By combining the GM (1,1) gray model with the variable-size sliding window technique, the model could dynamically update the latest monitoring data and significantly improve the prediction accuracy. The study also innovatively designed two data preprocessing mechanisms, i.e., equidistant and exponential. The model was validated at Kansai International Airport and Xiamen Xiang'an International Airport with an over 20% improvement in accuracy compared to the traditional gray method.

Using on-site examination, testing, and numerical simulation, Yang et al. (contribution 11) systematically analyzed the damage mechanism in the concrete components of a high-pile beam plate wharf. A full-scale 3D numerical model containing 40 rows of beams, 204 columns, and complex foundations was innovatively constructed. A high degree of consistency between the plastic damage distribution and the actual crack location was demonstrated by the model. Based on the analysis results, a graded crack repair strategy and load optimization scheme were proposed.

3. Conclusions

This Special Issue presents 11 groundbreaking research studies on the innovation and transformation of civil engineering materials and structures. The insights presented are expected to drive further progress and inspire future research in the field of civil engineering.

Funding: This research was funded by the National Natural Science Foundation of China (Grant No. 52208302), the Excellent Youth Science Foundation of Chongqing (CSTB2024NSCQ-JQX0006), and the Technology Project of the Transportation Department of Guizhou Province (Grant No. 2024-122-001).

Acknowledgments: We extend our gratitude to the authors who contributed their research to this Special Issue and to the reviewers for their meticulous evaluations. We also sincerely thank the editors for their dedication and perseverance in ensuring the success of this Special Issue.

Conflicts of Interest: The authors declare that they have no known competing financial interests or personal relationships that could have appeared to influence the work reported in this paper.

List of Contributions:

1. Cheng, C.; Chen, D.; Shao, S.; Na, R.; Cai, H.; Zhou, H.; Wu, B. Revealing the Impact of Depth and Surface Property Variations on Infrared Detection of Delamination in Concrete Structures Under Natural Environmental Conditions. *Buildings* **2024**, *15*, 10. https://doi.org/10.3390/buildings15010010.
2. Dong, R.; Chen, C.; Wang, Z. Visualization Analysis of Construction Robots Based on Knowledge Graph. *Buildings* **2024**, *15*, 6. https://doi.org/10.3390/buildings15010006.
3. Li, C.; Zhao, B.; Hao, D.; Gao, X.; Bian, H.; Zhang, X. A Numerical and Theoretical Investigation of the Flexural Behavior of Steel–Ultra-High-Performance Concrete Composite Slabs. *Buildings* **2025**, *15*, 166. https://doi.org/10.3390/buildings15020166.
4. Cai, H.; Cheng, C. Experimental Investigation of Infrared Detection of Debonding in Concrete-Filled Steel Tubes via Cooling-Based Excitation. *Buildings* **2025**, *15*, 465. https://doi.org/10.3390/buildings15030465.
5. Zhao, B.; Yang, J.; Qin, D.; Zou, Y.; Zhang, Z.; Zhang, K.; Leng, J. Flexural Response of UHPC Wet Joints Subjected to Vibration Load: Experimental and Theoretical Investigation. *Buildings* **2025**, *15*, 496. https://doi.org/10.3390/buildings15030496.
6. Zhu, C.; Du, C.; Qi, Y.; Jiang, Z.; Zhang, Z.; Yang, J.; Li, Y.; Cheng, J. Flexural Performance of RC Beams Strengthened with High-Strength Steel Wire Mesh and UHPC. *Buildings* **2025**, *15*, 589. https://doi.org/10.3390/buildings15040589.
7. Du, L.; Wu, D.; Wang, J.; Wang, S.; Zhao, B.; Tang, X. Experimental Study on Flexural Behavior of Retard-Bonded Prestressed UHPC Beams with Different Reinforcement Ratios. *Buildings* **2025**, *15*, 887. https://doi.org/10.3390/buildings15060887.

8. Yu, X.; Han, B.; Zhao, Y.; Deng, B.; Du, K.; Liu, H. A Study on the Calculations of the Bottom Void Range of an Underground Pipe Gallery Structure Under the Action of Ground Fissure Dislocations. *Buildings* **2025**, *15*, 920. https://doi.org/10.3390/buildings15060920.
9. Ran, H.; Li, L.; Wei, Y.; Xiao, P.; Yang, H. Study on the Underpinning Technology for Fixed Piers of Concrete Box Girder Bridges on Mountainous Expressways. *Buildings* **2025**, *15*, 1031. https://doi.org/10.3390/buildings15071031.
10. Ma, K.; Weng, H.; Luo, Z.; Sarajpoor, S.; Chen, Y. Dynamic Prediction Method for Ground Settlement of Reclaimed Airports Based on Grey System Theory. *Buildings* **2025**, *15*, 1034. https://doi.org/10.3390/buildings15071034.
11. Yang, C.; He, P.; Wang, S.; Wang, J.; Zhu, Z. Study on the Causes of Cracking in Concrete Components of a High-Pile Beam Plate Wharf. *Buildings* **2025**, *15*, 1352. https://doi.org/10.3390/buildings15081352.

References

1. Chen, L.; Huang, L.; Hua, J.; Chen, Z.; Wei, L.; Osman, A.I.; Fawzy, S.; Rooney, D.W.; Dong, L.; Yap, P.-S. Green construction for low-carbon cities: A review. *Environ. Chem. Lett.* **2023**, *21*, 1627–1657. [CrossRef]
2. Albuhairi, D.; Di Sarno, L. Low-carbon self-healing concrete: State-of-the-art, challenges and opportunities. *Buildings* **2022**, *12*, 1196. [CrossRef]
3. Drück, H.; Mathur, J.; Panthalookaran, V.; Sreekumar, V.M. Green buildings and sustainable engineering. In *Springer Transactions in Civil and Environmental Engineering*; Springer: Singapore, 2020.
4. Xue, X. Application and Environmental Impact Analysis of Green Building Materials in Civil Engineering. *Acad. J. Archit. Geotech. Eng.* **2024**, *6*, 12–17.
5. Valdez, B.; Schorr, M.; Quintero, M.; García, R.; Rosas, N. Effect of climate change on durability of engineering materials in hydraulic infrastructure: An overview. *Corros. Eng. Sci. Technol.* **2010**, *45*, 34–41. [CrossRef]
6. Mirza, S. Durability design of infrastructure and some related issues. *Can. J. Civ. Eng.* **2006**, *33*, 650–672. [CrossRef]
7. Mirza, S. Durability and sustainability of infrastructure—A state-of-the-art report. *Can. J. Civ. Eng.* **2006**, *33*, 639–649. [CrossRef]
8. Sánchez-Silva, M.; Gardoni, P.; Val, D.V.; Yang, D.Y.; Frangopol, D.M.; Limongelli, M.P.; Honfi, D.; Acuña, N.; Straub, D. Moving toward resilience and sustainability in the built environment. *Struct. Saf.* **2025**, *113*, 102449. [CrossRef]
9. Tiza, M.T.; Imoni, S.; Akande, E.O.; Mogbo, O.; Jiya, V.H.; Onuzulike, C. Revolutionizing Infrastructure Development: Exploring Cutting-Edge Advances in Civil Engineering Materials. *Recent Prog. Mater.* **2024**, *6*, 1–68. [CrossRef]
10. Najmi, A.; India, S. Innovative materials and techniques for sustainable building structures. *Int. J. Sci. Res. Eng. Manag.* **2024**, *8*, 1–5. [CrossRef]
11. Soliman, A.; Hafeez, G.; Erkmen, E.; Ganesan, R.; Ouf, M.; Hammad, A.; Eicker, U.; Moselhi, O. Innovative construction material technologies for sustainable and resilient civil infrastructure. *Mater. Today Proc.* **2022**, *60*, 365–372. [CrossRef]

Disclaimer/Publisher's Note: The statements, opinions and data contained in all publications are solely those of the individual author(s) and contributor(s) and not of MDPI and/or the editor(s). MDPI and/or the editor(s) disclaim responsibility for any injury to people or property resulting from any ideas, methods, instructions or products referred to in the content.

buildings

MDPI

Article

Revealing the Impact of Depth and Surface Property Variations on Infrared Detection of Delamination in Concrete Structures Under Natural Environmental Conditions †

Chongsheng Cheng [1,2,*], Dequan Chen [2], Shuai Shao [3,4], Ri Na [5], Haonan Cai [2], Hongwen Zhou [3,4] and Bo Wu [2]

1 State Key Laboratory of Mountain Bridge and Tunnel Engineering, Chongqing Jiaotong University, Chongqing 400074, China
2 School of Civil Engineering, Chongqing Jiaotong University, Chongqing 400074, China; dqchen1999@163.com (D.C.); 611210080005@mails.cqjtu.edu.cn (H.C.); bo.wu@cqjtu.edu.cn (B.W.)
3 Chongqing Geomatics and Remote Sensing Center, Chongqing 401120, China; xiachongbuyu@hotmail.com (S.S.); zhouhw_dlygzx@hotmail.com (H.Z.)
4 Key Laboratory of Monitoring, Evaluation and Early Warning of Territorial Spatial Planning Implementation, Ministry of Natural Resources, Chongqing 401120, China
5 Department of Civil and Environmental Engineering, University of Delaware, Newark, DE 19716, USA; nari@udel.edu
* Correspondence: ccheng@cqjtu.edu.cn
† This paper is an extended version of our paper published in Cheng, C.; Chen, D.; Cai, H.; Wang, S.; Wang, R.; Cheng, X. Investigation of the effect of surface condition variation on infrared imaging of concrete delamination. Bridge Maintenance, Safety, Management, Digitalization and Sustainability—Proceedings of the 12th International Conference on Bridge Maintenance, Safety and Management, IABMAS 2024, 2409–2416. Copenhagen, Denmark, 24–28 June 2024.

Abstract: Infrared thermography (IRT) is an effective nondestructive testing method for detecting delamination in concrete structures. However, erroneous data interpretation often diminishes its practical utility due to surface irregularities (e.g., color variations) during inspection. These "noisy conditions" alter the temperature distribution of the structure under solar heating cycles, making it challenging to quantify delamination based on the developed thermal contrast (ΔT). This study experimentally investigates the impact of different surface conditions (bare concrete vs. painted surfaces) on ΔT. Artificial delamination at varying depths was simulated and tested under natural environmental conditions, where the maximum ΔT values for shallow delamination were 9.40 °C (bare concrete), 7.35 °C (yellow paint), and 5.15 °C (white paint), respectively. This study measured and analyzed the absorptivity (bare concrete: 0.652, yellow paint: 0.538, and white paint: 0.369), emissivity, and the temperature difference (δT) between non-delaminated areas and air, revealing their correlation with ΔT variation. Based on the results, three typical scenarios are proposed to correlate δT with delamination detection. These findings contribute to a better understanding of the optimal detection window and present a new approach to quantifying delamination at different time windows. The conclusion also discusses the limitations of this study and future research directions.

Keywords: concrete delamination; infrared detection; surface condition variations; delamination depth

1. Introduction

Concrete structures are susceptible to various defects such as cracks, delaminations, spalling, and voids due to factors like steel reinforcement corrosion, thermal effects, continuous freeze–thaw cycles, and concrete shrinkage [1]. Compared to common issues like spalling or pitting [2], delamination defects are characterized by their high concealment and potential for extensive development [3,4]. If delaminations are not detected and addressed in a timely manner, they can evolve into large-scale open spalling, severely compromising structural safety and reducing service life [5,6]. In practice, delaminations often originate within the concrete and remain unnoticed until significant damage occurs [7]. Therefore, gaining a comprehensive understanding of the current state of the concrete, including voids, delaminations, and other defects, is crucial [8]. Early identification of delaminations in concrete structures is essential to prevent structural damage and to provide appropriate repair methods for potentially compromised structures, which is vital for the safe operation of the structure [4,9,10].

In recent years, numerous nondestructive testing methods have been employed to detect delamination in concrete structures. These methods include traditional approaches such as chain dragging and hammer sounding, as well as advanced techniques like ground-penetrating radar [11,12], ultrasonics [13,14], and impact echoes [15,16]. However, these methods require continuous contact or limited sensing distance with the concrete structure and entail significant time and resource expenditure for testing. In contrast, the primary advantage of infrared thermography (IRT) for detecting delamination in concrete structures lies in its ability to conduct large-area assessments in relatively short time frames and identify areas of the defect [17]. Additionally, as it does not necessitate close contact with the structure, this technology enhances the safety and efficiency of the inspection.

Current diagnostic methods based on infrared imaging primarily focus on surface temperature differentials to assess the temperature difference between defective and non-defective areas. The excitation methods for delamination infrared detection can be categorized into active and passive excitation. Active excitation involves researchers heating or cooling test specimens in a laboratory using a heat source such as a halogen lamp. On the other hand, passive excitation involves researchers conducting experiments outdoors without employing additional heat sources for heating or cooling the test specimens, such as solar radiation. In research utilizing solar radiation as the thermal excitation, the American Society for Testing and Materials (ASTM-D4788 [18]) standard recommends that the surface temperature differential of concrete structures should reach 0.5 °C after at least 4 h of sunlight exposure as the baseline criterion for detection. Kee et al. [19] and Omar and Nehdi [20] have evaluated this recommendation, suggesting that it is suitable for daytime inspections without shadow effects. Their research found that the maximum temperature difference can reach 7 °C and typically occurs around the 7th hour after sunrise. Washer et al. [21] conducted a study on structural surfaces of non-sunlit bridges. They found that a minimum environmental temperature change of 1.5 °C per hour is the lowest requirement for thermal imaging detection. Mac et al. [22] conducted comprehensive thermal imaging observations on a series of square artificial delaminated regions with various sizes and depths throughout the day. Their results revealed that during the daytime heating phase, the absolute temperature contrast ranged from 0.49 °C to 6.95 °C, while during the nighttime cooling phase, the temperature difference ranged from 0.32 °C to 3 °C. Pozzer et al. [23] fabricated three specimens containing nine artificial delaminations of different sizes and varying depths. They conducted experimental studies using solar radiation as a passive excitation method. They found that delaminations with larger areas at the same depth exhibited greater temperature thermal contrast, making them easier to detect, whereas delaminations with smaller areas were more challenging to identify. Tran [24] conducted experiments by fabricating 12 artificial delaminations of varying sizes and three different depths. The experimental results indicated that the absolute thermal contrast above the delaminations increased with the size of the delaminations and the duration of heating (both active and passive). Additionally, other studies have shown that the magnitude of thermal contrast is related to the time window [19,25], depth, and environmental conditions [26,27].

Since practical engineering applications primarily rely on sunlight as the thermal excitation method, determining the optimal detection time window is another crucial detection parameter. However, due to the high complexity of environmental factors and the uncertainty in forming delaminations, there is no consensus on the optimal detection time window in current research. Washer et al. [26] found that the optimal observation period is 5 to 9 h after sunrise. Meanwhile, Zhang et al. [28] found that the afternoon appears to be the optimal time for infrared detection during the summer. However, Yehia [29] indicated that delaminations with a diameter of less than 5 cm are more detectable from 10 a.m. to noon than from noon to 3 p.m. Furthermore, Tran [24] indicated that the optimal observation time for passive infrared radiation experiments on sunny days is from 11 a.m. to 2 p.m. These non-uniform observations suggest that determining the optimal detection time based solely on the specific time window of the maximum temperature difference, without considering geographical and climatic variations, is incomplete. More importantly, in the studies by Hiasa et al. [30,31], it was found that the temperature differential is highest at midday (resulting in stronger detectable signals). However, the signal noise is also more remarkable due to the influence of concrete surface texture (resulting in stronger interference signals). Conversely, although the signal strength is lower at night, the interference signals are weaker, making it a recommended primary time window for data collection. These results suggest that determining the optimal detection window enables detection personnel to quickly and accurately identify delamination in concrete structures.

Environmental factors such as air temperature and solar radiation are the primary factors affecting detection. In the study by Washer et al. [26], it was noted that there is a positive linear relationship between solar irradiance and temperature contrast, and a positive correlation was found between delamination depth and time after sunrise. In their further study [32], they incorporated solar intensity and air temperature into a finite element model (FEM) to predict temperature contrast. Watase et al. [25] employed a multiple regression model to predict surface temperatures of delaminated regions based on six parameters. Their analysis results indicated that air temperature was a significant influencing factor. Mac et al. [22] conducted a systematic experimental study to determine the relationship between the depth-to-diameter ratio and detectability. Mac et al. [33] studied infrared imaging detection on the concrete bridge deck surface under natural conditions. Their study found that detection reliability increases when the amplitude of environmental temperature changes throughout the day increases. Based on these findings, it is evident that the factors influencing the detectability of delaminations are the same as those influencing temperature contrast. Therefore, this study will investigate the factors influencing detection from the perspective of thermal contrast, which is defined as the temperature difference between the delaminated area and the non-delaminated area.

In infrared detection of delamination in concrete structures, the analysis of temperature differences on the concrete surface is used to identify delamination. Therefore, the surface temperature of the concrete structure is crucial in delamination infrared detection. In the study by Berg and Quinn [34], the temperature of white-painted roads with a reflectivity close to 0.55 (absorptivity close to 0.45) is almost the same as that of the surrounding environment. In contrast, the temperature of unpainted roads with a reflectivity close to 0.15 (absorptivity close to 0.85) is about 11 °C higher than the air temperature. Meanwhile, in the study by Santamouris [35], the temperature of asphalt reaches about 63 °C, while the temperature of white road surfaces is around 45 °C. In the study by Synnefa [36], the thermal behavior of different materials varies, primarily due to differences in their reflectivity, which affects their thermal behavior during the day, while their emissivity is the main factor influencing their thermal behavior at night. For concrete, its reflectivity and absorptivity add up to 1. These studies indicate that the surface temperature of concrete is influenced by its surface material properties (such as absorptivity and emissivity), which indirectly suggests that these properties will affect the infrared detection of delamination in concrete structures.

In actual concrete structures, there are material differences and various surface treatments (such as different colored paints, different surface polishing methods, etc.) on the concrete surface. In the study by Hiasa et al. [30], wooden sticks and gravel were placed on the surface of concrete specimens, and different-colored tapes were applied. The results indicated that wooden sticks affected the detection of delaminations, while the impact of gravel depended on its size. Additionally, different-colored tapes could be observed and distinguished using infrared thermography (IRT). The differences in surface properties can affect the effectiveness of infrared detection of delaminations in concrete structures. In the study by Ta et al. [9], two types of specimens were fabricated, each with 12 artificial delaminations of varying sizes and depths. One type of specimen was painted with green paint, while the other type remained untreated. Through experiments using square pulse thermography, it was found that, at the same size and depth of delaminations, the temperature difference was greater in the concrete coated with paint, indicating that delaminations in concrete with green paint were easier to detect. These results indicate that differences in surface properties can affect infrared delamination detection in concrete.

Cool-colored coatings are commonly used in concrete structures (such as rooftops, pavements, etc.) [36]. Therefore, this study applied white and yellow paints to alter the surface properties of concrete. Infrared thermography detections were conducted on specimens with different surface treatments (painted on both sides, with the middle part untreated). A comparative study was carried out to investigate the effect of surface property variations (absorptivity and emissivity variations) on delamination detection. In addition, this study also explored the impact of depth on detecting delamination under the same surface property. The effect of air temperature was considered to determine the optimal time window for infrared detection.

Based on relevant studies, paint is widely used in concrete structures, and the thermal behavior of surfaces with different colors varies. The absorptivity of these surfaces affects their thermal behavior during the day, while their emissivity influences their thermal behavior at night [36]. These differences in thermal behavior manifest as surface temperature variations, which are critical indicators for identifying delamination in infrared detection of concrete structures. Therefore, the color of the concrete surface impacts its infrared delamination detection, as demonstrated in existing studies. Although some researchers have investigated the influence of color on infrared detection, the impact of the properties behind these colors (absorptivity and emissivity) on detection remains unexplored. Thus, this study aims to investigate the effects of surface property variations and delamination depth on the infrared detection of interlayer delaminations in concrete structures under natural conditions. The thermal contrast of delaminations with varying colors and depths was compared and analyzed. Subsequently, the characteristics and detection capabilities at different time windows were discussed. This research bridges the gap in understanding how surface colors influence infrared detection in concrete structures and provides insights for detecting delaminations under surfaces of different colors. It further aids in identifying potential delaminations for maintenance, safeguarding public safety. The conclusion section also explores this study's limitations and suggests directions for future research.

2. Materials and Methods

2.1. Research Framework

The research framework for this study is illustrated in Figure 1. First, the surface emissivity of yellow paint and white paint was obtained using the two-reference-body method, and the surface absorptivity of bare concrete, yellow paint, and white paint was derived based on the principle of energy conservation. After obtaining the surface properties, three artificial delaminated concrete specimens with different burial depths were designed. Each specimen had three delaminations, and delamination detection experiments were conducted under natural environmental conditions. The thermal contrast curve of delamination over the day was segmented into three phases based on its different trends better to understand the thermal contrast development of delamination over time. Additionally, a correlation analysis was conducted to reveal the influence of surface properties on detection at different phases. Finally, further correlation analysis and curve fitting were performed on the temperature difference between the non-delaminated regions, air (δT), and the corresponding thermal contrast (ΔT).

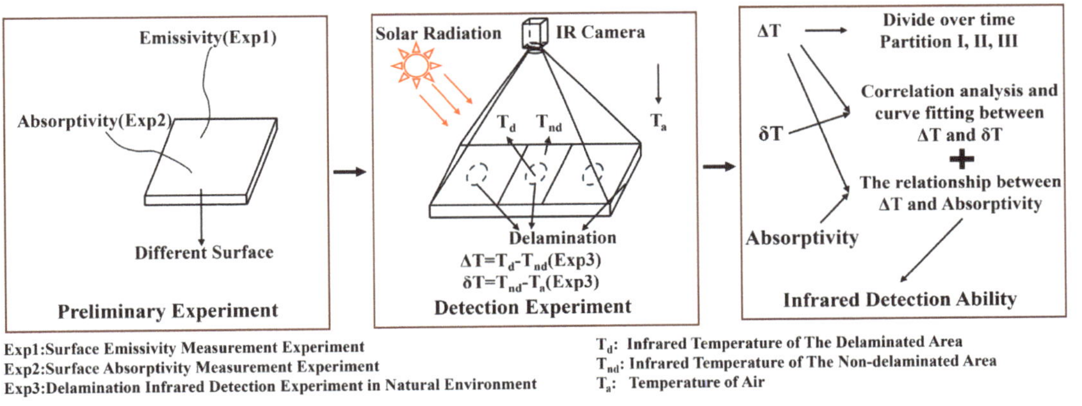

Figure 1. Research framework.

2.2. Experimental Design and Procedure [37]

Since the experiments needed to be carried out both day and night, the differences in surface properties of the concrete specimens could lead to variations in absorptivity and emissivity, potentially resulting in differences in surface thermal contrast. Therefore, three experiments were carried out: surface emissivity measurement, surface absorption measurement, and concrete delamination detection in the natural environment. The experiments were conducted at the Science City campus of Chongqing Jiaotong University in western China.

2.2.1. Surface Emissivity Measurement

The accuracy of surface emissivity affects the accuracy of surface temperature measurements [38], which in turn will affect the thermal radiation and thermal imaging of the surface of the specimen. The emissivity of a material's surface is a dimensionless quantity that characterizes the radiative capability of the material's surface and is an essential thermal property parameter [39]. Based on previous studies [39,40], the emissivity of the surface of dry, normal concrete was set to 0.95 in this study. This study designed a measurement experiment to determine the emissivity of yellow and white paints, both of which are acrylic paints. The emissivity measurement specimen (EMS) for this experiment was made by applying 1.35 kg/m^2 of white and yellow paint to the outer surface of a stainless-steel container, with 3M Scotch Super 88 (3M, Austin, TX, USA) insulation tape applied around the painted areas, as shown in the EMS in the lower right part of Figure 2. When measuring temperature with an infrared camera, 3M Scotch Super 88 insulation tape is often used as a reference body for emissivity, with emissivity value of 0.95 [41].

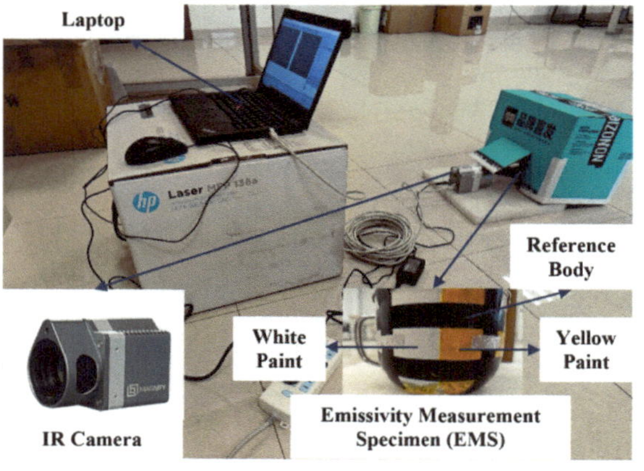

Figure 2. Emissivity measurement experiment: arrangement of experimental equipment and surface treatment of emissivity measurement specimen.

Figure 2 shows the instruments and experimental layout for the emissivity experiment. The EMS was filled with hot water at 100 °C and placed on the floor to cool naturally. A cardboard box with a small opening was made to cover the EMS, allowing the collection of infrared images through the opening using a scientific-grade thermal imager (MAG-F6, Shanghai Hugee Technology Co., Ltd., Shanghai, China, parameters shown in Table 1). Data were recorded every 15 min, capturing temperature readings ranging from 34 °C to 46 °C. To eliminate the influence of other factors such as ambient temperature and environmental reflection temperature [38], this experiment was performed in a controlled chamber. Then, the emissivity of the paint was estimated using the dual-reference method [38]. For detailed calculations, please refer to Section 2.3.1.

Table 1. Relevant parameters of MAG-F6.

Camera Type	MAG-F6
Detector type	Uncooled microbolometer
Noise Equivalent Temperature Difference (NETD)	<0.05 °C
Accuracy	±0.7 °C or 0.7%
Resolution	640 × 480 pixels
Spectral range	8~14 μm
Field of view	25° × 19°

2.2.2. Surface Absorptivity Measurement

Absorptivity is the ratio of the thermal radiation energy absorbed by an object to the total thermal radiation energy incident on the object, which could directly affect the temperature field distribution inside the concrete structure. The surface absorptivity can be calculated by using the principles of heat transfer and energy balance to establish a heat balance equation on the surface of the specimen [42]. A surface absorptivity measurement experiment was designed to measure the absorptivity of the yellow and white paint and their respective surfaces on the concrete specimen. The absorptivity measurement specimen (AMS) is depicted in Figure 3. Three concrete specimens with dimensions of 500 × 500 × 100 mm^3 (AMS1, AMS2, and AMS3) were fabricated for the experiment. AMS1, AMS2, and AMS3 represent concrete specimen surfaces without treatment, concrete specimen surfaces with yellow paint, and concrete specimen surfaces with white paint, respectively. And the concrete is standard C40 concrete, and the paints are consistent with that used in emissivity measurement experiment. A heat flux sensor (XM269C, greenteg AG, Rümlang ZH, Switzerland, parameters shown in Table 2) and four temperature sensors (PT100, Hangzhou Guizhong Technology Co., Ltd, Hangzhou, China, parameters shown in Table 3) were affixed to the surface of each specimen. These sensors were utilized to measure the heat flux at the center of each specimen and the temperatures in four directions on the surface of the specimen. Then, 1.35 kg/m^2 of yellow paint was applied to AMS2, and white paint was applied to AMS3.

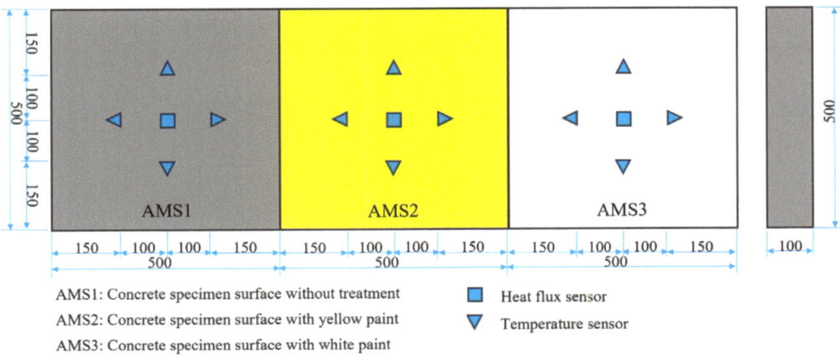

Figure 3. Absorptivity measurement specimen: dimensions of each specimen, surface treatment, and sensor arrangement (units: mm).

Table 2. Parameters of XM269C.

Name	Heat Flux Range (kW/m^2)	Sensitivity (μV/(W/m^2))	Temperature (°C)	Calibration Accuracy (%)
Measurement range	−150~150	1.5	−50~150	±3%

Table 3. Parameters of PT100.

Name	Temperature (°C)	Accuracy (°C)
Measurement range	−50~200	±0.15

Figure 4 shows the instruments and experimental setup for the absorption measurement of AMS. After transporting the AMS to the corresponding outdoor positions, a 5 mm thick glass cover was placed on its surface to prevent convective heat exchange between the AMS and the ambient atmosphere. A temperature sensor was installed in the gap between the glass plate and the specimen to measure the air temperature in the gap (the orange frame in Figure 4, T_a in Equation (3), Section 2.3.1), and another temperature sensor was attached to the inner surface of the glass to measure the surface temperature of the glass (the blue frame in Figure 4, T_2 in Equation (4), Section 2.3.1). To collect real-time weather data, a portable weather station (FT-BQX7, Shandong Fengtu Internet of Things

Technology Co., Ltd., Weifang, China, see Table 4) was set up near the specimens to collect meteorological conditions. The sampling frequency was set to 15 min, collecting information such as solar radiation (Q_t in Equation (2), Section 2.3.1). A similar glass plate was placed above the weather station to avoid discrepancies in the data compared to the specimens. Utilizing the heat flux sensor (XM269C, parameters as shown in Table 2) to measure the magnitude of heat flux (q_a in Equation (5), Section 2.3.1) passing through the specimen surface, measurements were conducted using a heat flux data acquisition instrument (DQ-100, Yi'an (Shanghai) Electronic Technology Co., Ltd., Shanghai, China), with data collected every minute. The temperature sensors (PT100, parameters as shown in Table 3) were connected to a multi-channel data logger (TP700, Shenzhen Topray Electronic Co., Ltd., Shenzhen, China, parameters as shown in Table 5) for collecting the temperature of each specimen surface (T_1 in Equation (4) and T_s in Equation (3)), the glass plate, and the air gap between the glass plate and the specimen, with data collected every minute. All data were collected during a period around noon when the sun was nearly perpendicular to the test site. Data from time periods with relatively stable solar radiation were selected for calculating the absorptivity, with detailed methods provided in Section 2.3.1.

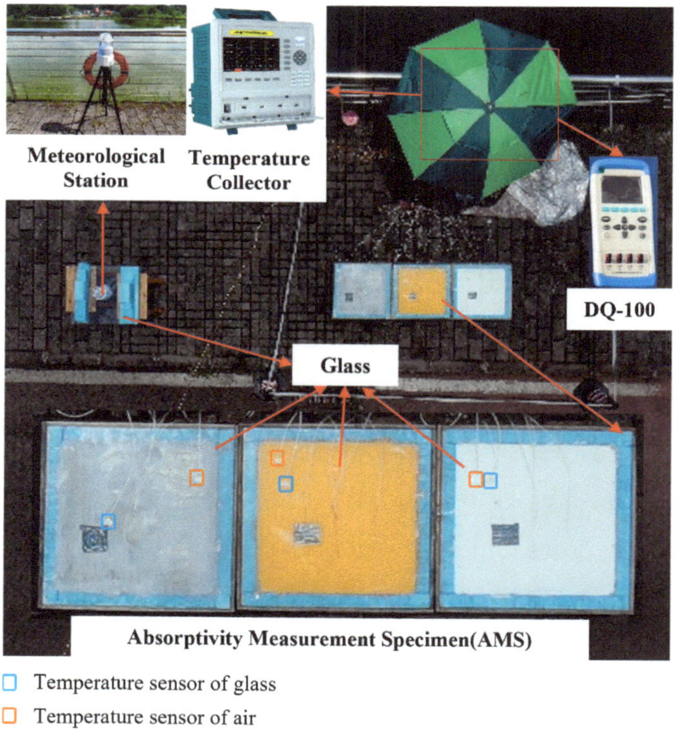

Figure 4. Absorptivity measurement experiment: arrangement of experimental equipment and placement of AMS.

Table 4. Parameters of FT-BQX7.

Name	Wind Speed (m/s)	Wind Direction (°)	Air Temperature (°C)	Solar Radiation (10^2 W/m^2)
Measurement range	0~60	0~360	−40~60	0~20
Error range	±0.1	±2	±0.3	<±3%

Table 5. Parameters of TP700.

Name	Number of Channels	Thermoresistance	Storage Interval
Range/Type	1~200	PT1000, PT100	1~19,999 s

2.2.3. Delamination Detection Experiment Under Natural Environment

As shown in Figure 5, three concrete specimens (CS1, CS2, and CS3), measuring 2100 mm × 900 mm × 100 mm, were designed for the delamination detection experiment under natural environmental conditions. The specimens were cast using the same specification of concrete as in the absorptivity measurement experiment. CS1, CS2, and CS3 represent concrete specimens with foam simulators' depths of 20 mm, 50 mm, and 80 mm, respectively. Inside each specimen, three foam simulators of delamination were embedded. These delaminations were elliptical in shape and identical in size, each enclosed within a 300 mm × 300 mm square, with a thickness of 10 mm. Each delamination was numbered, with C, Y, and W representing concrete, yellow paint, and white paint surfaces, respectively, and 1, 2, and 3 indicating delamination depths of 20 mm, 50 mm, and 80 mm. For example, CS1-C represents a delamination located 20 mm beneath the concrete surface. It is worth noting that the thermal conductivity of polystyrene foam is k = 0.027 W/(m·°C), which is similar to air, k = 0.024 W/(m·°C) [9]. For the concrete surface, white and yellow paints were brushed on both sides of the specimens at a rate of 1.35 m^2/kg, aimed at altering the surface properties of the concrete (such as emissivity and absorptivity). The paint used was the same type as that used in the aforementioned experiments. Each color of paint covered an area of 900 × 800 mm^2, leaving the middle section as bare concrete, with an area of 900 × 500 mm^2 (Figure 5). To reduce heat conduction in both directions, thermal insulation boards were installed around the specimens to prevent heat exchange between the concrete specimens (except for the surface and bottom) and the environment.

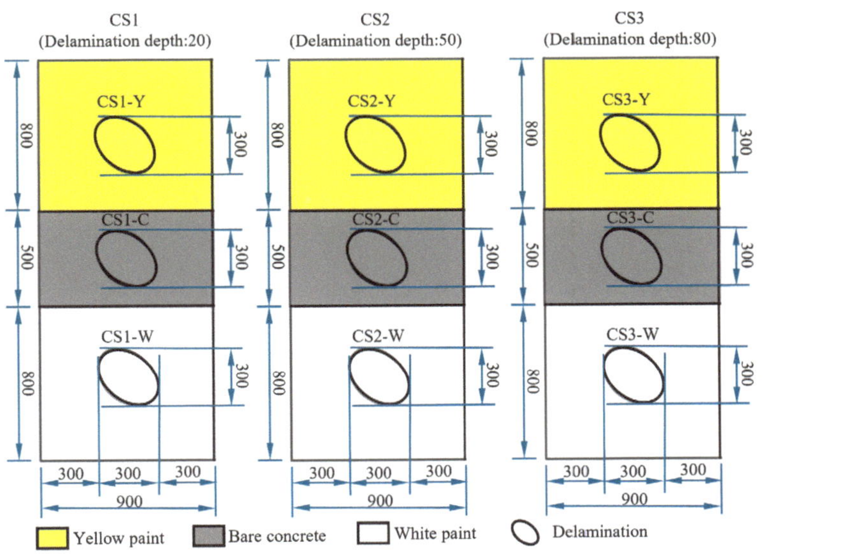

Figure 5. Delamination detection experiment: dimensions of each concrete specimen, surface treatment (units: mm), dimensions of each delamination and delamination arrangement (C, Y, and W representing concrete, yellow paint, and white paint).

Figure 6 illustrates the setup of equipment and the experimental site. In this experiment, solar radiation was utilized as the thermal excitation method for detection research. According to the American Society for Testing and Materials (ASTM-D4788) recommendation, the specimens were pre-exposed to sunlight for a minimum of 4 h before data collection commenced. The data collection equipment for meteorological data, temperature sensor data, and infrared images are described in Sections 2.2.1 and 2.2.2. Additionally, data were continuously collected over multiple day–night cycles to avoid the occurrence of incidental phenomena.

Figure 6. Delamination detection experiment: arrangement of experimental equipment and placement of concrete specimens.

2.3. Measurement and Evaluation Indicators

2.3.1. Surface Emissivity and Absorptivity Measurement Indicators

The experimental data related to EMS and AMS have been obtained in this study. This study estimated surface emissivity using the dual-reference method [38]. This approach involves using a blackbody and a high-reflectivity diffuser as references, maintaining the test object at the same temperature as the blackbody, while the diffuser temperature equals the background temperature. The radiation energy of the specimen, blackbody, and diffuser are measured using a thermal imager. The emissivity is then calculated using Equation (1) [38]:

$$\varepsilon_s = (T_r^n - T_u^n)/(T_0^n - T_u^n) \tag{1}$$

In the equation, ε_s represents the emissivity of the test object, T_r, T_0, and T_u denote the temperatures of the test object, blackbody, and background measured by the thermal imager (Table 4), respectively, and $n = 3.9889$.

The temperature data for the two paints were obtained from the thermal images extracted from multiple measurements in this experiment. Then, using Equation (1), multiple emissivities were calculated, and the average value was computed as the emissivity of the test object.

For absorptivity, this study utilized heat transfer principles and energy balance equations to establish a thermal balance equation for the surface of the specimens, from which the absorptivity of various materials on the surface of the specimens was calculated [42]. The relevant formula is as follows:

$$\alpha = Q_a/Q_t + Q_c/Q_t + Q_r/Q_t \tag{2}$$

In the equation, α represents absorptivity, Q_t denotes the total solar radiation on the surface of the concrete, Q_a signifies the heat absorbed by the concrete, Q_c represents the convective heat exchange between the concrete surface and the surrounding air, and Q_r

denotes the long-wave thermal radiation between the concrete surface and surrounding objects. The total solar radiation Q_t is measured using the portable weather station FT-BQX7.

The magnitude of convective heat transfer is generally calculated based on the temperature difference between the concrete surface and the surrounding air. In a state of complete stillness, natural convection heat exchange occurs between the concrete surface and the surrounding air. If the surface natural convection heat transfer coefficient is known, the convective heat transfer can be represented using Newton's cooling formula as follows:

$$Q_c = \alpha'(T_s - T_a)A \quad (3)$$

where α' is the natural convection heat transfer coefficient; T_s is the concrete surface temperature; T_a is the surrounding air temperature; and A is the surface area of the concrete exchanging heat with the air. According to the principles of heat transfer and engineering experience, α' is taken as 5.8 (W/m^2·K) [42].

Radiative heat transfer is another form of heat exchange, referring to the emission of primarily long-wave thermal radiation from an object due to its temperature. As long as the object's temperature is above absolute zero, it emits long-wave thermal radiation outward, exchanging heat with surrounding objects. The radiative heat transfer between the surfaces of any two objects per unit area can be calculated by the following equation:

$$Q_r = \sigma \cdot (T_1^4 - T_2^4)/(\frac{1}{\varepsilon_1} + \frac{1}{\varepsilon_2} - 1) \cdot \varphi_{12} \cdot A \quad (4)$$

In the equation, ε_1 is the surface emissivity of plane 1, ε_2 is the surface emissivity of plane 2, T_1 is the surface temperature of plane 1, T_2 is the surface temperature of plane 2, σ is the Stefan–Boltzmann constant, $\sigma = 5.67 \times 10 - 8$ W/(m^2·K^4), and φ_{12} is the view factor from plane 1 to plane 2. Here, the emissivity of concrete is taken as 0.95, and the emissivity of glass is taken as 0.83 (according to Chinese standard GB/T 2680-2021 [43]).

The heat absorbed by concrete can be measured using a heat flux sensor, and the equation is as follows:

$$Q_a = q_a \cdot A \quad (5)$$

In the equation, q_a is the data measured by the heat flow sensor, and A is the area of the AMS.

The relevant data obtained from the absorption rate experiment are used to substitute Equations (2)–(5) to obtain the corresponding absorption rates. The absorption rate's average value is calculated when the radiation intensity is relatively stable. The temperatures mentioned in the absorption rate are all obtained from the temperature sensor.

2.3.2. Evaluation Indicators for Delamination Detection Under Natural Conditions

In this study, we analyze the detectability by examining the temperature difference between the delaminated and non-delaminated regions. Delamination is considered detectable when it can be visually discerned from the infrared images (qualitative) or when its thermal contrast (ΔT) is not less than 0.5 °C (following ASTM D4788-03). Thermal contrast is calculated using Equation (6) [1,9,44]:

$$\Delta T = T_d - T_{nd} \quad (6)$$

In the equation, T_d, T_{nd}, and ΔT represent the average surface temperatures of the delaminated area (red box in Figure 7), non-delaminated area (blue box in Figure 7), and the temperature difference between the delaminated and non-delaminated areas, respectively. The areas within the yellow, gray, and purple rectangles represent regions of the concrete specimen coated with yellow paint, bare concrete, and coated with white paint, respectively. These average temperatures are extracted from the infrared images. When $\Delta T > 0$, it indicates that Td > Tnd, meaning the temperature in the delaminated area is higher than that in the non-delaminated area (referred to as the "heat island" area); conversely, if the temperature in the delaminated area is lower than that in the non-delaminated area (referred to as the "cold island" area), $\Delta T < 0$.

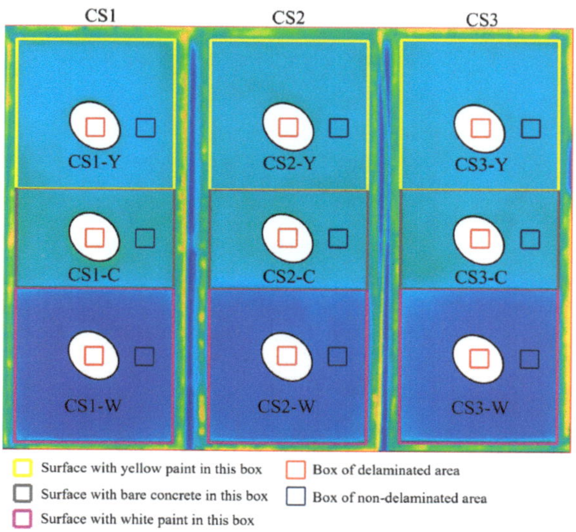

Figure 7. Sensitive areas of delaminated and non-delaminated areas.

3. Results and Discussions

3.1. Results of Absorptivity and Emissivity of Surface Materials

The emissivity was obtained using Equation (1) from Section 2.3.1. Figure 8a shows the box plot of emissivity for different materials, where the short lines on the top and bottom of the boxes represent the maximum and minimum values in the data set. The solid black line represents the mean value of the data. The mean emissivity value for each material falls within the box, and the variances are nearly 0, 0.00003, and 0.00004, respectively. The emissivity values for yellow paint and white paint were found to be 0.960 and 0.965, respectively. The emissivity of concrete, based on relevant studies [45], was taken as 0.950. The absorptivity was obtained using Equation (2) from Section 2.3.1. The average values were then derived from these calculations. Figure 8b shows the box plot of absorptivity for different materials, where the short lines on the top and bottom of the boxes represent the maximum and minimum values in the data set. The solid black line represents the mean value of the data. The mean absorptivity value for each material falls within the box, and the variances are nearly 0, 0.00008, 0.00014, and 0.00022, respectively. The absorptivity values for bare concrete, yellow paint, and white paint were found to be 0.652, 0.538, and 0.369, respectively. According to relevant studies [46], the absorptivity of normal concrete ranges from 0.57 to 0.66, yellow concrete ranges from 0.52 to 0.58, and white concrete ranges from 0.31 to 0.40, which are comparable to the measurement of this study.

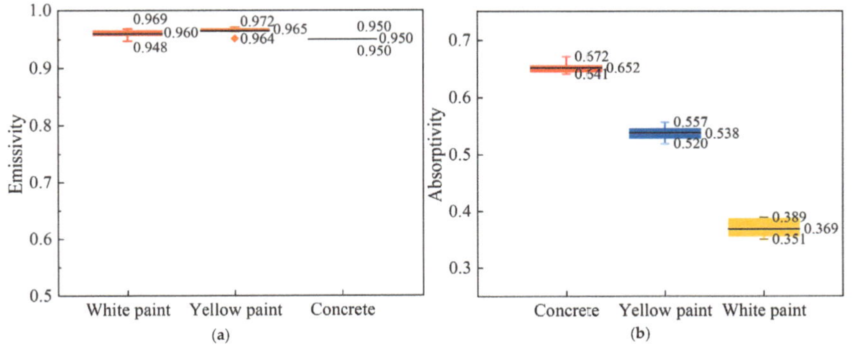

Figure 8. Box plot of surface emissivity and absorptivity; (**a**) box plot of emissivity for different materials; and (**b**) box plot of absorptivity for different materials.

3.2. Analysis of Variation in Delamination Thermal Contrast over Time

Figure 9 displays the thermal contrast distribution of CS1 with a delamination depth of 20 mm. The thermal contrast (ΔT) shown in the figure is obtained from the surface of CS1 using an infrared thermal imager. The curve is segmented into different zones based on different time intervals throughout the day, illustrating the process of calculating thermal contrast for various surface materials of CS1, including yellow paint, white paint, and exposed concrete.

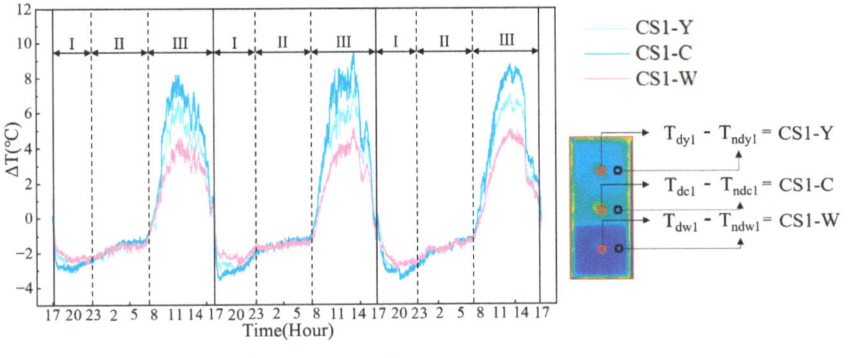

Figure 9. Partitioning of the thermal contrast diagram.

Zone I corresponds to the period when the radiative heat dissipation ceases, and there is intense internal heat conduction within the concrete specimen and heat exchange with the environment (approximately from 17:00 to 23:00). During this period, all three ΔT curves sharply decrease, followed by an increase, exhibiting significant gradients. The minimum values of ΔT are approximately −3.1 °C (CS1-C), −2.8 °C (CS1-Y), and −2.5 °C (CS1-W), respectively.

Zone II corresponds to a period of relatively stable internal heat conduction within the concrete specimen and heat exchange with the environment (approximately from 23:00 to the next day at 7:00). During this period, the three ΔT curves are nearly overlapping and gradually rise. ΔT increases gradually from around −2 °C to approximately −1 °C.

Zone III represents a period of rapid temperature changes within the concrete specimen after the appearance of radiation (approximately from 7:00 to 17:00). During this period, the three ΔT curves experience sharp increases, intense oscillations, and rapid declines. There are distinct gradients during the intense oscillations, and the maximum ΔT values are approximately 9 °C (CS1-C), 7 °C (CS1-Y), and 5 °C (CS1-W), respectively.

Partitioning the thermal contrast enables this study to more intuitively perceive the trends of the thermal contrast in different periods.

3.3. Analysis of the Influence of Surface Properties and Ambient Conditions on Detection

This section focuses on the influence of surface properties and ambient conditions on detection.

In Zone I, the median values of ΔT for white, yellow, and concrete are −2.25, −2.71, and −3.00, respectively. A linear fit of these values gives an Adj-R^2 of 0.9993, indicating a robust linear relationship. Additionally, numerically, ΔT increases with the increase in absorptivity. Moreover, the factors influencing detection are surface absorptivity and ambient conditions. Due to heating from solar radiation, the surface absorptivity varies for differently treated surfaces of the specimen, leading to variations in energy accumulation in this region. Additionally, due to the restriction caused by delamination, the thermal exchange between the specimen and the surroundings occurs after the radiation disappears, gradually causing the temperature in the delaminated area to be lower than that in the non-delaminated area, as shown in Figure 10a.

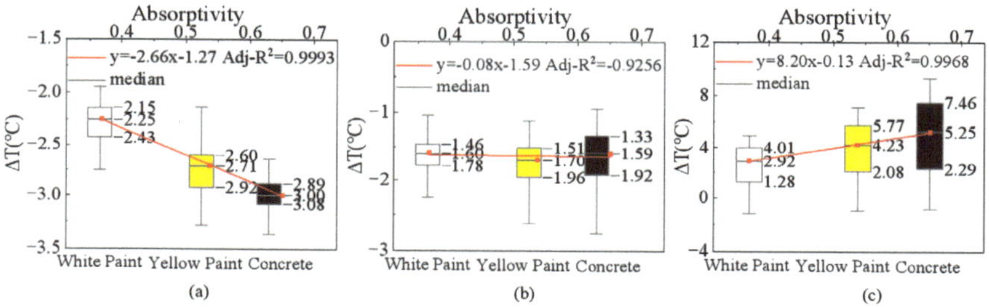

Figure 10. The relationship between ΔT and absorptivity (concrete, yellow paint, and white paint are 0.652, 0.538, and 0.369, respectively): (**a**) Zone I; (**b**) Zone II; and (**c**) Zone III.

In Zone II, the median values of ΔT for white, yellow, and concrete are −1.60, −1.70, and −1.59, respectively. A linear fit of these values gives an Adj-R^2 of −0.9256, indicating a poor correlation. At this point, ΔT is not related to absorptivity. Moreover, the factors influencing detection are surface emissivity and ambient conditions. After intense heat exchange, the specimen undergoes a relatively stable heat exchange with the surroundings. Due to the similar surface emissivity, the thermal contrast curves for different regions tend to converge or overlap, as shown in Figure 10b.

In Zone III, the median values of ΔT for white, yellow, and concrete in this region are 2.92, 4.23, and 5.25, respectively. A linear fit of these values gives an Adj-R^2 of 0.9968, indicating a strong linear relationship. Additionally, ΔT increases with the increase in absorptivity. Moreover, the factors influencing detection include surface absorptivity and ambient conditions. The specimen's surface rapidly accumulates energy from solar radiation upon the emergence of radiation. Due to differing absorptivity, energy accumulation varies among different regions. At this point, thermal exchange between the specimen and the environment can be disregarded. The presence of delamination restricts thermal conductivity, resulting in higher temperatures in delaminated areas compared to non-delaminated areas, as shown in Figure 10c.

3.4. The Characteristics of Infrared Detection for Concrete Delamination in Different Time Windows

3.4.1. Zone I (Approximately 17:00 to 23:00)

Figure 11a shows the plot depicting the temperature difference between the non-delaminated area and the air (δT) and thermal contrast (ΔT) in Zone I. In the figure, the scatter points of δT with respect to ΔT for each delamination form a coherent group based on depths and surface conditions. Therefore, the k-means clustering method [47] was applied to process the delamination data of different surface materials and depths in Figure 11a, where k is set as 1. Here, k represents the centroids of each delamination data cluster, the data sets' central positions, as shown in Figure 11b, and the points with the same color and shape as in Figure 11a.

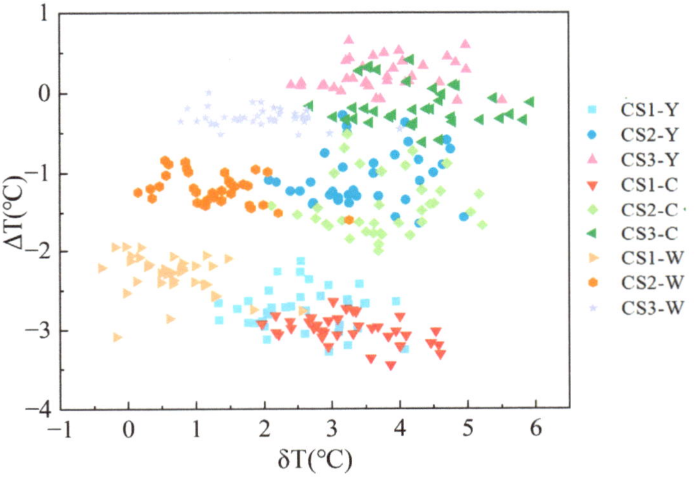

(a)

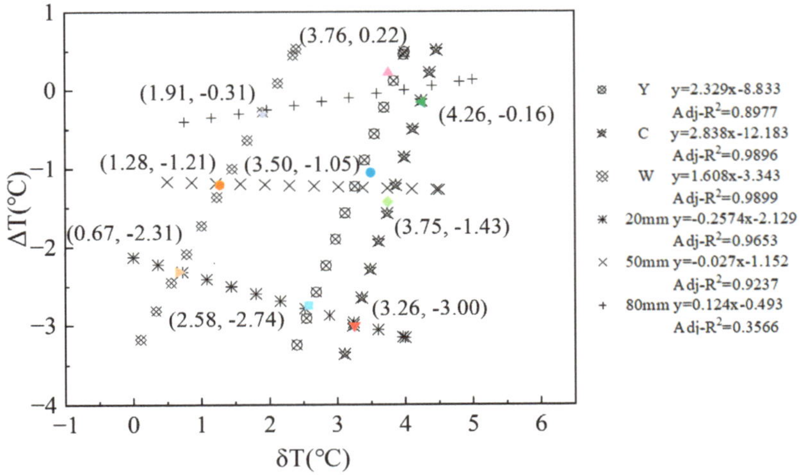

(b)

Figure 11. Zone I: the Map of thermal contrast (ΔT) and temperature difference between non-delaminated area and air (δT): (**a**) scatter plots of different delaminations and (**b**) fitting curves by depth or surface material.

The centroids were linearly fitted according to the same depth, resulting in three depth lines, as shown by the 20 mm, 50 mm, and 80 mm lines in Figure 11b. The value of ΔT is discussed based on its magnitude, without discussing its positive or negative sign. Across surface variations, the ΔT for a 20 mm depth delamination increases from 2.31 to 3.0, with an increase in δT from 0.67 to 3.26. Meanwhile, the linear correlation of the adjusted R^2 is strong (0.9653). For 50 mm depth delamination, ΔT remains almost unchanged (1.21, 1.05, and 1.43) with the increase in δT (from 1.28 to 3.75). Meanwhile, the linear correlation of the adjusted R2 is strong (0.9237). For 80 mm depth delamination, ΔT remains almost unchanged (0.31, 0.22, 0.16) with the increase in δT (from 1.28 to 3.75), but the linear correlation of the Adj-R^2 is weak (0.3566). The results show that shallow delamination developed thermal contrast (ΔT) more sensitive to the variation in surface absorptivity than the deeper delamination in this period. This finding indicates that when the surface absorptivity is given, the range of thermal contrast could be determined for shallow and middle depths delamination.

The centroids were linearly fitted to the same surface, resulting in three surface lines, as shown by the C, Y, and W lines in Figure 11b. The value of ΔT is discussed based on its magnitude, without discussing its positive or negative sign. For delamination under a concrete surface, ΔT decreases from 3.00 to 0.16 as depth changes, with δT increasing from 3.26 to 4.26, and the linear correlation is strong with an adjusted R^2 of 0.9896. For delamination under a yellow paint surface, ΔT decreases from 2.74 to 0.22, with δT increasing from 2.58 to 3.76, and the linear correlation is strong with an adjusted R^2 of 0.8977. For delamination under a white paint surface, ΔT decreases from 2.31 to 0.31 as δT increases from 0.67 to 1.91, and the linear correlation is strong with an adjusted R^2 of 0.9899. The results indicate that the thermal contrast (ΔT) is more sensitive to depth changes under concrete surfaces while less so under lighter-colored surfaces. This finding suggests that, given a specific depth, the thermal contrast range can be determined for concrete and other colored surfaces.

In summary, δT, depth, and surface material all influence the determination of ΔT. At the same depth or the same surface material, this study can determine ΔT uniquely through δT. However, there may be misjudgments when depth and surface material are considered simultaneously. For example, under the same δT, delaminations may coincide in concrete at a depth of 20 mm and under yellow paint at a depth of 50 mm, illustrating a limitation of this model. And during this period, delaminations with a depth of 80 mm have ΔT values ranging from −0.5 °C to 0.5 °C. According to the threshold set in this study, delaminations within this range are considered ineffective for detection within Zone I, while other delaminations can be detected effectively.

3.4.2. Zone II (Approximately from 23:00 to 07:00 the Next Day)

As shown in Figure 12, the plot depicts the temperature difference between the non-delaminated area and the air (δT) and thermal contrast (ΔT) in Zone II. Linear fittings were performed for each delamination, resulting in nine curves. Additionally, linear fittings were conducted for the delaminations at the same depth, producing three overall depth-related curves, totaling twelve. At the same depth, the slopes of each delamination curve are very similar (20 mm: −0.191, −0.278, −0.259; 50 mm: −0.122, −0.091, −0.117; 80 mm: −0.063, −0.026, −0.026), and the slopes of the delamination curves at the same depth are close to those of the corresponding overall curves. These results indicate that different absorptivities have little to no impact on delamination detection. However, at different depths, the slopes of the overall curves decrease as the depth increases.

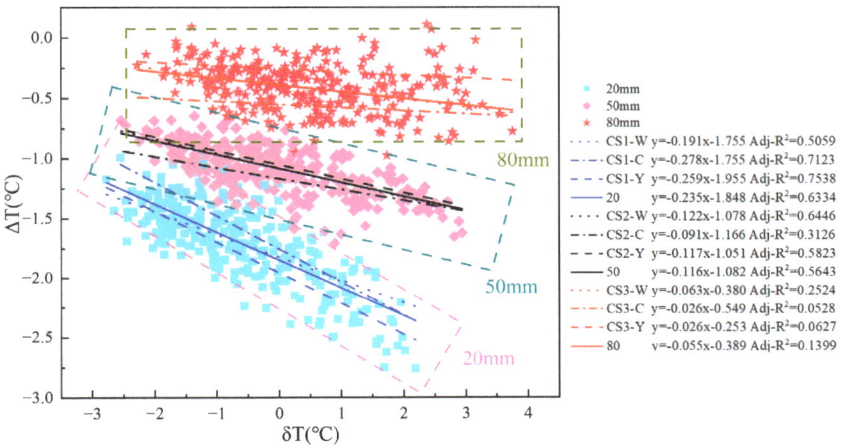

Figure 12. Zone II: the map of thermal contrast (ΔT) and temperature difference between non-delaminated area and air (δT).

As shown in Table 6, the variance analysis [48] was conducted on ΔT with the same depth and different surface treatment methods, and their correlation coefficients p were all less than 0.05, indicating a significant correlation between data with the same depth but different absorptivities. Under the same depth, the mean values (20 mm: −1.77, −1.65, −1.66; 50 mm: −1.04, −1.20, −1.02; 80 mm: −0.27, −0.58, −0.38) and median values (20 mm: −1.70, −1.59, −1.60; 50 mm: −1.04, −1.19, −0.99; 80 mm: −0.27, −0.58, −0.37) are relatively close, which further confirms their strong correlation. Indeed, the scatter plots for each surface material are very close, and except for the 80 mm delamination, ΔT increases with increasing δT. The results show that during this period, ΔT is unrelated to surface absorptivity but is related to depth and surface emissivity. This finding indicates that when depth and δT are determined, the range of ΔT can be determined.

Table 6. Correlation analysis of ΔT between different delamination layers.

Delamination Type	Mean (°C)	Median (°C)	Variance	*p*-Value
CS1-Y	−1.77	−1.70	0.121	
CS1-C	−1.65	−1.59	0.158	2.9×10^{-2}
CS1-W	−1.66	−1.60	0.084	
CS2-Y	−1.04	−1.04	0.037	
CS2-C	−1.20	−1.19	0.049	8.57×10^{-11}
CS2-W	−1.02	−0.99	0.026	
CS3-Y	−0.27	−0.27	0.025	
CS3-C	−0.58	−0.58	0.025	2.25×10^{-34}
CS3-W	−0.38	−0.37	0.021	

Under the same surface material, significant differences exist in the mean and median values for the same surface material at different depths. It shows that the skewness of their scatter plots gradually decreases with increasing depth, indicating that heat conduction within the delaminated areas of the concrete specimen is restricted to varying degrees with depth. Therefore, deeper delamination results in less restricted heat conduction, leading to closer temperatures between the delaminated and non-delaminated areas.

This study shows that changes in surface properties have little impact on the infrared detection of delamination in concrete structures during Zone II. In contrast, delamination depth still has a significant negative impact on detection. During this period, the ΔT of the painted delaminations with a depth of 80 mm ranged from −0.5 °C to 0 °C. However, some un-delaminated concrete delaminations without treatment had values greater than 0.5 °C. According to the threshold set in this study, delaminations falling within this range could not be effectively detected in Zone II, while all other delaminations could be detected. This study did not consider the size of the delaminations, which poses certain limitations. In their studies, researchers [22–24,49] have suggested that even relatively deep delaminations of sufficient size could be detected. Meanwhile, when δT is positive, this study can uniquely determine the depth or ΔT through δT alone.

3.4.3. Zone III (Approximately 7:00–17:00)

As shown in Figure 13, the plot depicts the temperature difference between the non-delaminated areas and the air (δT) and thermal contrast (ΔT) in Zone III. When the sun appears, the ambient temperature gradually rises. Although the temperature of the specimen also gradually increases, when the specimen's temperature is higher than the ambient temperature, the specimen radiates energy to the outside. This results in different ΔT values at the same δT, forming a behavior similar to a hysteresis curve. It was observed that the scatter plot resembles a hysteresis model. By using the hysteresis model, the relevant hysteresis curves were fitted accordingly [50]. The hysteresis curve rises rapidly to its maximum value at the same depth. Then, it slowly decreases to a specific value before rapidly dropping to its minimum value. It finally returns slowly to the origin, as shown in the upper half of Figure 13c. The degree of expansion of the hysteresis curve increases with the absorption rate, with the curves for the 20 mm deep delamination particularly noticeable. For instance, the ΔT for CS1-C rapidly rises from 0 °C to around 8 °C, then decreases to about 6.5 °C before sharply dropping to around −1.5 °C and finally returning to the origin. Similarly, the ΔT for CS1-Y quickly climbs from 0 °C to about 5 °C, decreases to around 4 °C, then drops to around −1 °C, and returns to the origin. For CS1-W, the ΔT rises swiftly from 0 °C to approximately 3.5 °C, then decreases to about 3 °C, drops to around −0.5 °C, and finally returns to the origin. This phenomenon could be because surfaces with higher absorption rates have a greater ability to absorb solar radiation, accumulating more heat. Additionally, due to the restriction of heat transfer in the delaminated areas, the temperature in the delaminated regions tends to be higher than that in the non-delaminated regions.

For the same surface material, the expansion of the hysteresis curve decreases with increasing depth. This is particularly evident for different depths of delamination on untreated concrete surfaces, as shown in the bottom of Figure 13c. For CS1-C, ΔT rapidly rises from 0 °C to about 8 °C, then drops to around 6.5 °C before quickly falling to approximately −1.5 °C and finally returning to the origin. For CS2-C, ΔT rises rapidly from 0 °C to about 4 °C, then gradually rises to around 4.5 °C before returning to the origin. For CS3-C, ΔT rises rapidly from around −0.5 °C to about 1.7 °C before gradually returning to the origin. This phenomenon could be because the heat conduction inside the concrete specimen is restricted in the delaminated regions. The degree of restriction varies with depth, meaning that deeper delaminations experience less restriction. As a result, the temperature difference between the delaminated and non-delaminated regions decreases as the depth increases. Therefore, this study concludes that in Zone III, both surface property variations and delamination depth negatively impact the detection of subsurface delamination in concrete.

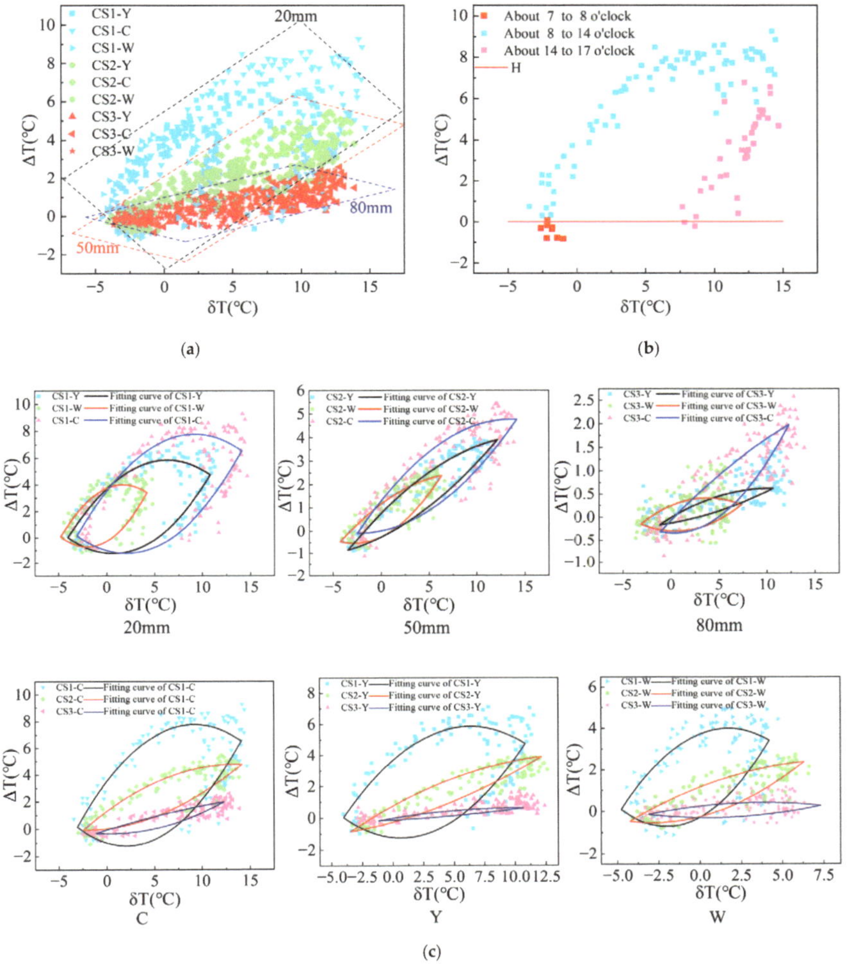

Figure 13. Zone III: (**a**) the map of thermal contrast (ΔT) and temperature difference between non-delaminated area and air (δT); (**b**) segment the CS1-C in Figure a by time for reference; and (**c**) hysteresis curve graph for the same depth or the same surface material.

From 7 to 8 o'clock, it is difficult to determine the delamination depth. However, from 8 to 14 o'clock, this study suggests that the delamination depth can be preliminarily determined using ΔT and δT and further refined based on the detection time. Similarly, the period from 14 to 17 o'clock is similar to the period from 8 to 14 o'clock. According to the threshold set in this study, delaminations at depths other than 80 mm can generally be effectively detected during this period. However, 80 mm deep delaminations require sufficiently large δT values to be detectable.

4. Conclusions

This study conducted experimental research and data analysis to reveal the effects of surface absorptivity and depth on the infrared delamination detection of concrete structures under natural environmental conditions. According to the findings, the ΔT curve can be divided into three characteristic zones based on its variation pattern based on the daily solar loading cycle. In Zone I, linear fitting of the centroids obtained from k-means clustering can distinguish the ΔT distribution under different absorptivities and depths. In Zone II, δT and depth can be used to determine the range of ΔT, and variance analysis indicates that the impact of absorptivity on detection is negligible during this period. Zone III has a hysteresis effect between ΔT and δT, where depth and absorptivity influence the hysteresis loop, effectively impacting ΔT. The shallower the depth, the larger the hysteresis loop; the more significant the absorptivity, the larger the hysteresis loop.

In summary, this study has found that depth has a negative impact on infrared detection of concrete delamination, while absorptivity has a positive impact on detection. For delamination with different depths and the same absorptivity, the effectiveness of infrared detection decreases with increasing depth, such as delamination under the surface of concrete, whose maximum ΔT values are 9.40 °C (20 mm), 5.54 °C (50 mm), and 2.63 °C (80 mm), respectively. For delamination with the same depth but different absorptivity, as the absorptivity increases, the effectiveness of infrared detection also increases. For example, for delamination with a depth of 20 mm, their maximum ΔT values are 5.15 °C (absorptivity of 0.369), 7.35 °C (absorptivity of 0.538), and 9.40 °C (absorptivity of 0.652), respectively. And the larger the δT, the easier it is to detect delamination. For infrared delamination detection of concrete structures with different surface colors (cool colors), this study provides a reference for further quantifying and assessing concrete delamination using δT among colors with similar absorptivity. This approach is particularly useful for regions with latitudes and climate conditions similar to those of Chongqing.

Although this study has revealed the effects of absorptivity and depth on the infrared detection of concrete structure delamination, the impact of delamination size, the region where the concrete structure is located, and the environment (climate, season) on infrared detection of delamination remains unclear, and the detectable limit of delamination depth is also yet to be determined. These will be the directions for future research.

Author Contributions: Conceptualization: C.C.; methodology: D.C.; validation: C.C.; formal analysis: D.C., S.S. and H.C.; resources: H.Z.; data curation: D.C.; writing—original draft preparation: C.C. and D.C.; writing—review and editing: C.C., D.C., S.S., R.N., H.C., H.Z. and B.W.; project administration: C.C.; funding acquisition: C.C. All authors have read and agreed to the published version of the manuscript.

Funding: This work was partially supported by the National Natural Science Foundation of China (52108267, 52478301), the Chongqing Natural Science Foundation of China (CSTB2022NSCQ-MSX1379 and CSTB2022TIAD-KPX0205), the Special Support Program of Chongqing Postdoctoral Research (2021XMT007), and the China Postdoctoral Science Foundation under Grant Number 2024T171100.

Data Availability Statement: The original contributions presented in this study are included in this article. Further inquiries can be directed to the corresponding authors.

Acknowledgments: The authors would like to thank the State Key Laboratory of Mountain Bridge and Tunnel Engineering for its invaluable support in the experimental phase of this research.

Conflicts of Interest: The authors declare no conflicts of interest.

References

1. Ahlborn, T.M.; Brooks, C.N. *Evaluation of Bridge Decks Using Non-Destructive Evaluation (NDE) at Near Highway Speeds for Effective Asset Management*; Michigan Technological University: Houghton, MI, USA, 2015.
2. Gheitasi, A.; Harris, D.K. Performance Assessment of Steel–Concrete Composite Bridges with Subsurface Deck Deterioration. *Structures* **2015**, *2*, 8–20. [CrossRef]
3. Zhang, G.; Harichandran, R.S.; Ramuhalli, P. Automatic Delamination Detection of Concrete Bridge Decks Using Impact Signals. *J. Bridge Eng.* **2012**, *17*, 951–954. [CrossRef]

4. Raja, B.N.K.; Miramini, S.; Duffield, C.; Sofi, M.; Mendis, P.; Zhang, L. The Influence of Ambient Environmental Conditions in Detecting Bridge Concrete Deck Delamination Using Infrared Thermography (IRT). *Struct. Control Health Monit.* **2020**, *27*, e2506. [CrossRef]
5. Pozzer, S.; Omidi, Z.; El Refai, A.; López, F.; Ibarra-Castanedo, C.; Maldague, X. Passive Infrared Thermography for Subsurface Delamination Detection in Concrete Infrastructure: Capabilities. *Constr. Build. Mater.* **2024**, *419*, 135542. [CrossRef]
6. Gucunski, N.; Imani, A.; Romero, F.; Nazarian, S.; Yuan, D.; Wiggenhauser, H.; Shokouhi, P.; Taffe, A.; Kutrubes, D. *Nondestructive Testing to Identify Concrete Bridge Deck Deterioration*; Transportation Research Board: Washington, DC, USA, 2012. ISBN 978-0-309-12933-6.
7. Huh, J.; Mac, V.; Tran, Q.; Lee, K.-Y.; Lee, J.-I.; Kang, C. Detectability of Delamination in Concrete Structure Using Active Infrared Thermography in Terms of Signal-to-Noise Ratio. *Appl. Sci.* **2018**, *8*, 1986. [CrossRef]
8. Milovanović, B.; Gaši, M.; Gumbarević, S. Principal Component Thermography for Defect Detection in Concrete. *Sensors* **2020**, *20*, 3891. [CrossRef] [PubMed]
9. Ta, Q.T.; Mac, V.H.; Huh, J.; Yim, H.J.; Tran, Q.H. Nondestructive Detection of Delamination in Painted Concrete Structures through Square Pulse Thermography. *J. Build. Eng.* **2023**, *70*, 106360. [CrossRef]
10. Raja, B.N.K.; Miramini, S.; Duffield, C.; Sofi, M.; Zhang, L. Infrared Thermography Detection of Delamination in Bottom of Concrete Bridge Decks. *Struct. Control Health Monit.* **2022**, *29*, e2886. [CrossRef]
11. Hu, D.; Wang, M.; Guo, R.; Li, S. Bridge Deck Condition Assessment Using GPR: System Configuration and Defects' Characterization. In Proceedings of the Construction Research Congress 2024, Des Moines, IA, USA, 20–23 March 2024; American Society of Civil Engineers: Reston, VA, USA, 2024; pp. 1210–1219.
12. Alsharqawi, M.; Zayed, T.; Shami, A. Ground Penetrating Radar-Based Deterioration Assessment of RC Bridge Decks. *Constr. Innov.* **2020**, *20*, 1–17. [CrossRef]
13. Li, M.; Anderson, N.; Sneed, L.; Maerz, N. Application of Ultrasonic Surface Wave Techniques for Concrete Bridge Deck Condition Assessment. *J. Appl. Geophys.* **2016**, *126*, 148–157. [CrossRef]
14. Kim, J.; Cho, Y.; Lee, J.; Kim, Y.H. Defect Detection and Characterization in Concrete Based on FEM and Ultrasonic Techniques. *Materials* **2022**, *15*, 8160. [CrossRef]
15. Scherr, J.F.; Grosse, C.U. Delamination Detection on a Concrete Bridge Deck Using Impact Echo Scanning. *Struct. Concr.* **2021**, *22*, 806–812. [CrossRef]
16. Hendricks, L.J.; Baxter, J.S.; Chou, Y.; Thomas, M.; Boekweg, E.; Guthrie, W.S.; Mazzeo, B.A. High-Speed Acoustic Impact-Echo Sounding of Concrete Bridge Decks. *J. Nondestruct. Eval.* **2020**, *39*, 58. [CrossRef]
17. Vemuri, S.H.; Atadero, R.A. Case Study on Rapid Scanning Techniques for Concrete Bridge Decks with Asphalt Overlay: Ground-Penetrating Radar and Infrared Thermography. *Pract. Period. Struct. Des. Constr.* **2017**, *22*, 05016004. [CrossRef]
18. *D4788-03*; Standard Test Method for Detecting Delaminations in Bridge Decks Using Infrared Thermography. ASTM International: West Conshohocken, PA, USA, 2003.
19. Kee, S.-H.; Oh, T.; Popovics, J.S.; Arndt, R.W.; Zhu, J. Nondestructive Bridge Deck Testing with Air-Coupled Impact-Echo and Infrared Thermography. *J. Bridge Eng.* **2012**, *17*, 928–939. [CrossRef]
20. Nehdi, M.; Omar, T. Application of Passive Infrared Thermography for the Detection of Defects in Concrete Bridge Elements. In Proceedings of the 2016 Conference of the Transportation Association of Canada, Toronto, ON, Canada, 25 September 2016.
21. Washer, G.; Fenwick, R.; Nelson, S.; Rumbayan, R. Guidelines for Thermographic Inspection of Concrete Bridge Components in Shaded Conditions. *Transp. Res. Rec. J. Transp. Res. Board* **2013**, *2360*, 13–20. [CrossRef]
22. Mac, V.H.; Tran, Q.H.; Huh, J.; Doan, N.S.; Kang, C.; Han, D. Detection of Delamination with Various Width-to-Depth Ratios in Concrete Bridge Deck Using Passive IRT: Limits and Applicability. *Materials* **2019**, *12*, 3996. [CrossRef]
23. Pozzer, S.; Dalla Rosa, F.; Pravia, Z.M.C.; Rezazadeh Azar, E.; Maldague, X. Long-Term Numerical Analysis of Subsurface Delamination Detection in Concrete Slabs via Infrared Thermography. *Appl. Sci.* **2021**, *11*, 4323. [CrossRef]
24. Tran, Q.H. Passive and Active Infrared Thermography Techniques in Nondestructive Evaluation for Concrete Bridge. *AIP Conf. Proc.* **2021**, *2420*, 050008.
25. Watase, A.; Birgul, R.; Hiasa, S.; Matsumoto, M.; Mitani, K.; Catbas, F.N. Practical Identification of Favorable Time Windows for Infrared Thermography for Concrete Bridge Evaluation. *Constr. Build. Mater.* **2015**, *101*, 1016–1030. [CrossRef]
26. Washer, G.; Fenwick, R.; Bolleni, N.; Harper, J. Effects of Environmental Variables on Infrared Imaging of Subsurface Features of Concrete Bridges. *Transp. Res. Rec. J. Transp. Res. Board* **2009**, *2108*, 107–114. [CrossRef]
27. Kirimtat, A.; Krejcar, O. A Review of Infrared Thermography for the Investigation of Building Envelopes: Advances and Prospects. *Energy Build.* **2018**, *176*, 390–406. [CrossRef]
28. Zhang, Q.; Ro, S.H.; Wan, Z.; Babanajad, S.; Braley, J.; Barri, K.; Alavi, A.H. Automated Unmanned Aerial Vehicle-Based Bridge Deck Delamination Detection and Quantification. *Transp. Res. Rec. J. Transp. Res. Board* **2023**, *2677*, 24–36. [CrossRef]
29. Yehia, S.; Abudayyeh, O.; Nabulsi, S.; Abdelqader, I. Detection of Common Defects in Concrete Bridge Decks Using Nondestructive Evaluation Techniques. *J. Bridge Eng.* **2007**, *12*, 215–225. [CrossRef]

30. Hiasa, S.; Birgul, R.; Matsumoto, M.; Necati Catbas, F. Experimental and Numerical Studies for Suitable Infrared Thermography Implementation on Concrete Bridge Decks. *Measurement* **2018**, *121*, 144–159. [CrossRef]
31. Hiasa, S.; Birgul, R.; Watase, A.; Matsumoto, M.; Mitani, K.; Catbas, F.N. A Review of Field Implementation of Infrared Thermography as a Non-Destructive Evaluation Technology. In Proceedings of the Computing in Civil and Building Engineering (2014), Orlando, FL, USA, 23–25 June 2014; American Society of Civil Engineers: Reston, VA, USA, 2014; pp. 1715–1722.
32. Rumbayan, R.; Washer, G.A. Modeling of Environmental Effects on Thermal Detection of Subsurface Damage in Concrete. *Res. Nondestruct. Eval.* **2014**, *25*, 235–252. [CrossRef]
33. Mac, V.H.; Huh, J.; Doan, N.S.; Lee, B.Y.; Haldar, A. A Comprehensive Study on Identification of Both Deck and Soffit Defects in Concrete Bridge Decks through Thermographic Investigation of Shaded Side under Natural Conditions. *Constr. Build. Mater.* **2021**, *303*, 124452. [CrossRef]
34. Berg, R.; Quinn, W. Use of Light Colored Surface to Reduce Seasonal Thaw Penetration beneath Embankments on Permafrost. In Proceedings of the Second International Symposium on Cold Regions Engineering, Fairbanks, Alaska, 12–14 August 1976; University of Alaska: Fairbanks, Alaska, 1978; pp. 86–99.
35. Santamouris, M. *Energy and Climate in the Urban Built Environment*; Santamouris, M., Ed.; Routledge: London, UK, 2013. ISBN 9781134257904.
36. Synnefa, A.; Santamouris, M.; Livada, I. A Study of the Thermal Performance of Reflective Coatings for the Urban Environment. *Sol. Energy* **2006**, *80*, 968–981. [CrossRef]
37. Cheng, C.; Chen, D.; Cai, H.; Wang, S.; Wang, R.; Cheng, X. Investigation of the effect of surface condition variation on infrared imaging of concrete delamination. In *Bridge Maintenance, Safety, Management, Digitalization and Sustainability. Proceedings of the 12th International Conference on Bridge Maintenance, Safety and Management, IABMAS 2024, 2409–2416, Copenhagen, Denmark, 24–28 June 2024*; CRC Press: Boca Raton, FL, USA, 2024.
38. Yang, L.; Yang, Z. *The Principles and Technology of Infrared Thermal Imaging Temperature Measurement*; Science Press: Beijing, China, 2012.
39. Vollmer, M.; Mllmann, K.-P. *Infrared Thermal Imaging: Fundamentals, Research and Applications*; John Wiley & Sons, Inc.: Hoboken, NJ, USA, 2018.
40. ACI. *Committee Report on Nondestructive Test Methods for Evaluation of Concrete in Structures*; American Concrete Institute: Farmington Hills, MI, USA, 2013.
41. López, G.; Basterra, L.A.; Acuña, L.; Casado, M. Determination of the Emissivity of Wood for Inspection by Infrared Thermography. *J. Nondestruct. Eval.* **2013**, *32*, 172–176. [CrossRef]
42. Liu, W.; Geng, Y. Experimental Study on Solar Radiation Absorption Rate of Concrete Surface. *n.a. Concr. Cem. Prod.* **2004**, *2004*, 8–11. [CrossRef]
43. *GB/T 2680-2021*; Glass in Building-Determination of Light Transmittance, Solar Direct Transmittance, Total Solar Energy Transmittance, Ultraviolet Transmittance and Related Glazing Factors. National Standards of the People's Republic of China: Beijing, China, 2021.
44. Tran, Q.H.; Han, D.; Kang, C.; Haldar, A.; Huh, J. Effects of Ambient Temperature and Relative Humidity on Subsurface Defect Detection in Concrete Structures by Active Thermal Imaging. *Sensors* **2017**, *17*, 1718. [CrossRef]
45. Cheng, C.; Shen, Z. Semi Real-Time Detection of Subsurface Consolidation Defects during Concrete Curing Stage. *Constr. Build. Mater.* **2021**, *270*, 121489. [CrossRef]
46. Marceau, M.L.; Vangeem, M.G. Solar Reflectance Values of Concrete. *Concr. Int.* **2008**, *30*, 52–58.
47. Lloyd, S. Least Squares Quantization in PCM. *IEEE Trans. Inf. Theory* **1982**, *28*, 129–137. [CrossRef]
48. Baraheni, M.; Tabatabaeian, A.; Amini, S.; Ghasemi, A.R. Parametric Analysis of Delamination in GFRP Composite Profiles by Performing Rotary Ultrasonic Drilling Approach: Experimental and Statistical Study. *Compos. B Eng.* **2019**, *172*, 612–620. [CrossRef]
49. Cotič, P.; Kolarič, D.; Bosiljkov, V.B.; Bosiljkov, V.; Jagličić, Z. Determination of the Applicability and Limits of Void and Delamination Detection in Concrete Structures Using Infrared Thermography. *NDT E Int.* **2015**, *74*, 87–93. [CrossRef]
50. Jiles, D.C.; Atherton, D.L. Theory of Ferromagnetic Hysteresis (Invited). *J. Appl. Phys.* **1984**, *55*, 2115–2120. [CrossRef]

Disclaimer/Publisher's Note: The statements, opinions and data contained in all publications are solely those of the individual author(s) and contributor(s) and not of MDPI and/or the editor(s). MDPI and/or the editor(s) disclaim responsibility for any injury to people or property resulting from any ideas, methods, instructions or products referred to in the content.

buildings

MDPI

Article

Visualization Analysis of Construction Robots Based on Knowledge Graph

Runrun Dong *, Cuixia Chen and Zihan Wang

College of Civil Engineering, Henan University of Technology, Zhengzhou 450001, China; 17861160536@163.com (C.C.); 16631090975@163.com (Z.W.)
* Correspondence: dongrunrun@haut.edu.cn; Tel.: +86-18623719217

Abstract: Construction robots are pivotal in advancing the construction industry towards intelligent upgrades. To further explore the current research landscape in this domain, the CNKI Chinese database and the Web of Science core database were employed as data sources. CiteSpace software (version 6.2R4) was utilized to visualize and the analyze relevant literature on construction robots from 2007 to 2024, generating pertinent maps. The findings reveal an annual increase in the number of publications concerning construction robots. An analysis of institutions and authors indicates closer collaboration among English institutions, while Chinese authors exhibit stronger cooperation. However, overall institutional and author collaboration remains limited and fragmented, with no prominent core group of authors emerging. Research hotspots in both the Chinese and English literature are largely aligned, focusing on intelligent construction, human-robot collaboration, and path planning. Notably, the Chinese literature emphasizes technical aspects, whereas the English literature is more application-oriented. Future trends in the field are likely to include human-robot collaboration, intelligent construction, robot vision technology, and the cultivation of specialized talent.

Keywords: construction robots; visual analysis; CiteSpace; research hotspot

1. Introduction

The construction of buildings is fundamental to human existence, representing a substantial component of the material production sector within the broader national economy [1]. However, several challenges are currently faced by the construction industry, even amidst its growth. In numerous countries, an aging population, alongside a decline in both the quantity and quality of the labor force, is presenting a significant challenge [2]. Simultaneously, increasing concerns are being raised regarding the efficacy of development methods, workforce productivity, quality of output, incidence of engineering safety accidents, and levels of energy consumption and emissions within the industry. An urgent transformation towards intelligent practices is needed to drive advancements with innovative construction models [3].

Scientific and technological innovation is currently highly valued across all sectors of society, with robots constituting a significant element of the intelligent construction industry due to their substantial research and application benefits. Construction robots have emerged as a central focus within this domain, contributing to enhanced construction efficiency and quality while reducing costs and risks. These advancements facilitate industrial upgrading, attract policy support, and bolster international competitiveness. In January 2023, the "Robot+" Application Action Implementation Plan was issued by

China's Ministry of Industry and Information Technology along with seventeen other departments. This plan explicitly advocates for the promotion of construction robots to expand their application scope. Additionally, it aims to synergize the development of intelligent construction with new building industrialization. The plan further details the implementation of "Robot+" application innovation and practice, the establishment of international and domestic exchange platforms, and the cultivation of a conducive environment for the widespread adoption of robotic applications [4].

This study was conducted to enhance the understanding of the current state of research in construction robots and to identify potential future research hotspots. Prior to 2007, the number of journal articles on construction robots in the CNKI Chinese database and the WOS core database was nearly nonexistent, indicating that related research was minimal and lacked reference value. The CiteSpace bibliometric method is employed in this study to conduct a metrological analysis of the research literature on construction robots from 2007 to 2024 in the CNKI Chinese database and the WOS core database. The objective is to analyze research institutions, authors, research vectors, hotspots, challenges, and trends within the field of construction robots.

2. Data Collection and Research Methods

In this study, CiteSpace software is utilized to quantify and examine the relevant literature within the domain of construction robots from the CNKI Chinese database and the WOS core database. The objective is to generate a series of knowledge maps to analyze publication volume, research focal points, research vectors, and emerging trends in this field.

2.1. Data Collection

An advanced search was conducted in the CNKI Chinese database, selecting "academic journals" as the literature type and using "Topic = construction robot" as the search term. Similarly, in the WOS Core Database, an advanced search was performed with the term "TS = construction robot" and a time range from 1 January 2007 to 30 June 2024. After de-duplication and the manual elimination of the unrelated literature, conference notices, interviews, and policy document interpretations, 710 valid documents were obtained from the CNKI database, while 2237 valid documents were retrieved from the WOS Core database. The findings in this paper are based on this dataset. Figure 1 illustrates the annual publication count in both databases, revealing a consistent upward trend in overall volume. Notably, a significant increase in article volume was observed starting in 2018. Based on the upward trend and article volume in the first half of 2024, it is reasonably inferred that CNKI will publish approximately 130 articles, while WOS will publish around 400 articles in 2024.

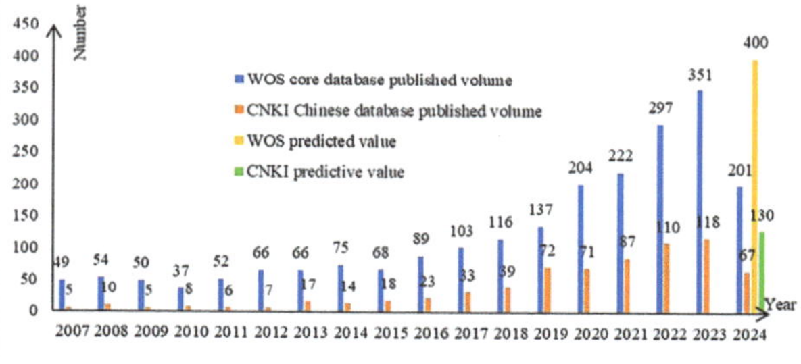

Figure 1. The annual number of construction robots-related publications from 2007 to 2024.

2.2. Research Methods

CiteSpace, a scientific bibliometric visualization and analysis software developed by Prof. Chaomei Chen of Drexel University on the Java platform, is employed in this study [5]. CiteSpace6.2R4 is utilized to enumerate the annual publication count in order to analyze trends in article volume, as well as the distribution and collaboration among authors and institutions, thereby elucidating research trends in this field. Keywords are analyzed through co-occurrence, clustering, and prominence to identify research hotspots, thematic veins, and trends. Based on this analysis, the representative literature is examined, and all results are comprehensively assessed to explore future development trends in the field of construction robots.

3. Results

3.1. Analysis of Institutional Cooperation Network

The comparative analysis of the literature revealed notable discrepancies in publication output among the top ten organizations in each database, as shown in Table 1. While the CNKI Chinese database documented a modest total of 15 articles, the WOS Core Database demonstrated a significantly greater output, with 72 articles—nearly five times that of the CNKI. Furthermore, the tenth-ranked organization in the WOS database published more than ten articles, which is beyond what the leading organization in the CNKI database achieved. This highlights the considerably smaller number of articles in the CNKI database compared to the WOS Core Database.

Network density is a measure defined as the ratio of the actual number of relationships to the theoretical maximum number of relationships within a network. In a directed network comprising n nodes, where the actual number of relationships is m, and the theoretical maximum is $n(n-1)$, the network's density is expressed as $\frac{m}{n(n-1)}$, with a maximum achievable value of 1. An increase in density indicates a higher level of cooperation among organizations as the limit of 1 is approached [6]. The Institutional Co-occurrence Network can illuminate the authoritative standing of a particular research domain, analyze the characteristics of inter-institutional collaboration, and offer a new perspective for assessing the academic influence of research institutions. Figure 2 illustrates the Institutional Co-occurrence Network within the CNKI Chinese database. The prominence of fonts representing the School of Architecture and Urban Planning of Tongji University and Bright Dream Robotics signifies a substantial number of published articles, suggesting notable engagement in the field of construction robots and demonstrating strong research capabilities. The network depicted in Figure 2 consists of 329 nodes and 111 connecting lines, resulting in a network density of 0.0021, which is significantly lower than 1. This low density indicates a scarcity of cooperation among institutions, particularly between universities and enterprises, and underscores a lack of awareness regarding the necessity for such collaboration. Consequently, it is recommended that research on construction robots should be directed towards enhancing institutional collaboration and developing a university-enterprise linkage model to further progress in this field.

The institutional co-occurrence network map of the WOS core database is depicted in Figure 3. The network consists of 423 nodes and 693 connecting lines, with a network density of 0.0078. A relatively close collaboration between institutions is evident, with several prominent networks having formed around key institutions such as the Chinese Academy of Sciences, the Swiss Federal Institutes of Technology Domain, Harbin Institute of Technology, Zhejiang University, the French National Center for Scientific Research, and Shanghai Jiao Tong University. This configuration within the WOS core database illustrates the prevalence of extensive and intimate collaborations among these institutions.

Table 1. Top 10 institutions in database annual publication volume from 2007 to 2024.

	Institution	Count
CNKI Chinese database	School of Architecture and Urban Planning, Tongji University, Shanghai, China	15
	Bright Dream Robotics, Foshan, China	13
	Guangdong Bogarto Construction Technology Co. Foshan, China	8
	China Construction Eighth Engineering Bureau Co. Shanghai, China	6
	School of Electro-mechanical Engineering, Guangdong University of Technology, Guangzhou, China	5
	School of Geodesy and Geomatics, Wuhan University, Wuhan, China	5
	School of Mechanical and Aerospace Engineering, Jilin University, Changchun, China	4
	Wuhan Business University, Wuhan, China	4
	School of Robots Science and Engineering Northeastern University, Shenyang, China	4
	Guangdong Nonferrous Engineering Survey and Design Institute, Guangzhou, China	4
WOS core database	Chinese Academy of Sciences, Beijing, China	72
	University of Michigan System, Michigan, America	52
	Swiss Federal Institutes of Technology Domain, Zurich, Switzerland	38
	Harbin Institute of Technology, Harbin, China	37
	Zhejiang University, Hangzhou, China	32
	Centre National de la Recherche Scientfique, Paris, France	32
	Shanghai Jiao Tong University, Shanghai, China	28
	Tsinghua University, Beijing, China	27
	Hong Kong Polytechnic University, Hong Kong, China	27
	Beijing Institute of Technology, Beijing, China	24

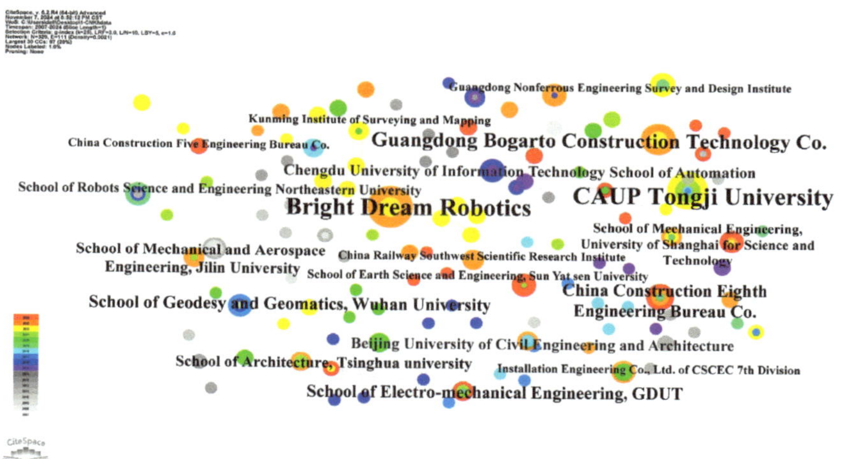

Figure 2. Institutional co-occurrence network map in CNKI Chinese database.

By analyzing the current state of inter-institutional cooperation, it is suggested that such collaboration needs to be further strengthened. Strengthening cooperation not only facilitates the integration of information, human resources, equipment, and technology to achieve resource sharing and complementarity of professional advantages for promoting collaborative innovation, but it also contributes to the cultivation and development of talent. This approach broadens individuals' horizons, enhances their comprehensive abilities, and fosters a sense of collaboration, which holds irreplaceable significance for the advancement of all involved parties and the overall progress in the field of construction robots.

Figure 3. Institutional co-occurrence network map in WOS core database.

3.2. *Analysis of Author Co-Occurrence Network*

Analyzing the relationships between authors gives a clear picture of patterns of academic collaboration within a field and can help identify key authors within that field. In examining the relationship between the number of scientists and the volume of scientific publications, the renowned American historian of science, Price, identified a distinctive pattern between the total number of scientists and the number of eminent ones. This pattern was discerned through the analysis of a substantial dataset, leading to the calculation of a coefficient of 0.749. This study applies Price's theory to analyze central authors, with the condition that the number of articles considered is at least N. Generally, when the articles authored by the "core authors" constitute 50% of the total number of articles, it can be inferred that a "core group of authors" has been established. The mathematical formula for determining core authors is as follows [7]:

$$N = 0.749\sqrt{\eta_{\max}} \tag{1}$$

In the formula: N is the number of publications that meets the minimum requirement for core authors; $\eta_{\max}$ is the number of publications of the author with the highest number of publications.

By utilizing CiteSpace software to analyze and enumerate the authors within the literature, the ten most prolific authors in terms of publication volume are identified, as presented in Table 2. In the CNKI Chinese database, the author has the highest publication count and has produced 15 articles, which corresponds to the maximum value of $\eta_{\max} = 15$. According to Formula (1), this author's output can be approximated as $N \approx 2.901$. In the WOS core database, the author with the highest publication count in the field of construction robotics has published 10 articles, which corresponds to the maximum value of $\eta_{\max} = 10$, resulting in a value of $N \approx 2.369$ according to Formula (1). Consequently, it can be concluded that authors who have published three or more articles are considered the core contributors in this field.

Table 2. Top 10 authors in databases annual publication volume from 2007 to 2024.

	Author	Institution	Count
CNKI Chinese database	Yuan, Feng	CAUP Tongji University, Shanghai, China	15
	Duan, Han	Guangdong Bogarto Construction Technology Co. Foshan, China	6
	Wang, Peng	School of Architecture, Tsinghua University, Beijing, China	4
	Li, Xiao	School of Electro-mechanical Engineering, GDUT, Guangzhou, China	4
	Lu, Songyao	Guangdong Nonferrous Engineering Survey and Design Institute, Guangzhou, China	4
	Chen, Linxin	Guangdong Bogarto Construction Technology Co.	4
	Chen, Gaohong	Bright Dream Robotics, Foshan, China	3
	Zhang, Junhua	Kunming Institute of Surveying and Mapping, Kunming, China	3
	Liu, Jinyue	HBUT School of Mechanical Engineering, Harbin, China	3
	Lu, Chunting	Installation Engineering Co., Ltd. of CSCEC 7th Division, Zhengzhou, China	3
WOS core database	Li, Heng	Hong Kong Polytechnic University, Hong Kong, China	10
	Kamat, Vineet R	University of Michigan, Dept Civil, Ann Arbor, Michigan, USA	9
	Du, Jing	Department of Civil and Coastal Engineering, University of Florida, Gainesville, Florida, USA	8
	Zhang, Tao	School of Economics and Management, Tongji University, Shanghai, China	8
	Menassa, Carol C	Department of Civil and Environmental Engineering, University of Michigan, Ann Arbor, Michigan, USA	8
	Zhou, Tianyu	Department of Civil and Coastal Engineering, University of Florida, Gainesville, Florida, USA	8
	Zhu, Qi	Department of Civil and Coastal Engineering, University of Florida, Gainesville, Florida, USA	6
	Kromoser, Benjamin	University of Natural Resources and Life Sciences, Green Civil Engineering Institute, Vienna, Austria	6
	Jebelli, Houtan	Department of Civil and Environmental Engineering, University of Illinois at Urbana-Campaign, Urbana and Champaign, Illinois, USA	6
	Menges, Achim	Institute for Computational Design and Structures ICD, University of Stuttgart, Stuttgart, Germany	6

The author co-occurrence network map, produced using CiteSpace, serves to quantify the extent of scholarly contribution and collaboration among researchers [8]. Figure 4 illustrates the co-occurrence network map of authors in the CNKI Chinese database. As evidenced in conjunction with Table 2, Yuan Feng, from the School of Architecture and Urban Planning at Tongji University, who focuses on intelligent building design and the construction of building robots, emerges as the most prolific author. In terms of collaborative groups, several have been established, with notable research groups centered around Yuan Feng and Duan Han of Guangdong Bogarto Architecture Technology Co. Overall, Figure 4 features 436 nodes and 311 connections, resulting in a network density of 0.0033. The number of connections is lower than the number of author nodes, and the low network density indicates a relatively dispersed group of researchers with limited collaboration. The author co-occurrence network map in the WOS core database is depicted in Figure 5, in conjunction with Table 2. This map consists of 566 nodes, 269 connecting lines, and a network density of 0.0017, forming a collaborative network centered on Kamat, Vineet R., and Menassa, Carol C., which includes Liu, Yizhi, Lee, Sang Hyun, as well as Zhou, Tianyu, and Du, Jing, as part of the core collaborative group.

Figure 4. Author co-occurrence network map in CNKI Chinese database.

Figure 5. Author co-occurrence network map in WOS core databases.

In conclusion, it can be observed that authors in the CNKI Chinese database engage in more frequent and extensive collaboration with each other. However, the level of interconnectivity among authors across both databases remains comparatively limited and dispersed. The analysis revealed that 23 core authors in the CNKI database account for 12.29% of the total number of articles, while 46 core authors in the WOS database represent 9.03% of the total articles. Notably, none of these authors constitute a core group, as defined by the proposed Formula (1).

3.3. Analysis of Hot Research Topics

Keywords are of great importance to the research topic of the article, reflecting the core of the research and the direction of the literature. They play a crucial role in revealing the development of the research topic [9].

3.3.1. Keyword Co-Occurrence

The data were imported into CiteSpace software, with the time interval set to one year to facilitate the analysis of keywords. Figure 6 presents the keyword co-occurrence map for the Chinese literature, which comprises 366 nodes and 447 connecting lines, resulting in a network density of 0.0067. The nodes represent keywords, while the connecting lines indicate the co-occurrence relationships between them. Table 3 illustrates the ten most frequently occurring keywords in the Chinese literature, where "year" denotes the year of the keyword's first appearance. A review of Figure 6 and Table 3 reveals that "robot" is the keyword with the highest frequency, accompanied by a centrality score of 0.44, indicating that robots are the most prominent research topic in this domain. Conversely, the prevalence of keywords such as deformation monitoring, intelligent construction, and artificial intelligence suggests a heightened interest in research related to construction robots. In addition to keywords with high frequency and centrality, terms like new engineering, talent cultivation, teaching reform, and curriculum system reflect the ongoing educational reforms being implemented by various universities in response to contemporary developments.

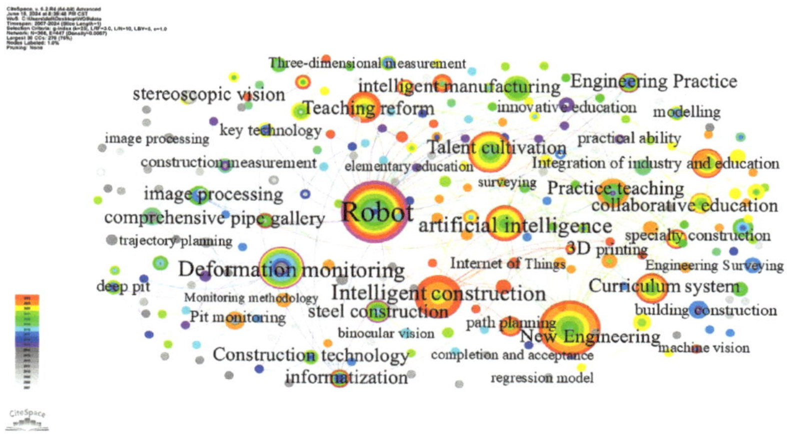

Figure 6. Keyword co-occurrence network map in the Chinese literature.

Table 3. Top 10 keywords in the Chinese literature frequency.

Keyword	Count	Centrality	Year
Robot	64	0.44	2007
New engineering	47	0.04	2019
Deformation monitoring	32	0.2	2008
Talent cultivation	31	0.02	2013
Intelligent construction	30	0.07	2021
Artificial intelligence	21	0.05	2019
Teaching reform	18	0.01	2010
Curriculum system	18	0.04	2019
Practice teaching	12	0.01	2013
Integration of industry and education	12	0.01	2019

Figure 7 illustrates the co-occurrence map for English keywords, consisting of 562 nodes and 2718 connecting lines, resulting in a network density of 0.0172. Table 4 presents the ten most frequently occurring keywords in the English literature. An examination of Figure 7, in conjunction with Table 4, reveals that the keywords "construction", "robot", and "mobile robots" demonstrate high centrality, which aligns with the prevalent research themes in this field. Furthermore, the substantial word frequency and centrality of keywords such as "design", "system", "model", "algorithm", "optimization", "path planning", "tracking", and "task analysis" reflect the research methodologies employed within this domain. Specifically, path planning, tracking, and task analysis are representative of the methodologies utilized in this field.

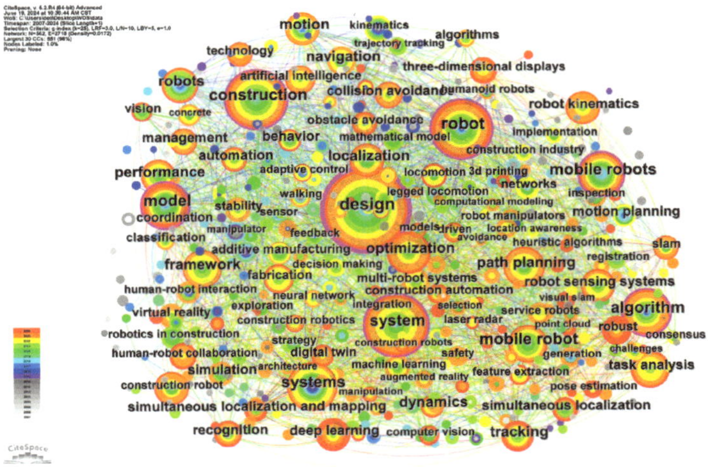

Figure 7. Keyword co-occurrence network map in the English literature.

Table 4. Top 10 keywords in the English literature frequency.

Keyword	Count	Centrality	Year
design	235	0.23	2007
construction	150	0.18	2007
system	146	0.12	2008
robot	134	0.17	2007
mobile robot	82	0.09	2008
model	79	0.13	2008
mobile robots	78	0.12	2007
systems	76	0.09	2011
algorithm	75	0.11	2007
optimization	67	0.06	2012

3.3.2. Keywords Cluster Analysis

Cluster analysis is a widely employed technique in the fields of statistical data analysis and knowledge discovery, utilized to uncover hidden themes within textual data. Through the application of cluster analysis, a substantial number of keywords can be classified into various research themes. This process allows for the categorization of keywords into related themes, thereby facilitating the identification of research trends and the interconnections within a given research field. Utilizing CiteSpace software, the literature was clustered to ascertain the relevance of the keywords. The ordinal number of a smaller cluster signifies its size, where larger and more cohesive clusters suggest greater activity and potential as current or future hotspots [10]. Figure 8 illustrates the keyword clustering timeline mapping of the Chinese literature, comprising nine clusters. It is evident that the clustering scale of robot engineering, path planning, and the related literature is larger. Similarly, Figure 9 presents the keyword clustering timeline mapping of the English literature, which also indicates that clusters pertaining to humanoid robots, path planning, and task analysis are notably larger. From this analysis, it can be initially inferred that path planning emerges as a significant research focus and a prominent topic within the field of construction robots.

The Chinese keyword clustering timeline mapping produced a total of 366 nodes and 447 connecting lines, resulting in a network density of 0.0067. The findings included a Modularity Q of 0.7609 and a Weighted Mean Silhouette S of 0.9678. A *Q* value greater than 0.3 indicates that the delineation of the structure is significant and that the clustering results are optimal. Similarly, an *S* value greater than 0.5 suggests that the clustering results are reasonable [11]. As shown in Figure 8, the English keyword clustering timeline mapping consists of 562 nodes. The *Q* value, recorded at 0.3891, surpasses 0.3, signifying that the discerned structure is significant and the clustering outcome is optimal. Furthermore, the Weighted Mean Silhouette is 0.7372, which demonstrates that the clustering is efficient and convincing when the *S* value is at 0.7. This implies that the homogeneity of this clustering mapping is high, and the overall clustering quality is excellent.

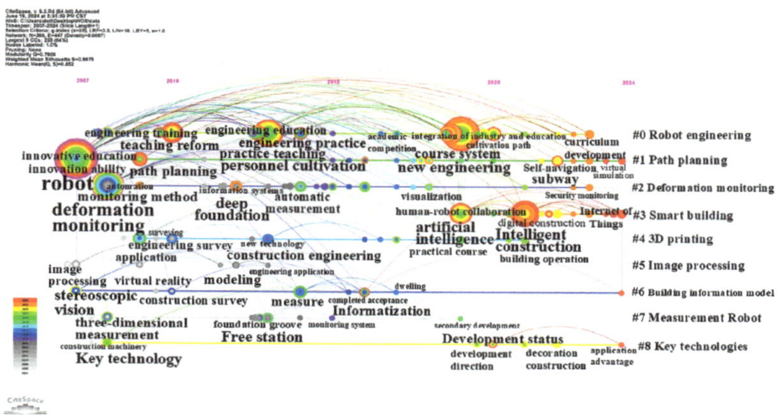

Figure 8. Keywords clustering time map in Chinese literature.

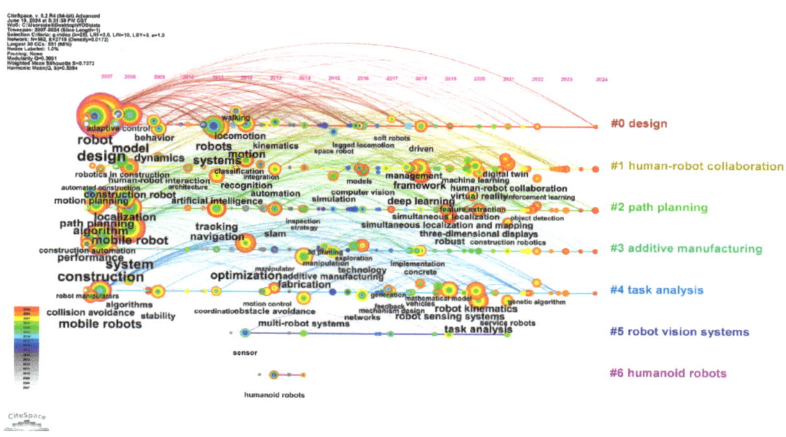

Figure 9. Keywords clustering time map in the English literature.

As illustrated in Figure 8, the keyword clustering timeline mapping of Chinese literature indicates that robot engineering, path planning, deformation monitoring, and intelligent construction are the current research focal points within the domain of construction robots. Research related to robots was initially introduced in the field of construction engineering in 2007 and has continually evolved to the present day. Teaching reform was first introduced in 2010, with the subsequent reform in 2013 placing greater emphasis on practical teaching in talent cultivation. In 2019, the concept of "new engineering" was proposed, leading to a transformation in talent cultivation modes and pathways. The focus shifted towards the integration of industry and education, emphasizing the learning of artificial intelligence, intelligent construction, robot engineering, and related knowledge areas. Additionally, innovative applications such as digital construction were promoted. As illustrated in Figure 9, the English keyword clustering timeline mapping reveals that design and human-robot collaboration have emerged as significant research foci. Since 2007, research on construction robots in English literature has predominantly concentrated on these areas, along with path planning, task analysis, and robot vision systems.

3.3.3. Analysis of Research Hotspot

A timeline mapping of keyword clustering, examined alongside the theme of construction robots, reveals that current research in this field is concentrated on several key areas, including path planning, deformation monitoring, intelligent construction, 3D printing, human-robot collaboration, task analysis, and robot vision systems. The objective of this article is to review and analyze representative papers related to these topics.

1. Path planning

The term "path planning" refers to the process of employing automated systems and algorithms to determine optimal movement paths. In the realm of construction robots, path planning is of paramount importance as it relates to how robots navigate and execute tasks with efficiency and safety on construction sites. A hybrid genetic ant colony algorithm was proposed by Zhang et al. [12], which integrates pheromones of genetic algorithm chromosomes and ant colony algorithm search paths to study construction robot pathways. Wang et al. [13] developed a navigation map based on a BIM model, combined with an optimized A* algorithm, to achieve optimal global path planning for construction robots, enabling real-time obstacle avoidance. Fernando Torres et al. [14] introduced a Lidar-based plain valley path method that is immune to OSM errors, thereby compensating for the low local accuracy of OSM and facilitating seamless obstacle avoidance and task execution by construction robots. Yang et al. [15] enhanced the traditional RRT algorithm using BIM navigation maps, integrating it with the DWA algorithm to increase obstacle avoidance flexibility for moving construction robots. Gao et al. [16] proposed an autonomous exploration method for mobile robots utilizing graph-based SLAM with 2D laser technology, with experimental validation demonstrating its efficacy in robot mapping and exploration tasks. Huang et al. [17] generalized the emerging edge computing paradigm to multi-robot SLAM by combining a multi-robot laser SLAM system for efficient co-processing in the cloud. Liu et al. [18] introduced a semantic-assisted Lidar tightly-coupled SLAM method to reduce interference from dynamic objects and environments, extending its applicability to complex 3D dynamic environments. Concurrently, Li et al. [19] employed a method integrating Euclidean space and angle thresholds to enhance the efficiency and accuracy of the SLAM algorithm point cloud processing, verifying the algorithm's feasibility and reliability in real-world scenarios. Current research in path planning predominantly focuses on three critical areas: enhancing localization accuracy, advancing autonomous navigation, and mitigating the impacts of dynamic environments on robot performance. The aim of this research is to improve the precision and resilience of construction robots in their operational activities.

2. Deformation monitoring

Deformation monitoring technology is employed not only to ensure the safety and stability of structures but also to facilitate the advancement of construction robotics. Zhou et al. [20] introduced a deep foundation pit monitoring and early warning system that utilizes a Leica TPS1200+ measuring robot. This system collects data via a VB platform, GeoCOM interface, and remote wireless transmission to enable communication control between the robot and a computer. Data processing is subsequently conducted using wavelet analysis. Zhang et al. [21] proposed a real-time detection method based on Grubbs' criterion, which is explored as a type of historical data analysis. Zheng et al. [22] developed software for automated pit deformation detection using the Android operating system, Bluetooth communication, and a measurement robot as its hardware foundation. This software satisfies the requirements for convenience, accuracy, and timeliness in pit deformation detection. Zhang et al. [23] investigated the bonding properties of UHPC to stone under various interface treatment methods to enhance building stability through material selection. Ji et al. [24] proposed an integrated approach combining GNSS technology, inclination sensors, and high-precision measuring robots for the horizontal deformation monitoring of ultra-high-rise buildings. Research in deformation monitoring technology aims to improve the precision and responsiveness of monitoring processes. Currently, these techniques are predominantly applied in the monitoring of pits and the deformation of high-rise structures to ensure the safety of construction projects.

3. Intelligent construction

The advancement of construction robots represents a promising development within the realm of intelligent construction. The increasing demand for intelligent construction has spurred innovation in the design, manufacturing, and control technologies associated with construction robots. Yuan et al. [25,26] introduced a digital construction technology system tailored for batch customization in prefabricated buildings, additionally proposing a digital building and intelligent design and construction process specifically for bricks. Duan et al. [27] identified construction robots as core components for optimizing quality and enabling the full-cycle digital formation of the building construction process, further emphasizing their role as key drivers in the advancement of intelligent construction. Zhao et al. [28] advocated for the integration of BIM (Building Information Modeling), detailed design, automated processing, and 3D simulation techniques to enhance the efficiency and quality of designing and constructing complex-shaped buildings. Gao [29] proposed an intelligent construction method for box house structures utilizing BIM visualization programming alongside the RT-Star algorithm. Wang et al. [30] highlighted the digital model of spatial and temporal information of buildings, which evolves with project progress, as the digital infrastructure critical to construction, forecasting that future digital building engineering will inherently possess characteristics of intelligence, platform integration, and industrialization. In addressing the construction industry's aspiration to create architecture that seamlessly integrates humans, the built environment, and technology, as well as the necessity for highly skilled technical professionals, Li et al. [31] proposed a model for cultivating talent in new engineering disciplines within an intelligent construction context. Currently, research in the field of intelligent construction is flourishing, predominantly concentrating on three areas: ongoing technological innovation, expanding application scenarios, and enhanced talent development and team-building efforts.

4. 3D printing

Currently, 3D printing technology is utilized for the rapid and precise fabrication of building components. The integration of construction robots with 3D printing technology enables the automated and intelligent production of these components, thereby improving the efficiency and quality of building construction. A comprehensive overview of the progressive integration of 3D printing in construction engineering was provided by Zuo et al. [32], who proposed a technical framework for the advancement of 3D printing technology, specifically in the context of ultra-high-rise buildings. Liu [33] proposed a

3D printing technology system for filler construction consisting of two components: a 3D printing assembly line robot operation system and a 3D printing technology adjustment system for filler construction. Hong et al. [34] suggested the use of an enhanced Q-learning technique to align the information decision-making process with the optimal robot state as identified through machine learning. This approach aims to facilitate the generation of optimal 3D printing robot paths while minimizing computational time requirements. Zeng et al. [35] introduced a novel method for the intelligent detection of porosity in 3D-printed concrete, combining a target detection algorithm with a lightweight, intelligent approach to address the limitations of existing techniques, such as dependence on manual labor, extended detection times, and high costs. At present, 3D printing is primarily employed in architecture for architectural modeling, meeting the specific requirements of the construction process. Technology is continuously advancing, with the aim of reducing time consumption, labor costs, and expenses.

5. Human-robot collaboration

The model of human-robot collaboration emphasizes the necessity of close interaction between humans and robots in the workplace, highlighting the importance of joint efforts to achieve specific work objectives. The effective implementation of human-robot collaboration not only enhances the efficiency and quality of building construction but also facilitates the transformation and upgrading of the construction industry. In their study, Huang et al. [36] employed a reinforcement learning (RL)-based construction robot to engage in active learning of automated control through environmental interactions. This approach allows the robot to operate based on its perceived state and objectives without human intervention, thereby reducing personnel workload. Additionally, Zhu et al. [37] proposed deep reinforcement learning (DRL)-based optimization methods that are directly applicable to various robot-assisted construction scenarios. Liang [38,39] and his team introduced a model of human-robot interaction employing the 12PL-DT-VR system, where humans are responsible for high-level task planning and workflow supervision while robots execute physical tasks. This model facilitates telecollaboration between construction workers and robots. Concurrently, the team proposed a novel method for categorizing collaborative human-robot work, indicating a promising direction for future research in this field. Yang et al. [40] suggested a framework designed to serve as an organizational tool to support future research and exploration in human-robot collaboration by interlinking the domains of systems, design, and human-centeredness. Liu et al. [41] introduced a method to enhance the safety of human-robot collaboration by focusing on the worker and using electroencephalography to capture brainwaves. This method enables the robot to acquire relevant task information and execute it accordingly. Current research in human-robot collaboration primarily focuses on reducing the physical burden on workers while ensuring construction quality and safety. It is anticipated that advancing intelligent construction management will further enhance work efficiency.

6. Task analysis

Task analysis serves as a critical tool in advancing the construction industry toward enhanced efficiency, safety, and sustainability across various key domains. In research conducted by Bakdi et al. [42], a dual Kinect camera vision system was proposed to furnish depth information to robots. This system, coupled with image processing technology, accurately simulates the surrounding environment. Genetic algorithms are then employed to devise the optimal path for avoiding object collisions, while adaptive fuzzy logic is utilized to control the robot's speed, thus realizing the planning and execution of the mobile robot's optimal path and enhancing its overall performance. The impact of digital fabrication on the construction industry was investigated by De Soto et al. [43], who analyzed both the cost and time requirements for the on-site construction of robotically fabricated complex concrete walls and evaluated the implications for construction productivity. Hartmann et al. [44] presented a system designed to parallelize complex task and motion planning problems using an iterative approach. In this system, smaller subproblems are iteratively solved until the desired solution is obtained. Optimization methods are integrated with a sampling-based bi-directional spatial-temporal path planner to address constraints and enable simultaneous collaboration among multiple robots. Elmakis et al. [45] introduced an innovative method to enhance the flexibility of ground-based unmanned robots. This method involves employing unmanned aerial vehicles for precise aerial localization and mapping, which enables robots to undertake site preparation tasks beyond mere sensing activities. The field of task analysis now encompasses a wide spectrum of facets related to the advancement of construction robotics technology. These facets include but are not limited to, path analysis, exploration of human-robot collaboration, data acquisition and analysis, environmental sensing and navigation, and the automation of construction tasks.

7. Robot vision systems

The integration of robot vision systems into construction robots holds significant potential to enhance automation and intelligence within the construction process, thereby improving accuracy safety, and facilitating more effective construction management and progress monitoring. A novel vision-guided path planning system has been proposed by Pinto, A.M. et al. [46], utilizing existing cable-driven robots. This system optimizes the robot's trajectory by accounting for the robot's position, the positions of targets and obstacles, as well as the interactions between obstacles, cables, and surrounding scenarios. In a recent study, Navid Kayhani [47] and colleagues introduced a cost-effective, adaptable, lightweight visual-inertial localization methodology for unmanned aerial vehicles. This approach employs a basic inertial measurement unit sensor configuration and a single-lens camera utilizing AprilTags. Mei [48] proposed a dynamic RGB-D visual SLAM dense map construction method that employs pyramidal L-K with multi-view geometric constraints. This method not only enhances the accuracy of camera position estimation and facilitates the construction of dense maps following dynamic object processing but also improves system stability and environmental reconstruction accuracy. Liu et al. [49] introduced a vision-based robot-assisted component installation system designed to address the challenges of on-site assembly of prefabricated components in building construction. Following initial alignment using a crane, two robots collaborate to precisely adjust the position and orientation of the components. Feng et al. [50] developed a planar rebar tying robot and proposed an innovative two-stage recognition method that combines a depth-of-look camera and an industrial camera to capture image information of the tying area. A review of the literature indicates that research on vision systems for robots is predominantly focused on enhancing the flexibility and accuracy of environmental perception, particularly in practical applications within the domain of construction robots.

3.4. Analysis of Research Trends

The use of keyword emergence mapping in CiteSpace software serves as a powerful tool for analyzing research hotspots within a specific field. This method effectively highlights keywords that appear suddenly and increase rapidly in frequency over a defined period—referred to as emergent words. Such emergent words signify new research hotspots or directions that experience a swift rise in attention, providing timely insights into emerging trends and prominent topics in the field. When an emergent word appears recently with high intensity and prolonged duration, it indicates that the research topic is in a stage of rapid development and may continue to attract significant attention in the near future, with the potential to become a future research hotspot or key area of development.

By employing CiteSpace software for keyword emergence analysis, Chinese (Figure 10a) and English (Figure 10b) keyword emergence maps were successfully generated. These maps were utilized to conduct an in-depth analysis of the historical development of research within this field and to identify the research hotspots that have emerged over time. The analysis revealed that the research landscape can be categorized into two distinct stages, each distinguished by a specific set of research trends and priorities.

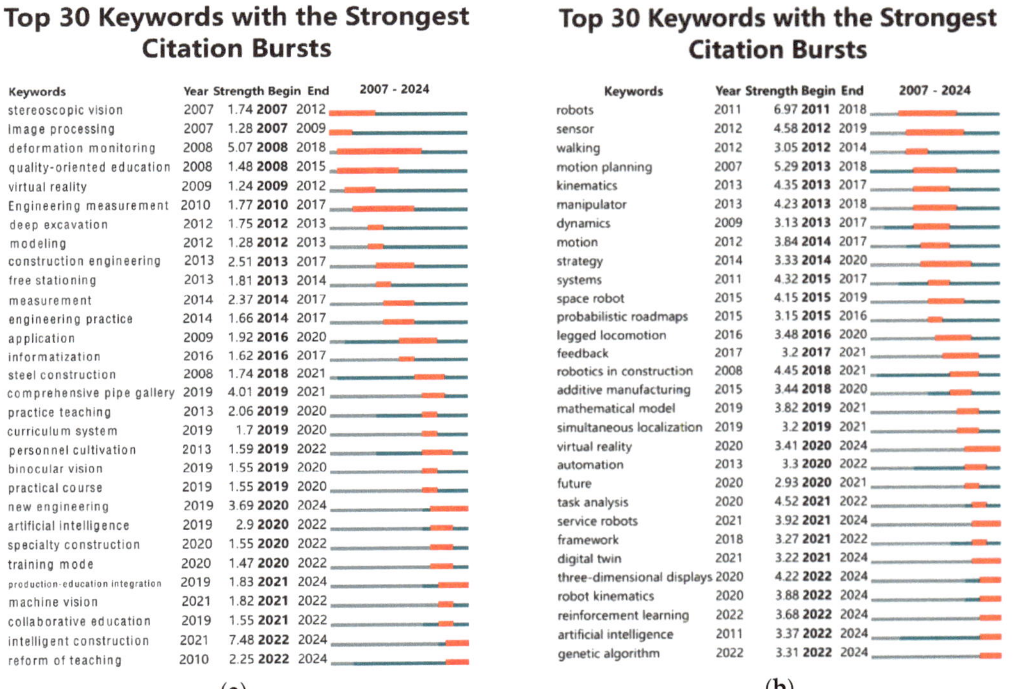

(a) (b)

Figure 10. (**a**) Emergence mapping of Keywords in the Chinese literature; (**b**) Emergence mapping of keywords in the foreign literature.

In summary, the period from 2007 to 2016 is characterized as the initial phase of field exploration. During this time, the predominant research areas included modeling, deformation monitoring, sensors, motion planning, manipulators, strategies, space robots, engineering practice, and information technology. The subsequent period from 2017 to 2024 marks the latest phase of frontier research. The focus during this phase encompasses but is not limited to, areas such as intelligent construction, talent training, new engineering, machine vision, automation, task analysis, three-dimensional displays, robot kinematics, and artificial intelligence.

1. Keyword terms exhibiting strong emergence intensity include "intelligent construction" and "robots". The application of construction robots is contingent upon the overarching framework of intelligent construction. Robots can perform tasks automatically, either by following a preset program or by operating autonomously based on principles developed through artificial intelligence technology. In the 1980s, the Shimizu Corporation of Japan pioneered the development of the world's first construction robot, the SSR-I refractory coating robot [51]. Since that time, countries have increasingly prioritized research and development in the field of construction robots.

The application of robotic technology is now considered a pivotal area of focus within the construction industry and represents a significant trend in current research.

2. The term "deformation monitoring" has the longest documented history of use, having been practiced for a decade. The primary objective of deformation monitoring is to ensure the safety and stability of engineering structures while providing a scientific foundation for the construction and maintenance of these structures. In the current context of smart construction, deformation monitoring has become more intelligent and automated [52]. Construction robots possess the potential to facilitate real-time monitoring and early warning systems for buildings, forming an integral part of the deformation monitoring process. To better align with evolving detection requirements, construction robots must undergo continuous enhancements to optimize the performance of their sensors, measurement equipment, and other associated components. Deformation monitoring constitutes a pivotal aspect of an intelligent construction system and serves as a critical driver for the intelligent transformation and modernization of the entire construction industry.
3. The genetic algorithm is a term that has recently emerged, rooted in the principles of genetics. It optimizes problem-solving gradually by simulating the natural processes of selection, crossover, and mutation. The introduction of genetic algorithms has significantly enhanced the performance and efficiency of construction robot systems. These algorithms are applied not only to optimize the robot's path planning but also to refine the design parameters of the robots.

When combined with the timeline chart, emergent words can clearly illustrate the emergence time of keywords, their developmental history, and their appearance at different stages. By analyzing these time-series data, the evolutionary trend of research hotspots can be observed, enabling an understanding of which topics are gradually gaining attention and which are losing prominence. This analysis also allows for predictions regarding the direction of future research. In conclusion, future research on construction robots can be divided into two main areas: firstly, the incorporation of genetic algorithms and other technologies to enhance the performance and practicality of construction robots in all aspects, and secondly, addressing potential issues that may arise from the practical application of construction robots within the industry. This is aimed at fostering the collaborative advancement of all aspects of the construction robot sector and promoting the broader application of these technologies.

4. Discussion

To facilitate a deeper and more efficient understanding of the foundational overview and research progress in the field of construction robots, this paper offers a visualization and analysis of 2947 documents related to construction robots sourced from the CNKI Chinese database and the WOS core database. This analysis was conducted using CiteSpace software, which enabled an examination of temporal distribution, authorship distribution, organizational distribution, keyword co-occurrence, and keyword clustering timelines. The findings are detailed below:

1. Over the past 18 years, the number of research articles pertaining to construction robots has steadily increased, with a particularly marked surge in publications commencing in 2018. Before 2013, the volume of publications in this field was nearly negligible. This trend indicates a gradual expansion in the scope of research and a growing sophistication in the content being explored. Moreover, it underscores the pivotal role of construction robots as essential tools for the intelligent upgrading of the construction industry, highlighting their significance as a prominent area of current research within this sector.

2. The analysis of institutions and authors reveals that cooperation within the field of construction robots is limited and fragmented, lacking a cohesive research system with a core group of authors. Consequently, enhancing communication networks among various research institutions and author teams is of considerable significance. Such improvements are essential to advancing the in-depth development and broad application of construction robot technology.
3. A comprehensive keyword analysis indicates that current research hotspots in the field of construction robots are predominantly centered on path planning, deformation monitoring, vision systems, image processing, and other technological innovations in research and development. Significant advancements have been achieved in these areas in recent years. However, attention has recently started to shift towards the practical applications of robots in building construction, encompassing considerations such as environmental constraints, safety, technological maturity, and human-robot collaboration. These issues related to practical applications are also subjects of ongoing research.

A comprehensive analysis of the literature on construction robots reveals certain limitations in the current research. Technically, challenges persist in multi-robot cooperative operations and human-robot collaboration. The interaction and integration across various disciplines are insufficient, hindering the formation of effective knowledge-sharing and collaborative innovation mechanisms. From the perspective of talent and collaboration, a notable scarcity of professionals exists in construction robots, complicating efforts to cultivate expertise. Additionally, the frequency and depth of cross-organizational and cross-team collaborations are limited, impeding the development of a comprehensive, multi-level network that integrates industry, academia, research, and application. Furthermore, practical applications of construction robots face constraints. Most current robots exhibit single-functionality and lack precise adaptability for specific construction tasks. Since construction operations often demand high levels of precision, flexibility, and multifunctional integration, this mismatch restricts the functionality and applicability of construction robots in real-world engineering practices.

5. Conclusions

In conclusion, significant advancements have been witnessed in the field of construction robotics in recent times. Current research focuses on several pivotal areas, including human-robot collaboration, path planning, task analysis, robot vision technology, and the reform of university education and professional training. Future developments at the technical level may involve the integration of various technologies into the study of construction robots. Ensuring the safe coordination between humans and machines remains a central focus, particularly concerning the application of construction robots and the management of human-machine interactions.

To promote the long-term development of the construction robots' field, future research should prioritize the acceleration of core technology research and development. This objective may be achieved by enhancing communication among various institutions and authors, as well as by improving personnel training and fostering team-building. Moreover, it would be advantageous to encourage colleges, universities, research institutes, and enterprises to establish interdisciplinary and cross-field scientific research and cooperation teams. Additionally, proactive exploration and application of novel materials and processes should be encouraged.

Author Contributions: Conceptualization, R.D., C.C. and Z.W.; methodology, R.D.; software, C.C. and Z.W.; validation, R.D. and C.C. formal analysis, C.C.; investigation, C.C.; resources, C.C; data curation, C.C. and Z.W.; writing—original draft preparation, C.C.; writing—review and editing, R.D.; visualization, C.C. and Z.W.; supervision, R.D.; project administration, R.D.; funding acquisition, R.D. All authors have read and agreed to the published version of the manuscript.

Funding: This research received no external funding.

Data Availability Statement: The original contributions presented in this study are included in the article. Further inquiries can be directed to the corresponding author(s).

Conflicts of Interest: The authors declare no conflicts of interest. The funders had no role in the design of the study; in the collection, analyses, or interpretation of data; in the writing of the manuscript; or in the decision to publish the results.

References

1. Tian, C.; Geng, J.; Wang, L. Supporting Role and Trend Forecast of Construction Industry in China's Economy. *Constr. Econ.* **2016**, *37*, 10–15.
2. Chen, T.; Zhang, L.; Ma, X. Exploring the Opportunities and Challenges of Cultivating Intelligent Construction Professionals in the Context of Artificial Intelligence. *Educ. Prog.* **2023**, *13*, 4563–4569.
3. Chen, C.; Li, X.; Qiu, Z.; Yao, W.; Zhu, H. Research Progress of Construction Robots. *J. Archit. Civ. Eng.* **2022**, *39*, 58–70.
4. Zang, X. Implementation program of "Robot +" application action issued. *Shanghai Educ.* **2023**, *07*, 5.
5. Chen, Y.; Chen, C.; Liu, Z.; Hu, Z.; Wang, X. The methodology function of Cite Space mapping knowledge domains. *Stud. Sci. Sci.* **2015**, *33*, 242–253.
6. Ye, J.; Fanyang, Y.; Wang, J.; Meng, S.; Tang, D. A Literature Review of Green Building Policies: Perspectives from Bibliometric Analysis. *Buildings* **2024**, *14*, 2607. [CrossRef]
7. Liu, D.; Zhang, H. Research Hotspots and Frontiers of Internet Public Opinion in Chinese Colleges and Universities. *Inf. Sci.* **2022**, *40*, 176–185.
8. Zheng, H.; Kan, X.; Chen, S.; Yuan, H. The Evolution Path and Hotspot Analysis of the Research on Aging Adaptation of Old Residential Community. *Constr. Econ.* **2023**, *44*, 72–80.
9. Bi, Q. Document Clustering Analysis on Semi-Supervised-Related Medical Literatures. Master's Thesis, Lanzhou University, Lanzhou, China, 2023.
10. Xie, C.; Feng, T.; Hu, L. A Bibliometric Analysis of the Outdoor Thermal Environment Based on CiteSpace. *Buildings* **2024**, *14*, 1384. [CrossRef]
11. En, X.; Liu, Y. Visualization analysis of high-speed railway research based on CiteSpace. *Transp. Policy* **2020**, *85*, 1–17.
12. Zhang, R.; Ren, R.; Yuan, B.; Luo, G.; Bi, L. Construction path optimization method of assembly building robot. *Comput. Eng. Des.* **2021**, *42*, 3516–3524.
13. Wang, F.; Li, T.; Liu, J.; Zhao, H. Research on Autonomous Path Planning and Obstacle Avoidance of Building Robot Based on BIM. *Comput. Eng. Appl.* **2020**, *56*, 224–230.
14. Muñoz-Bañón, M.; Velasco-sanchez, E.; Candelas, F. Openstreetmap-based autonomous navigation with lidar naive-valley-path obstacle avoidance. *IEEE Trans. Intell. Transp. Syst.* **2022**, *23*, 24428–24438. [CrossRef]
15. Yang, Z.; Zhuang, Y.; Chen, Y. Path planning of a building robot based on BIM and an improved RRT algorithm. *Exp. Technol. Manag.* **2024**, *41*, 31–42.
16. Gao, H.; Zhang, X.; Wen, J.; Yuan, J.; Fang, Y. Autonomous indoor exploration via polygon map construction and graph-based SLAM using directional endpoint features. *IEEE Trans. Autom. Sci. Eng.* **2018**, *16*, 1531–1542. [CrossRef]
17. Huang, P.; Zeng, L.; Chen, X. Edge robotics: Edge-computing-accelerated multirobot simultaneous localization and map. *IEEE Internet Things J.* **2022**, *9*, 14087–14102. [CrossRef]
18. Liu, P.; Bi, Y.; Shi, J. Semantic-Assisted LIDAR Tightly Coupled SLAM for dynamic environments. *IEEE Access* **2024**, *12*, 34042–34053. [CrossRef]
19. Li, S.; He, R.; Guan, H. A 3D LiDAR-Inertial Tightly-Coupled SLAM for Mobile Robots on Indoor Environment. *IEEE Access* **2024**, *12*, 29596–29606. [CrossRef]
20. Zhou, E.; Liu, X.; Qing, Z. Research and Implementation of Monitoring and Early Warning System of Deep Foundation Pit. *Chin. J. Undergr. Space Eng.* **2013**, *9*, 204–210.
21. Zhang, S.; Huang, S.; Yang, X. Research on real-time gross error detection method based on geo robot. *Bull. Surv. Mapp.* **2021**, 153–155+161. [CrossRef]
22. Zheng, H.; Fu, K. Design and Implementation of Foundation Pit Deformation Automatic Monitoring Software Based on Android. *Geospat. Inf.* **2022**, *20*, 120–123.

23. Zhang, Z.; Pang, K.; Xu, L.; Zou, Y.; Yang, J.; Wang, C. The bond properties between UHPC and stone under different interface treatment methods. *Constr. Build. Mater.* **2023**, *365*, 130092. [CrossRef]
24. Ji, Y.; Shi, K.; Dong, J. Research on Accuracy Improvement Method of Horizontal Deformation Intelligent Monitoring for Super High-rise Buildings. *Constr. Technol.* **2023**, *52*, 20–24+79.
25. Yuan, F.; Chai, H.; Zhang, X. Exploration of small-batch customization production mode of timber architecture based architectural robot. *Build. Struct.* **2018**, *48*, 39–43+55.
26. Yuan, F.; Chai, H. Digital construction technology system of assembled buildings for mass customization-Taking the innovation of assembled wooden buildings as an example. *New Build.* **2022**, *04*, 9–14.
27. Duan, H.; Zhang, F.; Chen, G. The Practice and Development of Intelligent Construction Driven by Construction Robots. *Constr. Econ.* **2022**, *43*, 5–12.
28. Zhao, P.; Zhang, M.; Zhang, W. Application of BIM parametric design in complex buildings with different shapes. *Build. Struct.* **2023**, *53*, 2060–2062.
29. Gao, Y.; Shu, J.; Yu, K. Research on robotic construction of light weight structures using BIM visual programming. *J. Build. Struct.* **2022**, *43*, 296–304.
30. Wang, P.; Xiao, J.; Xiao, X. Application and Prospect of Digital Technology in Building Construction. *J. Tongji Univ. (Nat. Sci.)* **2024**, *52*, 1068–1078.
31. Li, D.; Jiang, B.; Gu, W. The transformation and talent training mode of architectural engineering technology for intelligent construction. *Archit. Sci.* **2024**, *40*, 196.
32. Zuo, Z.; Huang, Y.; Zang, L. 3D printing technology and feasibility analysis of its application in super high-rise buildings. *Build. Struct.* **2022**, *52*, 95–101.
33. Liu, T. Tsinghua University State Key Laboratory of Hydro science and Engineering. 3D printing of large filled construction projects. *J. Tsinghua Univ. (Sci. Technol.)* **2022**, *62*, 1281–1291.
34. Hong, T.; Gao, X. Improve Path Generation Method of 3D Printing Robot under Q Learning. *Comput. Simul.* **2023**, *40*, 417–421.
35. Zeng, N.; Ma, Z.; Song, L. A porosity detection method for 3D printed concrete interface with self-attention mechanism. *Chin. J. Eng. Des.* **2024**, 1–10. Available online: https://kns.cnki.net/kcms/detail/33.1288.TH.20240527.1349.006.html (accessed on 12 December 2024).
36. Huang, L.; Zhu, Z.; Zou, Z. To imitate or not to imitate: Boosting reinforcement learning-based construction robotic control for long-horizon tasks using virtual demonstrations. *Autom. Constr.* **2023**, *146*, 104691. [CrossRef]
37. Zha, A.; Dai, T.; Xu, G. Deep reinforcement learning for real-time assembly planning in robot-based prefabricated construction. *IEEE Trans. Autom. Sci. Eng.* **2023**, *20*, 1515–1526. [CrossRef]
38. Liang, C.J.; Wang, X.; Kamat, V. Human–robot collaboration in construction: Classification and research trends. *J. Constr. Eng. Manag.* **2021**, *147*, 03121006. [CrossRef]
39. Wang, X.; Liang, C.; Menassa, C. Interactive and immersive process-level digital twin for collaborative human–robot construction work. *J. Comput. Civ. Eng.* **2021**, *35*, 04021023. [CrossRef]
40. Yang, X.; Amtsberg, F.; Sedlmair, M. Challenges and potential for human–robot collaboration in timber prefabrication. *Autom. Constr.* **2024**, *160*, 105333. [CrossRef]
41. Liu, Y.; Habibnezhad, M.; Jebelli, H. Brainwave-driven human-robot collaboration in construction. *Autom. Constr.* **2021**, *124*, 103556. [CrossRef]
42. Bakdi, A.; Hentout, A.; Boutami, H. Optimal path planning and execution for mobile robots using genetic algorithm and adaptive fuzzy-logic control. *Robot. Auton. Syst.* **2017**, *89*, 95–109. [CrossRef]
43. De Soto, B.G.; Agusti-juan, G.; Hunhevicz, J. Productivity of digital fabrication in construction: Cost and time analysis of a robotically built wall. *Autom. Constr.* **2018**, *92*, 297–311. [CrossRef]
44. Hartmann, V.; Orthey, A.; Driess, D. Long-horizon multi-robot rearrangement planning for construction assembly. *IEEE Trans. Robot.* **2022**, *39*, 239–252. [CrossRef]
45. Elmakis, O.; Shaked, T.; Degani, A. Vision-based uav-ugv collaboration for autonomous construction site preparation. *IEEE Access* **2022**, *10*, 51209–51220. [CrossRef]
46. Pinto, A.; Moreira, E.; Lima, J. A cable-driven robot for architectural constructions: A visual-guided approach for motion control and path-planning. *Auton. Robot.* **2017**, *41*, 1487–1499. [CrossRef]
47. Kavhani, N.; Zhao, W.; Mccabe, B. Tag-based visual-inertial localization of unmanned aerial vehicles in indoor construction environments using an on-manifold extended Kalman filter. *Autom. Constr.* **2022**, *135*, 104112.
48. Mei, J.; Zuo, T.; Song, D. Highly dynamic visual SLAM dense map construction based on indoor environments. *IEEE Access* **2024**, *12*, 38717–38731. [CrossRef]
49. Liu, C.; Wu, J.; Jiang, X. Automatic assembly of prefabricated components based on vision-guided robot. *Autom. Constr.* **2024**, *162*, 105385. [CrossRef]

50. Feng, R.; Jia, Y.; Wang, T. Research on the system design and target recognition method of the rebar-tying robot. *Buildings* **2024**, *14*, 838. [CrossRef]
51. Zhou, L.; Liu, S.; Yang, J. Comparative study on the driving path of competitiveness of technical standards in artificial intelligence industry at home and abroad. *Sci. Technol. Prog. Countermeas.* **2024**, 1–11. Available online: https://kns.cnki.net/kcms/detail/42.1224.G3.20241017.1046.004.html (accessed on 12 December 2024).
52. Zhang, W.; Song, Y.; Zhang, N. Application of intelligent construction in safe construction management of building projects. *Eng. Constr. Des.* **2023**, *13*, 263–265. [CrossRef]

Disclaimer/Publisher's Note: The statements, opinions and data contained in all publications are solely those of the individual author(s) and contributor(s) and not of MDPI and/or the editor(s). MDPI and/or the editor(s) disclaim responsibility for any injury to people or property resulting from any ideas, methods, instructions or products referred to in the content.

buildings

MDPI

Article

A Numerical and Theoretical Investigation of the Flexural Behavior of Steel–Ultra-High-Performance Concrete Composite Slabs

Changshui Li [1], Boyi Zhao [2], Dawei Hao [1], Xiaolong Gao [2,*], Hao Bian [1] and Xuanzheng Zhang [2]

[1] China Construction Seventh Engineering Division, Co., Ltd., Zhengzhou 450004, China; lichangshui@cscec.com (C.L.); haodw@cscec.com (D.H.); bianhao7@cscec.com (H.B.)
[2] College of Civil Engineering, Henan University of Technology, Zhengzhou 450001, China; iszhaoby@stu.haut.edu.cn (B.Z.); 2023930795@stu.haut.edu.cn (X.Z.)
* Correspondence: gxl0709@haut.edu.cn

Abstract: The steel–Ultra-High-Performance concrete (UHPC) composite slab is a new type of structure made of steel and UHPC connected by pegs, and its flexural mechanical properties and related design methods need to be further investigated. Firstly, a detailed numerical model of the steel UHPC composite slab is established and validated based on previous flexural behavior experimental research. Secondly, the flexural failure mechanisms of steel–UHPC composite slabs are clarified through finite element analysis. Under positive bending moments, when the degree of shear connection is lower than 100%, the ultimate load capacity of the composite slabs is determined by the shear capacity of the pegs. On the contrary, there are no significant changes in the load-carrying capacity of all the specimens, but there is a slight increase in stiffness. Under negative bending moments, the load-bearing capacity, stiffness, and crack resistance of the composite slab are improved as the degree of shear connection and reinforcement ratio increase. Finally, the method used to calculate the flexural capacity of steel–UHPC composite plates under positive and negative bending moments with high accuracy is proposed based on the analytical results. This paper provides a theoretical basis for the design of flexural performance of steel–UHPC composite slab.

Keywords: ultra-high-performance concrete (UHPC); steel–UHPC composite slab; flexural behavior; numerical analysis; calculation method

1. Introduction

Ultra-High-Performance Concrete (UHPC) is a novel eco-friendly cement-based composite material designed based on the principles of optimal particle packing, a water-to-binder ratio of less than 0.25, and fiber reinforcement. It is characterized by its ultra-high mechanical strength, enhanced toughness, superior durability, and excellent workability [1–4]. The interfacial bond strength between the UHPC matrix and steel fibers is significantly enhanced, allowing the UHPC matrix to maintain high tensile strength even after cracking. This results in the realization of metal-like tensile behavior and strain-hardening characteristics, which significantly improve the toughness and ductility of UHPC materials [5,6]. Compared to normal concrete (NC) and conventional high-performance concrete (HPC), the durability of UHPC, including resistance to chloride ion penetration and carbonation, is significantly enhanced by its high densification, which enables it to fully adapt to the demanding engineering requirements of harsh environments, such as corrosive and marine conditions [7,8]. Meanwhile, considering its excellent mechanical properties and ease of construction, UHPC is effectively utilized in the repair and strengthening of existing damaged or aging structures with good interfacial connection properties [9,10]. The mechanical performance and service life of the original structures are significantly enhanced [11–13]. More importantly, UHPC can be used in special scenarios such as large-span bridges, ultra-high-rise buildings and large-span heavy-duty industrial plants, etc., to minimize the amount of material under the condition of meeting the structural design requirements due to its high strength [14,15]. Traditional reinforced concrete slabs and steel–normal concrete slabs are relatively heavy. In composite cable-stayed bridges, the thickness of reinforced concrete slabs is typically greater than 250 mm, with their self-weight accounting for more than 70% of the weight of the bridge deck system [16]. Replacing the concrete layer in steel–concrete composite structures with UHPC can significantly improve the mechanical properties of the structure [17,18]. The new structure with thinner thickness and smaller mass can reduce structural deadweight and realize greater economic benefits. Due to the excellent tensile properties of UHPC, steel–UHPC composite slabs can effectively solve the problem of easy cracking in the negative moment zone of steel–concrete combined girder bridges and further improve the use of steel–concrete combined girder bridges in terms of span and range of application [19].

The application of steel–UHPC composite structures to bridge decks was first proposed by Shao et al. in 2012. Based on practical engineering needs, they conducted full-scale model static load tests. The results demonstrated that the novel steel–concrete composite bridge deck effectively reduced stresses in the deck structure and enhanced its stiffness, fundamentally mitigating the risk of fatigue cracking in the deck and extending the fatigue life of steel bridge decks [20]. In recent years, studies on steel–UHPC composite structures have focused on fatigue testing of steel–UHPC composite slabs [21–23], validating the shear behavior of peg connections [24–26], and optimizing the treatment of wet joints [27–29]. The reliability of the pegs is the basis for the perfect synergy between the steel–UHPC composite slabs, while the diameter size of the pegs and the separation spacing will have an impact on the working condition of the composite plates when subjected to bending loads. Hu et al. [30] validated the finite element model for the shear performance of pegs in steel–UHPC composite slabs. They discussed the effects of welded reinforcements, peg diameter, and spacing on the failure modes, shear capacity, and load-slip behavior of the peg connectors. The results indicated that when the peg diameters are 13 mm, 16 mm, and 19 mm, the group peg effect should be considered, and a shear-slip model for the pegs was established. Lu et al. [31] found that appropriately increasing the embedment depth and diameter of pegs can enhance their ultimate tensile capacity. However, it should be noted that changes in the failure mode may result in no increase or even a reduction in the load-carrying capacity. The bonded interface at the wet joint of steel–UHPC composite slabs is a weak point in the cooperative performance, affecting the structural safety. Ren et al. [28] designed seven steel–UHPC composite slabs and conducted four-point bending tests. The test parameters included reinforcement type, lapping details, joint width, and filling materials. The experimental results showed that all specimens experienced shear failure, with critical shear cracks forming within the shear span. Finally, a cohesive-friction hybrid model was developed. Concrete is known for its excellent compressive strength but exhibits relatively low tensile strength, which often results in the presence of cracks during service conditions [32]. Luo et al. [33] studied the flexural cracking behavior of steel–UHPC lightweight bridge deck systems by designing 40 steel–UHPC composite slabs and 8 steel–UHPC composite beams for transverse and longitudinal cracking tests. The results indicated that increasing the reinforcement ratio, reducing the UHPC cover thickness, decreasing the spacing of peg shear connectors, and thickening the UHPC layer effectively mitigated the initiation and propagation of cracks in the UHPC layer. Building on this, Wang et al. [34] proposed a theoretical formula for calculating the maximum crack width on the bottom surface of the UHPC layer through a series of experimental studies. This research provides theoretical guidance for the development of steel–UHPC composite structures. Zou et al. [35] proposed a hollow steel–UHPC composite bridge deck system without shear connectors, composed of hollow steel tubes and UHPC, to reduce the self-weight of the deck and enhance its crack resistance and durability. The results demonstrated that the embedded steel tubes and external UHPC achieved excellent composite performance. Additionally, a theoretical formula for predicting the flexural capacity of hollow steel–UHPC composite bridge decks was developed. Li et al. [36] proposed a steel–UHPC composite bridge deck system without longitudinal ribs. Compared to the orthotropic steel–UHPC composite bridge deck system, increasing the thickness of the UHPC layer alone was insufficient to effectively compensate for the stiffness reduction caused by the removal of U-shaped ribs. By reducing the longitudinal support spacing of the bridge deck system, steel consumption could be reduced by 28% and weld length by 80%, significantly lowering the risk of fatigue cracking in the deck. Li et al. [37] conducted flexural experiments on steel–UHPC composite slabs with PBL shear connectors. The results indicated that PBL connectors effectively ensured the integrated force transfer between the steel plate and the UHPC slab, providing the composite slab with excellent ductility under negative bending moments. Xiao et al. [38] investigated the influence of the number and type of shear connectors on the flexural performance of full-scale steel–concrete composite slabs. Their findings revealed that steel–UHPC composite slabs exhibited flexural failure, whereas steel–concrete composite slabs experienced punching shear failure. Among the composite slabs, specimens with sufficient perforated bonding rib shear connectors demonstrated the best flexural performance. In addition, machine learning and artificial neural network approaches are used to construct models of the intrinsic relationships of materials and to study the mechanical properties of composite structures [39–43]. Currently, domestic and international research on steel–UHPC composite structures is primarily conducted within the realm of bridge engineering. Much of this research focuses on meeting the design requirements for steel–UHPC lightweight composite bridge deck engineering, leading to a narrow scope that lacks broad representativeness. Studies on the effects of loading conditions, the degree of shear connection of the pegs and the reinforcement rate on the mechanical properties of steel–UHPC composite panels are still insufficient and lack depth.

In this paper, numerical simulation and theoretical analysis are used to investigate the bending force behavior of steel–UHPC composite slabs. Based on the determined UHPC principal model and damage evolution equations, the finite element numerical model of this paper is validated based on the bending performance tests of eight steel–UHPC composite slabs conducted by Gao et al. [44,45]. On this basis, the effects of parameters such as bending moments, shear connection degree, and reinforcement ratio on the flexural performance of composite slabs are investigated using numerical analysis methods. The failure mechanisms of steel–UHPC composite slabs under positive and negative bending moments are further clarified. Finally, based on the numerical analysis results, an out-of-plane load-bearing capacity calculation method for steel–UHPC composite slabs under positive and negative bending moments is established and compared with the experimental results.

2. Establishment of 3D FE Model

2.1. Cell Type and Mesh

In this paper, the finite element simulation of the steel–UHPC composite slab is carried out in ABAQUS 2020/Standard. The steel–UHPC composite slab is composed of UHPC slab, pegs, a steel plate, and steel reinforcement. Among them, the UHPC slab, pegs, and steel plate are simulated using the solid element C3D8R in ABAQUS. The reinforcement is simulated using truss element T3D2 and connected to the UHPC slab through embedding. The contact relationship between the pegs and the steel plate is the way the pegs are bound to the steel plate, while the contact relationship between the steel reinforcement and the UHPC slab is achieved through the steel reinforcement mesh built into the UHPC. In order to reduce the computational time, a 1/4 finite element model of the steel–UHPC composite slab is built for computational analysis due to its inherent biaxial symmetry. The finite element analysis results are significantly affected by the cell mesh size in the model. In order to ensure an accurate result, the meshing rules of each model in this paper are determined after checking the calculation. The finite element model meshing schematic is shown in Figure 1. The mesh size of the pegs is 4 mm, and the mesh size at the location of the peg holes in the UHPC slab is consistent with the outer surface of the pegs, increasing from 4 to 10 mm with the pegs as the center, in which the mesh size of the UHPC between the two pegs is 10 mm. The grid of the contact surface of the steel plate and the UHPC slab is consistent with the grid of the UHPC, with the grid size ranging from 4 mm to 10 mm, and the thickness of the steel plate is divided into three layers.

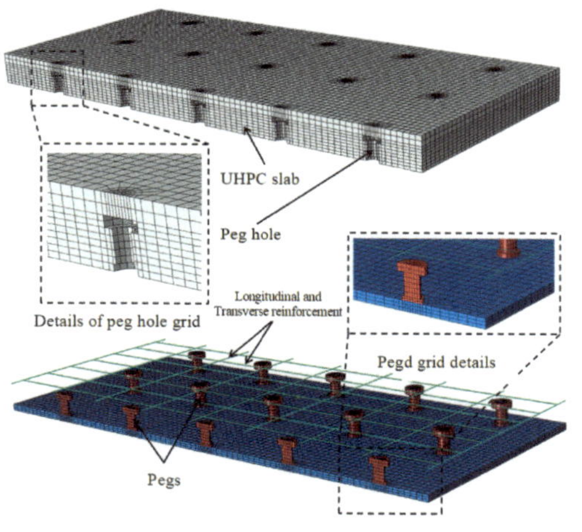

Figure 1. Schemes follow the same formatting.

2.2. Boundary Constraints and Loading Conditions

In order to avoid stress concentration, rigid pads with the same width as the width of the support and loaded beam in the test are set at the loading point and the support location, and the loads are applied to the upper rigid pads in the form of vertical displacements through the reference point. The hard contact mode in the normal direction and the penalty friction with a 0.25 friction coefficient in the tangential direction are used to define the contact relationship between the rigid pads and the composite slab at the bearing position [46,47]. Displacement constraints in three translational directions are imposed on the lower surface of the rigid pad at the support location. The finite element model contains two symmetrical surfaces, and the translational degrees of freedom in the normal direction of the symmetry surfaces and the rotational degrees of freedom in the other two directions are constrained.

The contact behavior of peg–UHPC and slab–UHPC is defined by means of Surface to Surface in ABAQUS. The surfaces of the steel plate and the pegs are defined as master surfaces, and the surface of the UHPC is defined as a slave surface. The normal direction of the contact surface is defined as "hard contact", the tangential direction is defined by "penalty function", and the friction coefficient is taken as 0.4 [48]. The final finite element model is shown in Figure 2.

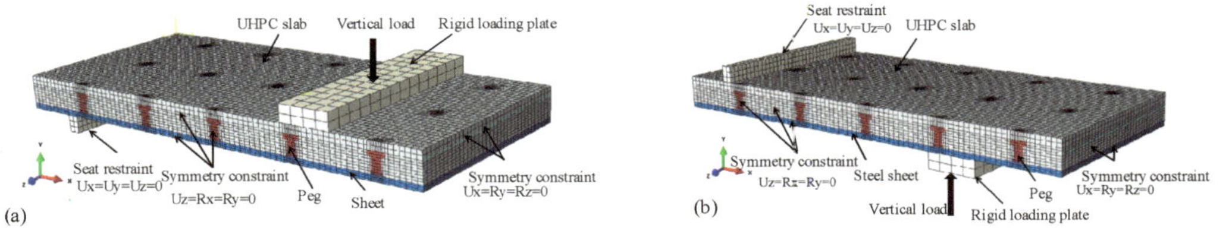

Figure 2. Finite element model of steel–UHPC composite slab: (**a**) Positive bending moment; (**b**) negative bending moments.

2.3. Definition of Material Properties

2.3.1. UHPC

The concrete damage plastic (CDP) model in ABAQUS has good applicability and convergence and is widely used in various applications such as unidirectional loading, cyclic loading and dynamic loading. The CDP model is used to define the constitutive relationships of UHPC in this study, in which the required axial tensile and axial compressive stress–strain relationships and damage variable–inelastic strain relationships are shown in Table 1, and the expressions of the UHPC axial tensile, and axial compressive constitutive relationships (σ_t) are shown in Equation (1) and the expression for the stress–seam width relationship corresponding to the softened section of UHPC is shown in Equation (2). The axial compressive stress–strain curves of UHPC and the corresponding damage variables and the axial tensile stress–seam width curves and the corresponding damage variables are shown in Figures 3 and 4, respectively [49].

Table 1. Mechanical properties of UHPC materials.

Tensile Strength f_{te} (MPa)	Ultimate Tensile Strength f_{tu} (MPa)	Ultimate Tensile Strain ε_{ut} (10^{-6})	Compressive Strength f_{cu} (MPa)	Axial Compressive Strength f_c (MPa)
7.83	9.42	2447	144.1	136

Note: The abbreviation f_{te} refers to the elastic tensile strength, f_{tu} refers to the ultimate tensile strength, ε_{ut} refers to the ultimate tensile strain, f_{cu} refers to the cube compressive strength, f_c refers to axial compressive strength, and the axial compressive strength is 0.95 times the compressive strength [50].

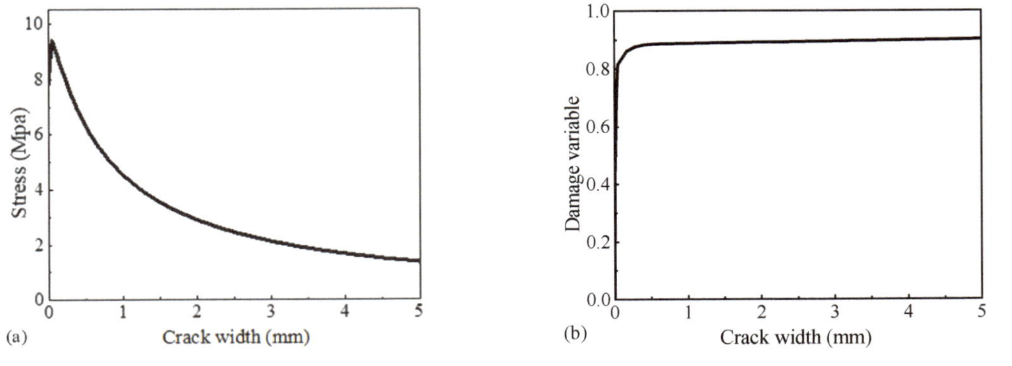

Figure 3. Axial tensile stress–damage relationship of UHPC: (**a**) Axial tensile stress–crack width curve; (**b**) axial tensile damage variables.

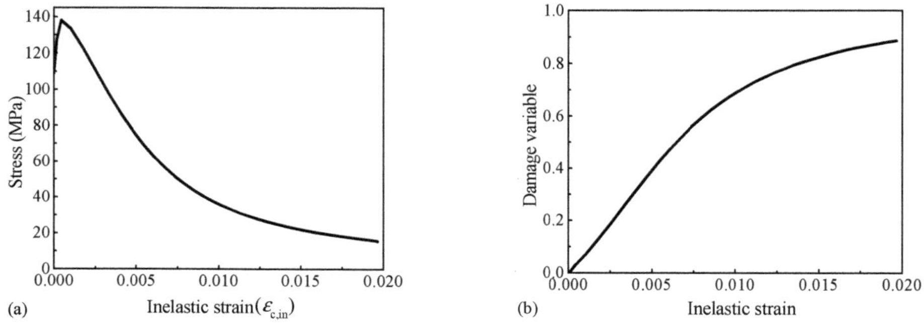

Figure 4. Axial compressive tensile stress–damage relationship of UHPC: (**a**) Axial compressive stress–strain curve; (**b**) axial compressive damage variables.

$$\sigma_t(w) = f_{te} + \frac{f_{tu} - f_{te}}{w_0} w \quad (0 < w < w_0) \tag{1}$$

where w refers to the crack width and w_0 refers to the ultimate tensile strength corresponding to a crack width of 0.05 mm.

$$\sigma_t(w) = f_{tu} \frac{1}{p_1(\frac{w}{w_0} - 1)^{p_2} + \frac{w}{w_0}} \quad (w_0 < w < w_u) \tag{2}$$

where p_1 and p_2 are parameters fitted based on the experimental results, −0.982 and 0.996, respectively.

2.3.2. Constitutive Relationship of the Pegs

The triple-fold model shown in Figure 5a is used to define the stress–strain relationship of the pegs [51–53]. According to the experimental test results, the measured tensile bearing capacity of the pegs under the damage state is 92.5 kN (corresponding to an engineering stress of 460.3 MPa), and the yield strength f_y of the pegs is calculated to be 400 Mpa, while the ultimate strength f_u is 460.3 Mpa, the yield strain ε_y is 0.199%, and the ultimate strain ε_u should be taken as 12% [54]. The basic mechanical properties of the peg are shown in Table 2. It should be noted that the measured ultimate strength of the pegs is the engineering stress, while the stress–strain relationship of the material entered in ABAQUS refers to the real stress–strain relationship curve (Figure 5b). It is necessary to convert the engineering stress–strain curve determined by the above parameters to the real stress–strain curve, and the conversion relationship is shown in Equations (3) and (4).

$$\sigma = \sigma_{nom}(1 + \varepsilon_{nom}) \tag{3}$$

$$\varepsilon = \ln(1 + \varepsilon_{nom}) \tag{4}$$

where σ and σ_{nom} refer to the true and engineering stresses, respectively; ε and ε_{nom} refer to the true and engineering strains, respectively.

Table 2. Material mechanical properties of steel and pegs.

Material Type	f_y (MPa)	f_u (MPa)	E_s (GPa)
Peg	400	430	201
Steel	357	462	201
Steel reinforcement	406 Mpa	539	207

Note: The abbreviation f_y refers to the yield strength, f_u refers to the ultimate strength, and E_s refers to the modulus of elasticity.

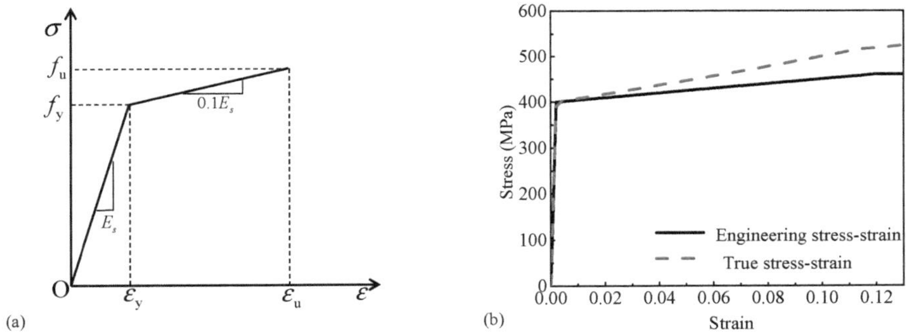

Figure 5. Definition of pegs' stress-strain relationship: (**a**) Trifold model; (**b**) stress-strain curves.

The flexible damage model in the ductile metal damage model is used to describe the damage process of the pegs, and the parameters include equivalent plastic strain, stress triaxiality, and strain rate. According to the Dominic Kruszewski et al. [51], the relationship between equivalent plastic strain and stress triaxiality (ε_{pl}) is shown in Equation (5).

$$\varepsilon_{pl} = \varepsilon_u e^{-\beta(\theta - 1/3)} \quad (5)$$

where ε_u is the ultimate strain; β is the characteristic parameter of the material, which is taken as 1.5; and θ is the stress triaxiality, which ranges from −0.33 to 2.0.

The plastic displacement approach is used to define the damage evolution process of the pegs in this study, where the relationship between the damage variables (D_i) and the plastic displacements (u_i^{pl}) can be determined by calculations based on Equations (6) and (7) [51]. Due to the lack of bolster stress-strain measured data, the definition of the bolster damage evolution process in this paper is determined according to [51], and the corresponding flexible damage model and its damage evolution process are shown in Figure 6.

$$D_i = \alpha_D [1 - \frac{\sigma_i}{f_u}] \quad (6)$$

$$u_i^{pl} = u_f^{pl} \frac{\varepsilon_i^{pl} - \varepsilon_u}{\varepsilon_f^{pl} - \varepsilon_u} \quad (7)$$

where α_D is the correction factor, σ_i is the stress in the descending section, u_f^{pl} is the fracture displacement taken as 1.8 mm, and ε_f^{pl} is the plastic strain at the time of peg fracture.

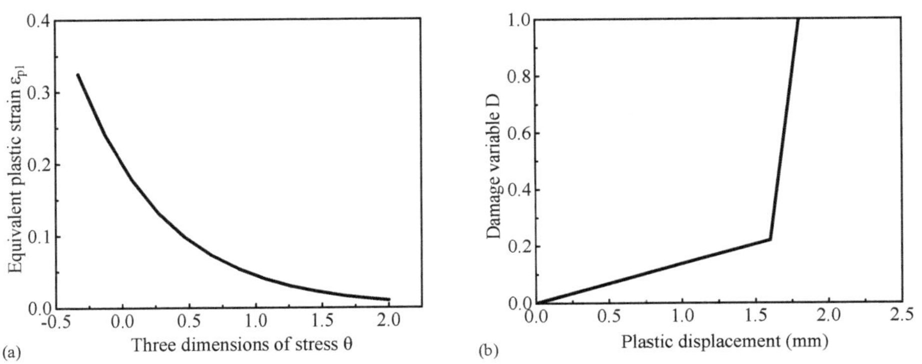

Figure 6. Definition of ductile damage model of pegs: (**a**) Flexible damage model; (**b**) stress-strain curves.

2.3.3. Steel and Steel Reinforcement Constitutive Relation

The isotropic model is used to simulate the steel and steel reinforcement, and the yield criterion adopts the default Mises yield condition in ABAQUS. The stress–strain relationship between the steel plate and reinforcement is defined using a bilinear model as shown in Figure 7, and the basic mechanical properties of the materials are shown in Table 2.

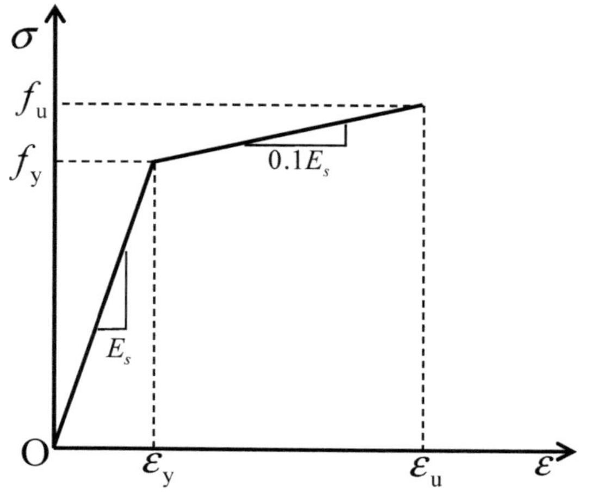

Figure 7. Bilinear model.

3. Validation of the Finite Element Model

In this section, the results of the previous experimental studies are compared with and used to correct the results of the finite element analysis in order to verify the accuracy and validity of the finite element model.

3.1. Experimental Generalization

According to the author's previous research [44], the basic design parameters and constructional diagrams of the tests are shown in Table 3 and Figure 8, respectively, and the test loading device and the main test results are shown in Figure 9 and Table 4.

Table 3. Basic design parameters of steel–UHPC composite slab specimens.

Specimen Number	s_a (mm)	s_b (mm)	s_c (mm)	s_d (mm)	r_a (mm)	r_b (mm)	ρ (%)	Loading Method
P-SUCS-3.1%-150	150	150	125	50	50	100	3.1	P
P-SUCS-3.1%-200	200	200	100	50	50	100	3.1	P
P-SUCS-3.1%-250	250	250	175	100	50	100	3.1	P
N-SUCS-3.1%-150	150	150	125	50	50	100	3.1	N
N-SUCS-3.1%-200	200	200	100	50	50	100	3.1	N
N-SUCS-3.1%-250	250	250	175	100	50	100	3.1	N
N-SUCS-2.0%-200	200	200	100	50	80	100	2.0	N
N-SUCS-3.8%-200	200	200	100	50	40	100	3.8	N

Note: s_a and s_b refer to the longitudinal and transverse spacing of the pegs. S_c and s_d refer to the longitudinal and transverse edge spacing of the pegs. R_a and r_b refer to the spacing of the transverse and longitudinal steel reinforcement. *P* refers to the steel reinforcement ratio of longitudinal steel reinforcement in UHPC. The specimens are named P and N for positive and negative moment loading, respectively. SUCS is the acronym used for the steel–UHPC composite slab and the numbers at the end represent the spacing of the pegs.

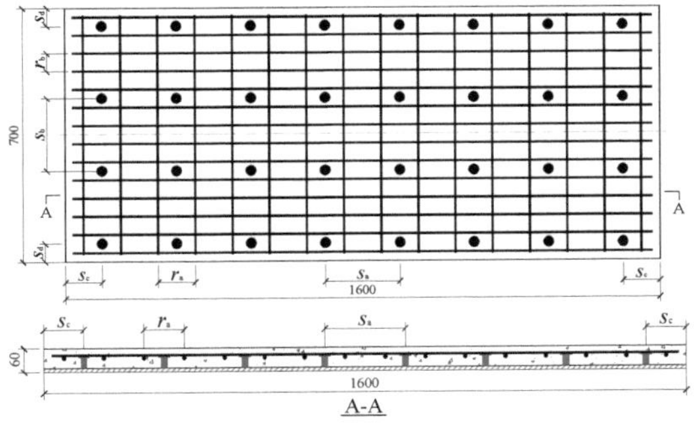

Figure 8. Dimensions of composite slab specimens.

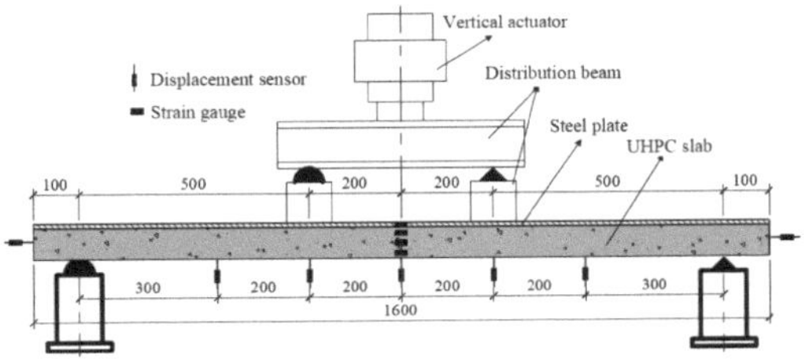

Figure 9. Schematic diagram of loading device.

Table 4. Test results of steel–UHPC composite slab specimens.

Specimen Number	P_{cr} (kN)	δ_{cr} (mm)	P_y (kN)	δ_y (mm)	$P_{u,Test}$ (kN)	δ_u (mm)	$\varepsilon_{y\text{-}s}$ (10^{-6})	$\varepsilon_{u\text{-}s}$ (10^{-6})
P-SUCS-150-3.1%	235.1	15.7	271.7	20.82	320.2	27.94	3097	4606
P-SUCS-200-3.1%	170.1	11.23	261.8	26.37	285.7	34.19	1772	3373
P-SUCS-250-3.1%	94.68	8.42	-	-	174.9	27.58	1542	3915
N-SUCS-3.1%-150	29.3	1.96	113.0	23.60	119.3	42.65	-	-
N-SUCS-3.1%-200	24.4	1.24	100.1	25.49	112.6	64.27	-	-
N-SUCS-3.1%-250	20.1	0.95	99.2	30.91	110.5	62.69	-	-
N-SUCS-2.0%-200	23.6	1.22	81.1	22.84	86.8	48.26	-	-
N-SUCS-3.8%-200	23.4	1.27	114.4	28.66	128.6	57.20	-	-

Note: P_{cr} and δ_{cr} represent the load and deflection when the crack width reaches 0.05 mm, P_y and δ_y represent the load and deflection when the steel plate yields, $P_{u,Test}$ and δ_u represent the ultimate load and deflection of the specimen, and $\varepsilon_{y\text{-}s}$ and $\varepsilon_{u\text{-}s}$ represent the maximum tensile strain of the steel plate and the maximum compressive strain of the UHPC, respectively, at the ultimate load.

3.2. Numerical Analysis of Steel–UHPC Composite Slabs Under Positive Bending Moments

3.2.1. Load–Deflection Curve

Under the action of positive bending moment, the load–deflection curves obtained from numerical analysis of steel–UHPC composite slabs are compared with the test results [44] as shown in Figure 10a–c. As shown in the figures, the load–deflection curves obtained via the numerical analysis can basically match the test results in the rising section. Table 5 lists the finite element calculation results and test results of the ultimate load capacity of the composite slab. The degree of shear connection between spigot and UHPC can be calculated according to the method of calculating the degree of shear connection of steel-mixed composite beams (ψ) [55], and the calculation formula is shown in Equation (8).

$$\psi = nP_{peg}/F_{min} \tag{8}$$

where n is the number of pegs in the shear span section of the steel–UHPC composite slab specimen. P_{peg} is the shear ultimate bearing capacity of the pegs. F_{min} is the smaller value of the tensile or compressive bearing capacity of the steel plate or UHPC.

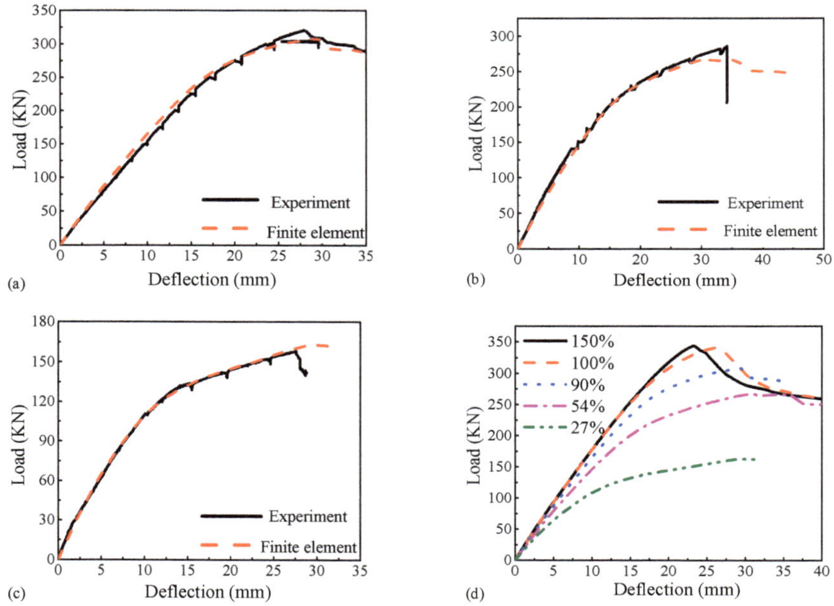

Figure 10. Finite element calculation results of steel–UHPC composite slabs: (**a**) Specimen P-SUCS-3.1%-150; (**b**) specimen P-SUCS-3.1%-200; (**c**) specimen P-SUCS-3.1%-250; (**d**) different shear connection degree.

Table 5. Ultimate bearing capacity of composite slabs under positive bending moments.

Specimen Number	Shear Connection Degree	$P_{u,Test}$ (kN)	$P_{u,FEA}$ (kN)	$P_{u,Test}/P_{u,FEA}$
P-SUCS-3.1%-150	90%	320.2	310.7	1.03
P-SUCS-3.1%-200	54%	285.7	267.1	1.07
P-SUCS-3.1%-250	27%	159.1	162.9	0.96
P-SUCS-3.1%-150-FEA1	100%	-	340.5	-
P-SUCS-3.1%-150-FEA2	150%	-	343.5	-

Note: $P_{u,FEA}$ refers to the ultimate bearing capacity value of the finite element analysis results.

The shear capacity of the peg connections (P_u) in UHPC is determined according to Equation (9), which was established in a previous study [45], and the calculation formula is shown in Equation (9).

$$P_u = A_{sc}f + \eta f_c d_{wc} l_{wc} \tag{9}$$

where A_{sc} refers to the cross-sectional area of the peg, f refers to the design value of the tensile strength of the peg, and d_{wc} and l_{wc} refer to the diameters and heights of the peg weld, respectively. η refers to the correction factor for the size of the weld, which is taken as 1.5 [38], whereas Luo suggested a correction factor of 2.5. After attempting the calculations, 2.0 was selected as the correction factor in this paper and was used to accurately assess the shear capacity of the peg connections [33].

As shown in Table 5, it can be seen that the average ratio between the test results and the finite element calculation results is 1.03, which is indicative of a high accuracy. A comparison of the curves for different degrees of peg connections is given in Figure 10d, where the specimens with 100% and 150% degree of connection are obtained by scaling up the yield and ultimate strengths of the pegs with equal proportions on the basis of P-SUCS-3.1%-150. According to Figure 10d and Table 5, for the specimens obtained after the degree of shear connection reaches 100%, the final damage state depends on the ultimate strengths of the steel plate and the UHPC. Therefore, the increase in the degree of shear connection will no longer cause a change in the ultimate load capacity of the specimen. Before 80% of the ultimate load capacity is reached, the peg connections of each specimen are in an elastic state and there is no significant difference in the flexural stiffness of each specimen. For specimens with less than 100% shear connection, the ultimate load capacity of the specimen depends on the ultimate shear capacity of the connection. A reduction in the degree of shear connection will result in a significant decrease in both the ultimate load-carrying capacity and the stiffness of the specimen.

3.2.2. Analysis of Damage Mechanism of Specimen

The finite element models of steel–UHPC composite slabs with shear connection degrees of 54% and 100% are subjected to ultimate loads, as depicted in Figures 11 and 12. As shown in Figure 11, the peg root of specimen P-SUCS-3.1%-200, which has a shear connection degree of 54%, exhibits an equivalent plastic strain ranging from 0.09 to 0.18. Significant plastic deformation leads to visible damage in the UHPC at the pegs' root. When the specimen reaches the ultimate load, significant compression damage occurs on the top surface of UHPC, with a thickness range of about 10 mm and a compression damage variable below 0.83, consistent with the failure mode observed in the test. From Figure 12, it can be observed that for the model with a shear connection degree of 100%, the pegs exhibit significant plastic deformation at the ultimate load. However, the equivalent plastic strain at the base of the pegs, ranging from 0.06 to 0.11, is lower than that of the pegs at the base of the composite slab specimen with a shear connection degree of 54%. As shown in Figure 12a, there is a noticeable increase in the compressive damage area on the upper surface of the UHPC board, extending downward to a depth of approximately 17 mm, while fully exerting its compressive strength.

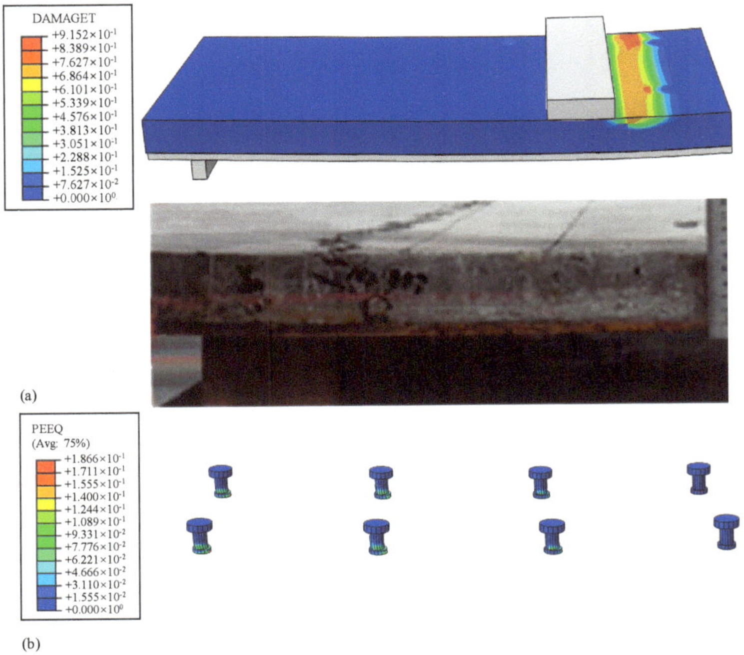

Figure 11. *Cont.*

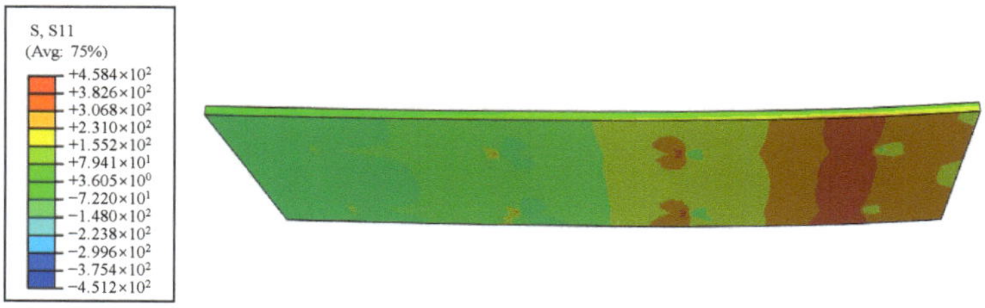

(c)

Figure 11. Analysis results of composite slabs with 54% shear connection degree: (**a**) Pressure damage contours for UHPC slab; (**b**) plot of equivalent plastic strain distribution of pegs; (**c**) longitudinal stress contours of steel plate.

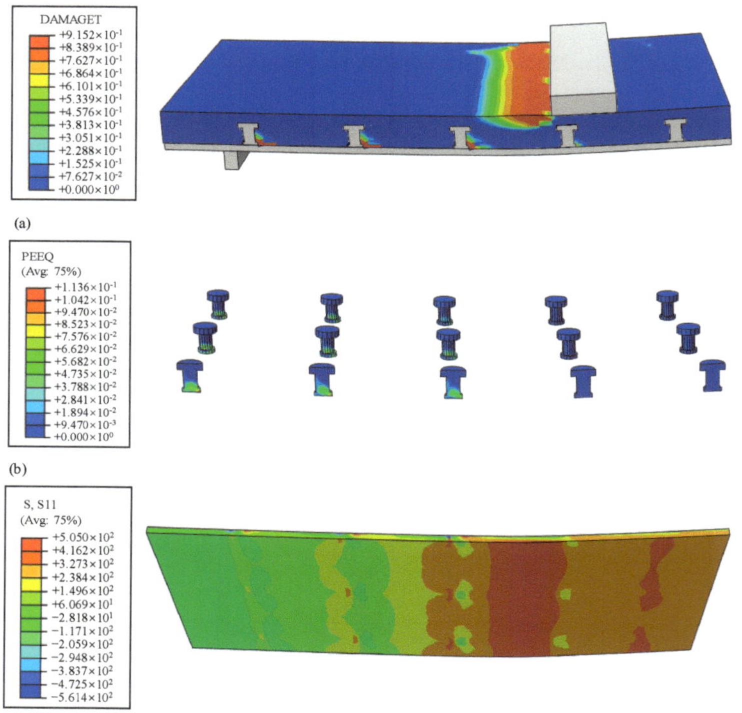

(c)

Figure 12. Analysis results of composite slab with 100% shear connection degree: (**a**) Pressure damage contours for UHPC slab; (**b**) plot of equivalent plastic strain distribution of pegs; (**c**) longitudinal stress contours in a steel plate.

The longitudinal stress distributions of steel plates with varying degrees of shear connection are presented in Figures 11c and 12c. In comparison, when the specimen reaches the ultimate load, the composite slab with a shear connection degree of 54% exhibits tensile stress on the lower surface of the steel plate in the pure bending region, with a corresponding tensile stress of 386.96 MPa. Meanwhile, the upper surface is in a compressed state, with a corresponding compressive stress of −85.93 Mpa. For the composite slab with a shear connection degree of 100%, the entire cross-section of the steel plate in the pure bending region is under tension. The tensile stresses corresponding to the lower and upper surfaces of the steel plate are 471.4 Mpa and 68.1 Mpa, respectively.

In summary, under the action of positive bending moment, insufficient shear connection of the pegs hinders the complete transfer of shear forces between the steel plate and the UHPC slab, as it is limited by the shear capacity of the connectors. This compromises the collaborative performance of the steel plate and UHPC slab, preventing the full utilization of their material properties.

3.3. Numerical Analysis of Steel–UHPC Composite Slabs Under Negative Bending Moments

3.3.1. Load–Deflection Curves

Under the action of a negative bending moment, the load–deflection curves obtained through finite element calculations are compared with the test results, as shown in Figure 13a–e. The comparison reveals that the load–deflection curve from finite element calculations is generally consistent with the test results, and the characteristic points of the curve align well with the test outcomes. Table 6 lists the finite element calculation results and test results for the ultimate load-carrying capacity of the composite slab. It can be observed that the average ratio between the test results and the finite element calculation results is 1.01, indicating high accuracy. The degree of shear connection listed in Table 6 is calculated according to Equation (8). However, under negative bending moments, the steel plate is compressed, while the UHPC slab is undergoing tension. Therefore, F_{min} in Equation (8) should take the smaller value between the compressive capacity of the steel plate and the tensile capacity of the UHPC slab. When calculating the tensile capacity of the UHPC slab, the contribution of the reinforcement to the overall capacity should be considered.

Table 6. Ultimate bearing capacity of composite slabs under negative bending moments.

Specimen Number	Degree of Shear Connection	$P_{u,Test}$ (kN)	$P_{u,FEA}$ (kN)	$P_{u,Test}/P_{u,FEA}$
N-SUCS-3.1%-150	250%	116.3	118.1	0.98
N-SUCS-3.1%-200	200%	112.6	110.9	1.02
N-SUCS-3.1%-250	100%	110.5	108.2	1.02
N-SUCS-2.0%-200	200%	86.8	87.4	0.99
N-SUCS-3.8%-200	200%	128.2	127.1	1.01
N-SUCS-3.1%-250-FEA1	50%	-	102.9	-
N-SUCS-3.1%-250-FEA2	0%	-	79.6	-
N-SUCS-4.5%-200	200%	-	136.2	-
N-SUCS-6.1%-200	200%	-	158.1	-

The comparison results of the load–deflection curves for steel–UHPC composite slabs with different shear connection degrees under negative bending moments are presented in Figure 13f. It can be observed that when the degree of shear connection is 100% or greater, the shear capacity of the connector surpasses the tensile strength of the UHPC layer that incorporates the steel reinforcement. By further increasing the degree of shear connection, the flexural stiffness of the specimen will undergo a significant improvement; bending stiffness increases by 34.73% when the degree of shear connection is increased from 100% to 250%. However, the ultimate bearing capacity of the specimen remains largely unchanged, as it is constrained by the tensile strength of the UHPC and the reinforcement. When the shear connection is insufficient, as indicated in Table 6 and Figure 13f, reducing the degree of shear connection from 100% to 50% leads to a 4.8% decrease in the ultimate load capacity of the specimen. Moreover, the flexural stiffness of the specimen is significantly reduced by 30.45%. The main reason for this is that as the shear connection degree of the composite slab decreases, the restraining effect of the steel plate on the deformation of the UHPC weakens. This leads to faster development of crack widths in the UHPC, resulting in a significant reduction in the stiffness of the specimen.

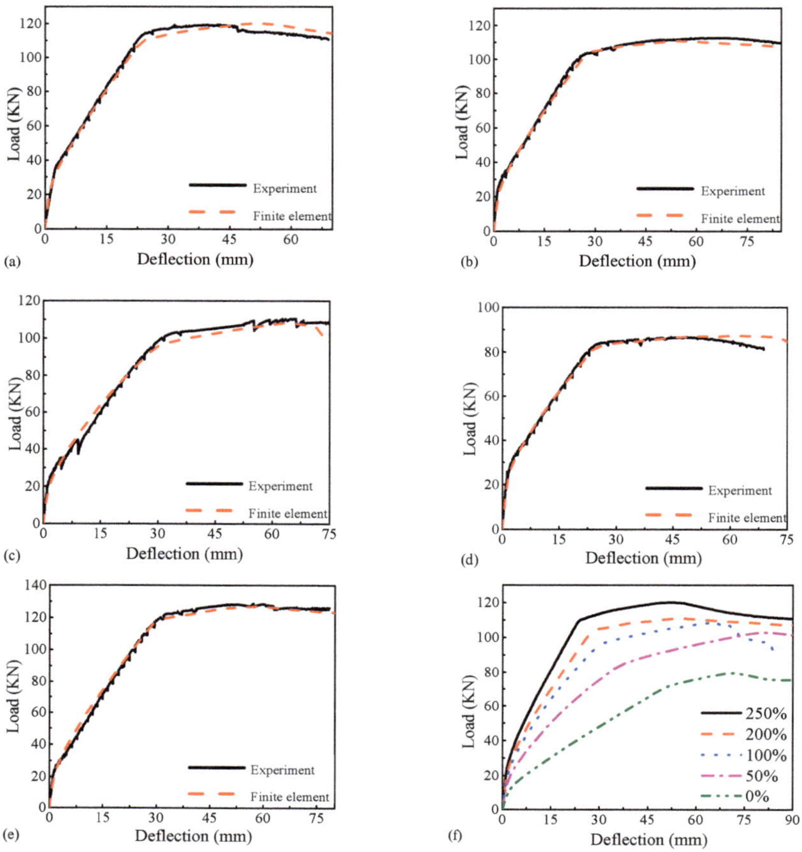

Figure 13. Finite element calculation results of steel–UHPC composite slabs: (**a**) Specimen N-SUCS-3.1%-150; (**b**) Specimen N-SUCS-3.1%-200; (**c**) Specimen N-SUCS-3.1%-250; (**d**) Specimen N-SUCS-2.0%-200; (**e**) Specimen N-SUCS-3.8%-200; (**f**) different shear connection degree.

The load–deflection curves of steel–UHPC composite slabs with different reinforcement ratios, obtained from finite element analysis, are shown in Figure 14. It can be observed that under negative bending moments, both the flexural stiffness and ultimate load capacity of the specimen increase significantly with the reinforcement ratio. This indicates that increasing the reinforcement ratio further enhances the load-bearing capacity and crack control ability of the UHPC slab.

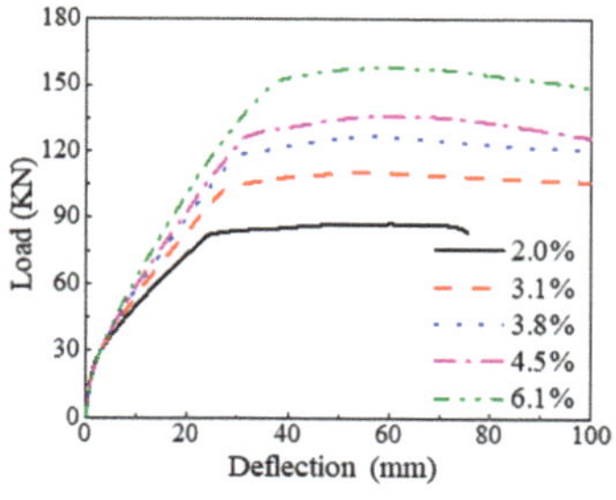

Figure 14. Analysis results of composite slabs with different reinforcement ratios.

3.3.2. Analysis of Specimen Damage Mechanism

The tensile damage contours of the UHPC slab and the stress contours of the steel plate and pegs in the finite element model for specimens N-SUCS-3.1%-150 and N-SUCS-3.1%-250 at ultimate load are depicted in Figures 15 and 16, respectively. It is evident that, for specimen N-SUCS-3.1%-150, none of the pegs in the shear span section yield during ultimate load due to the high degree of shear connection, allowing the steel plate and UHPC slab to achieve cooperative deformation. Furthermore, at ultimate load, the UHPC slab experiences full-section tension, while the steel plate at mid-span experiences compression on the lower surface and tension on the upper surface, with the neutral axis of the cross-section lying within the steel plate's thickness. In contrast, for the comparative specimen N-SUCS-3.1%-250, since the degree of shear connection is 100%, all pegs in the shear span section have significantly yielded at ultimate load. The increased interface slip after the pegs have yielded reduces the synergistic performance of the steel plate and UHPC slab, resulting in some of the lower edges of the UHPC slab near the loading point, on the side close to the pure bending section, being in a compressed state. Simultaneously, none of the stresses at the lower edges of the steel plate reach the yield state, indicating that the full strength of the steel plate is not utilized.

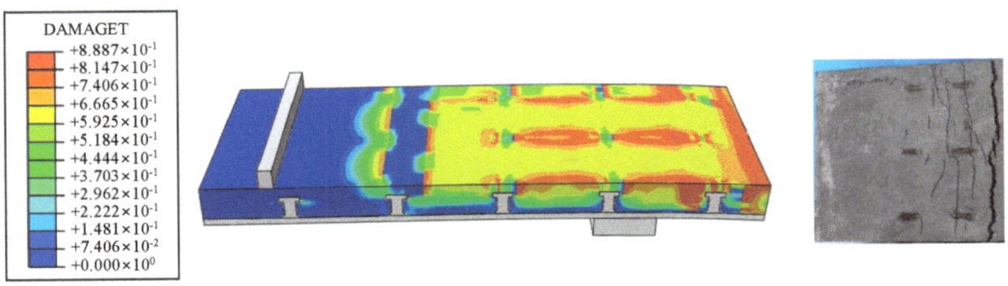

(a)

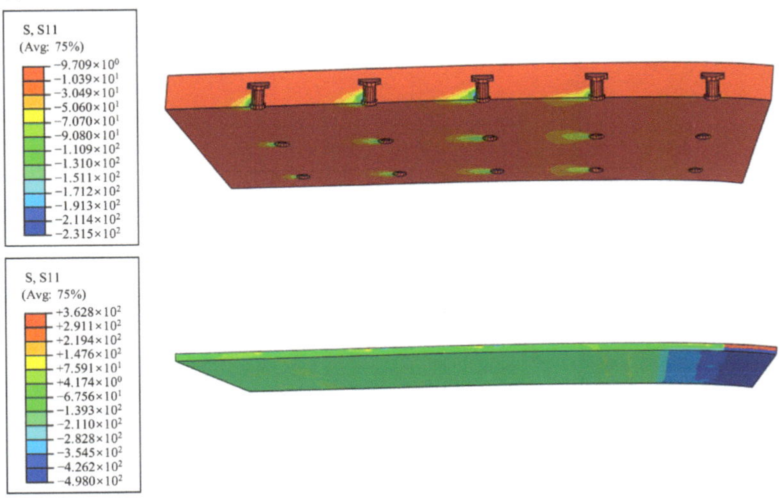

(b)

Figure 15. *Cont.*

(c)

Figure 15. Specimen N-SUCS-3.1%-150: (**a**) Tensile damage contours of UHPC slab; (**b**) longitudinal stress contours of UHPC slab and steel plate; (**c**) stress contours of the pegs.

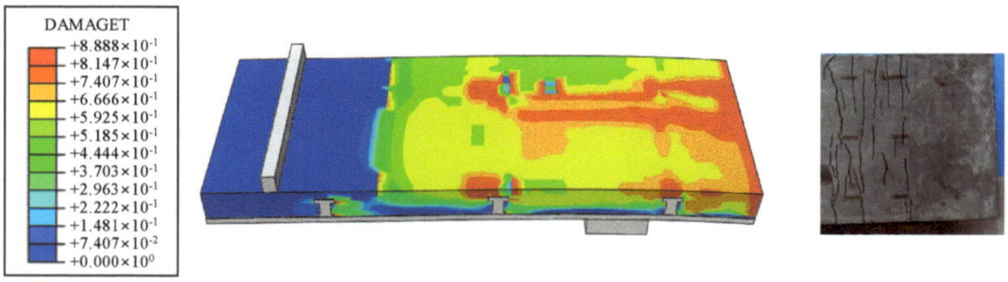

(a)

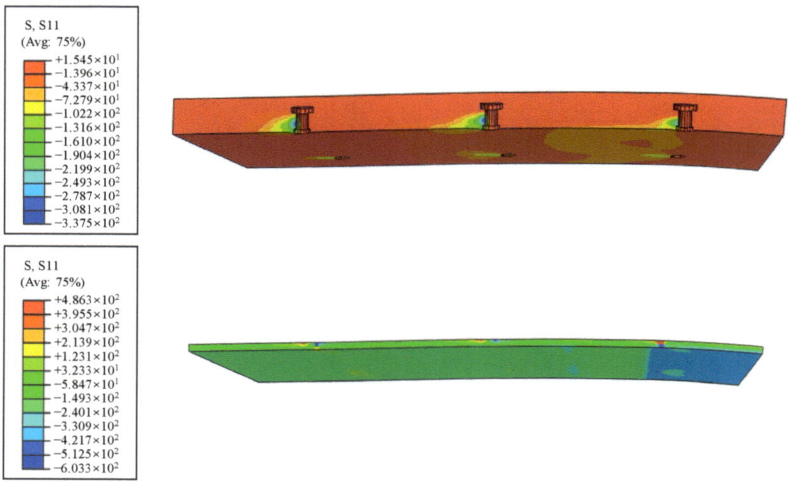

(b)

(c)

Figure 16. Specimen N-SUCS-3.1%-250: (**a**) Tensile damage contours of UHPC slab; (**b**) longitudinal stress contours of UHPC slab and steel plate; (**c**) stress contours of the pegs.

4. Calculation Method for Load-Bearing Capacity

4.1. Calculation Method for Flexural Capacity Under Positive Bending Moment

According to the results of the test and the numerical simulation, under the action of positive bending moment, when a fully shear-connected composite slab is subjected to ultimate load, the upper surface of the UHPC reaches its ultimate compressive stress and fails in compression, while other locations in the compression zone do not reach their ultimate compressive strength. The entire cross-section of the plate is in tension near the lower surface and yields. When the incompletely connected composite slab is under ultimate load, the UHPC slab undergoes compressive failure. The stress in the steel plate is dependent on the degree of shear connection in the composite slab. When the degree of shear connection is low, during the application of ultimate load, significant interface slip arises. Consequently, both the steel plate and the UHPC slab display a plastic neutral axis within the height of their cross-sections. To facilitate formula derivation and simplify calculations, given that the thickness of the steel plate in steel–UHPC composite structures is generally thinner compared to the thickness of the UHPC, the following assumptions are made in order to calculate the bearing capacity of steel–UHPC composite slabs under positive bending moments: (1) the tensile zone of the steel plate reaches the yield strength, ignoring the role of the compressive zone of the steel plate; (2) the distribution of the stress in the compressive zone of the UHPC is an inverted triangle, and the maximum compressive stress is the compressive strength of the UHPC, corresponding to the cross-section of the UHPC. The maximum compressive stress is the compressive strength of UHPC, and the corresponding cross-section stress distribution model is shown in Figure 17.

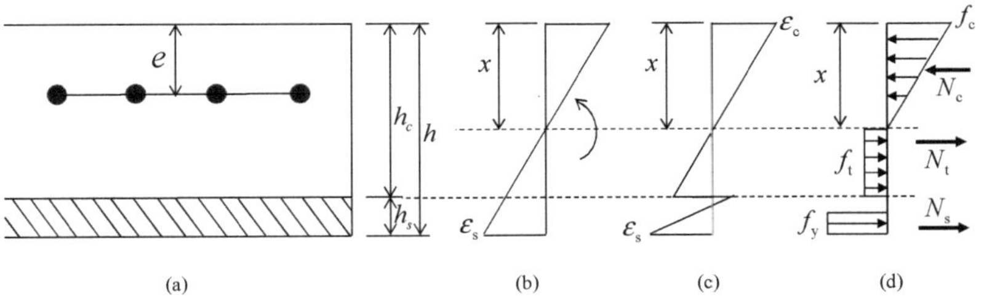

Figure 17. Calculation model of ultimate load under positive bending moment: (**a**) Composite plate cross-section calculation sketch; (**b**) Cross-sectional Strain Distribution-Full Shear Connection; (**c**) Cross-sectional strain distribution-Incomplete shear connection; (**d**) Cross-sectional stress distribution.

According to Figure 17d, the following equation can be obtained based on the axial force balance of the UHPC layer cross-section:

$$N_c - N_t - F_d = 0 \tag{10}$$

$$0.5 f_c b x - f_t b (h_c - x) - F_d = 0 \tag{11}$$

where f_c refers to the compressive strength of UHPC and f_t refers to the tensile strength of UHPC. The positive moment calculation takes the elastic tensile strength of UHPC, which, according to the results of the axial tensile test, is 7.42 Mpa; b refers to the width of the composite slab; x refers to the height of the neutralization axis; h_c refers to the thickness of the UHPC slab; h_s refers to the thickness of the steel plate; N_c, N_t, and N_s refer to the axial forces in the compressive region of the UHPC, the tensile region of the UHPC, and the tensile region of the steel plate; F_d refers to the shear-bearing capacity of the peg connections in the shear span section; and $F_d \leq \min(f_c b h_c, f_y b h_s)$.

From the axial force balance of the steel plate, we obtain the following calculation:

$$N_s = F_d = f_y b x_b \tag{12}$$

where x_b is the height of the tensile zone of the steel plate.

The equilibrium equation for the bending moment of the cross-section is as follows:

$$M = \frac{1}{3} f_c b x^2 + \frac{1}{2} f_t b (h_c - x)^2 + F_d (h - x) - \frac{1}{2 f_y b} {F_d}^2 \tag{13}$$

According to the boundary conditions of the four-point bending steel–UHPC composite slab, the relationship between the bending moment of the pure bending section cross-section and the vertical load can be expressed as follows:

$$p = \frac{2M}{a_0} \tag{14}$$

where a_0 is the length of the shear span section of the composite slab specimen, 500 mm.

Table 7 lists the comparisons of the ultimate bearing capacity of each steel–UHPC composite slab under a positive bending moment calculated according to Equations (10)–(14) based on the results of tests and finite element calculations, and it can be seen that the calculation results of the proposed formula can match the test results better, which can be used as a reference for designers.

Table 7. Theoretical calculation results of flexural bearing capacity under positive bending moments.

Specimen Number	$P_{u,Test}$ (kN)	$P_{u,FEA}$ (kN)	$P_{u,Cal}$ (kN)	$P_{u,Test}/P_{u,Cal}$	$P_{u,FEA}/P_{u,Cal}$
P-SUCS-3.1%-150	320.2	-	314.5	1.02	-
P-SUCS-3.1%-200	285.7	-	272.6	1.05	-
P-SUCS-3.1%-250	159.1	-	166.4	0.96	-
P-SUCS-3.1%-150-FEA1	-	340.5	334.6	-	1.02
P-SUCS-3.1%-150-FEA2	-	343.5	334.6	-	1.03

Note: $P_{u,Cal}$ refers to the calculated ultimate load.

4.2. Calculation Method of Flexural Capacity Under Negative Bending Moment

4.2.1. Theoretical Calculation of Cracking Load of Steel–UHPC Composite Slab

Under negative bending moments, the UHPC layers in steel–UHPC composite slabs are in a state of tension, posing a risk of cracking that could degrade structural durability. To ensure compliance with durability requirements and serviceability limit states specified in relevant codes, during structural design, it is essential to keep the maximum tensile stress in the UHPC layer within its design cracking strength, also referred to as the elastic ultimate strength, in the negative moment region [56]. In UHPC structures, the cracking load is generally defined as the load that corresponds to a maximum crack width of 0.05 mm in the UHPC layer. The UHPC used in this study is all strain-strengthening UHPC. Therefore, when the maximum crack width on the surface of the UHPC slab reaches 0.05 mm under negative bending moments, the corresponding surface strain of the UHPC exceeds its elastic ultimate tensile strain. When the UHPC slab reaches its cracking load, the neutral axis of the composite slab's cross-section is located within the UHPC slab, and the stress and strain distributions across the composite slab section are depicted in Figure 18.

The cracking strain of UHPC is assumed to be ε_{tc}, and according to the flat section assumption, the strain in the steel reinforcement is shown in Equation (15). The elastic ultimate tensile strain of UHPC (ε_{te}) is shown in Equation (16). The strains on the upper (ε_{tsc}) and lower surfaces of the steel plate (ε_{bsc}) are shown in Equations (17) and (18).

$$\varepsilon_{st} = \frac{\varepsilon_{tc}(x-e)}{x} \tag{15}$$

$$\varepsilon_{te} = \frac{\varepsilon_c x_1}{x} \tag{16}$$

$$\varepsilon_{tsc} = \frac{\varepsilon_c(h_c - x)}{x} \tag{17}$$

$$\varepsilon_{bsc} = \frac{\varepsilon_c(h - x)}{x} \tag{18}$$

where x refers to the height of the tensile zone of the UHPC slab, e refers to the thickness of the protective layer of the steel reinforcement, and x_1 refers to the height of the UHPC slab where the strain remains below the elastic ultimate tensile strain of UHPC.

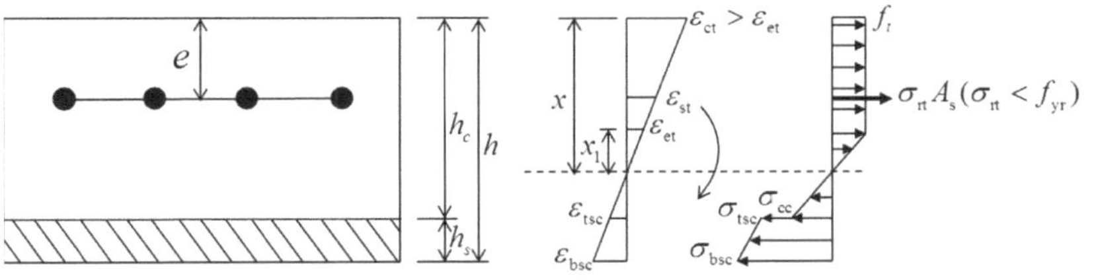

Figure 18. Calculation model of cracking load under negative moment action.

The axial force equilibrium condition of the cross-section is shown in Equation (19). The axial forces borne by the UHPC slab, steel reinforcement, and steel plate are shown in Equations (20)–(22).

$$N_c + N_r + N_s = 0 \tag{19}$$

$$N_c = f_{tu}b(x - x_1) + 0.5 f_{tu} x_1 b - 0.5 E_c \varepsilon_{tsc} b(h_c - x) \tag{20}$$

$$N_r = E_{rs}\varepsilon_{st}A_s \tag{21}$$

$$N_s = 0.5 E_s h_s b(\varepsilon_{tsc} + \varepsilon_{bsc}) \tag{22}$$

where N_c, N_r and N_s refer to the axial forces borne by the UHPC, steel reinforcement, and steel beams, respectively. E_{rs} and E_s refer to the modulus of elasticity of the steel reinforcement and steel plate, respectively, and A_s refers to the cross-sectional area of the steel reinforcement.

According to the cross-section moment equilibrium condition, the cracking moment of the steel–UHPC composite slab is shown in Equation (23). The bending moments borne by the UHPC slab, steel reinforcement, and steel plate are shown in Equations (24)–(26).

$$M_{cr} = M_c + M_r + M_s \tag{23}$$

$$M_c = 0.5 f_{tu} b(x^2 - x_1^2) + \frac{1}{3} f_{tu} b x_1^2 + \frac{1}{3} E_c b \varepsilon_{tsc}(h_c - x)^2 \tag{24}$$

$$M_r = E_{rs}\varepsilon_{st}A_s(x - e) \tag{25}$$

$$M_s = E_s h_s b[(\frac{1}{3}h_s + \frac{1}{2}h_c - \frac{1}{2}x)\varepsilon_{tsc} + (\frac{1}{6}h_s + \frac{1}{2}h_c - \frac{1}{2}x)\varepsilon_{bsc}] \tag{26}$$

where M_c, M_r, and M_s refer to the bending moments borne by the UHPC, steel reinforcement and steel beams, respectively, and M_{cr} refers to the section cracking moment caused by external loads. The above derivation process demonstrates that if the cracking strain ε_{tc} is known, the height x of the tensile zone of the UHPC can be determined using Equation (19). Subsequently, the cracking moment of the steel–UHPC composite slab can be calculated by applying Equation (23).

To determine the cracking strain ε_{tc} of UHPC in steel–UHPC composite slabs subjected to negative bending moments, the cracking loads of the composite slabs are calculated using the aforementioned formula across a range of cracking strains from 0.08% to 0.15%. The outcomes of these calculations are illustrated in Figure 19. By comparing the load values recorded during testing when cracks reached 0.05 mm with the cracking loads corresponding to various cracking strains depicted in Figure 19, the cracking strains of UHPC in each specimen were determined. Table 8 lists the cracking strains ε_{tc} of UHPC in the composite slab specimens, as determined by this method. It can be observed that, under negative bending moments, the cracking strain of UHPC in steel–UHPC composite slabs falls within a range of 0.099% to 0.139%, which is less than the ultimate tensile strain of UHPC measured in axial tensile tests. This is likely due to the uneven distribution of steel fibers resulting from the influence of pegs and steel reinforcement during the pouring of UHPC, subsequently leading to a reduction in the mechanical properties of UHPC.

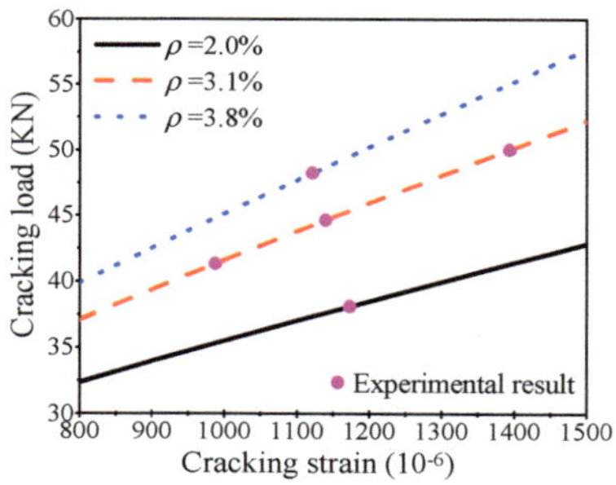

Figure 19. Corresponding relationship between cracking strain and cracking load.

Table 8. Theoretical calculation results of cracking strain of steel–UHPC composite slab.

Specimen Number	$P_{cr,Test}$ (kN)	ε_{tc} (%)	ε_{bst} (%)	ε_{bsc} (%)	$\varepsilon_{bst}/\varepsilon_{bsc}$
N-SUCS-3.1%-150	50	0.139	0.0273	0.0271	1.01
N-SUCS-3.1%-200	44.6	0.114	0.0251	0.0254	0.99
N-SUCS-3.1%-250	41.3	0.099	0.0230	0.0236	0.97
N-SUCS-2.0%-200	38.1	0.117	0.0222	0.0220	1.01
N-SUCS-3.8%-200	48.2	0.112	0.0290	0.0285	1.02

Note: $P_{cr,Test}$ refers to the cracking load obtained from the test, ε_{bst} refers to the experimental lower surface strain of the steel plate, and ε_{bsc} refers to the calculated lower surface strain of the steel plate.

To verify the accuracy of the formula presented in this paper for calculating the cracking strain of UHPC, the strain on the lower surface of the steel plate under the cracking load, calculated using Equation (18), is compared with the measured strain of the steel plate. The comparison results, listed in Table 8, indicate that the calculated results are relatively close to the test results, with the error kept within 3%.

4.2.2. Theoretical Calculation of Ultimate Load of Steel–UHPC Composite Slab

According to the results of the test and finite element analysis, under the action of a negative moment in an incompletely connected composite slab, the degree of constraint that the steel plate exerts on the bending deformation of the UHPC decreases, leading to premature cracking of the UHPC and a reduction in the stiffness and durability of the composite slab. Considering that achieving complete connection of the composite slab under negative moments is relatively feasible in practical engineering, this paper only derives the ultimate load capacity of the composite slab under negative moments and complete connection. For specimens with complete shear connection, at the ultimate load, the neutral axis of the section is located within the thickness of the steel plate. The lower surface of the steel plate yields under compression, while the UHPC experiences tension across the entire section. With reference to the Swiss specification MCS-EPFL [57], the height of the tensile zone of the reinforced UHPC slab is set at 0.9 h_c. Combining the results of the test and finite element analysis, the ultimate load-carrying capacity of the composite slab is calculated using the method applicable to steel–UHPC composite slabs under negative moments. The following basic assumptions are applied when calculating the ultimate load-carrying capacity: (1) the tensile and compressive zones of the steel plate have reached the yield strength; (2) neglecting the slip between the steel plate and the reinforced UHPC slab, the composite slab conforms to the flat cross-section assumption, and the corresponding cross-section stress distribution model is shown in Figure 20.

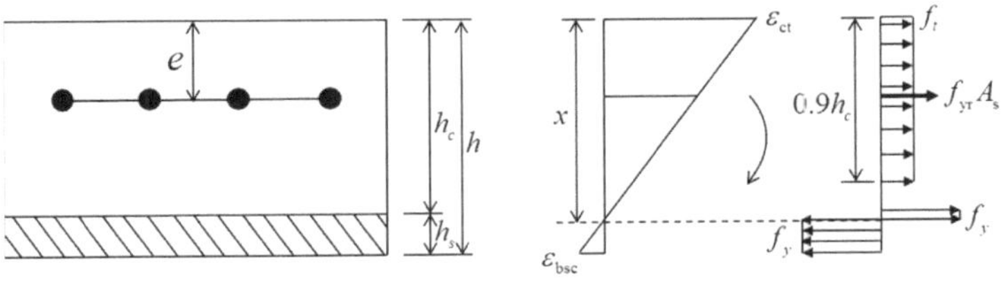

Figure 20. Calculation model of ultimate load under negative bending moments.

The axial force and bending moment equilibrium conditions are shown in Equations (27) and (28).

$$0.9f_t bh_c + f_{yr}A_s + f_y b(x - h_c) = f_y b(h - x) \tag{27}$$

$$M = 0.9f_t bh_c(x - 0.45h_c) + f_{yr}A_s(x - e) + 0.5f_y b[(x - h_c)^2 + (x - h)^2] \tag{28}$$

where A_s refers to the cross-sectional area of the steel reinforcement and f_{yr} refers to the tensile strength of the reinforcement. According to Equation (14), the ultimate bending moment corresponding to the ultimate load can be further calculated for each specimen and cross-section. The calculation results are presented in Table 9. It can be observed that the formula proposed in this paper, which calculates the ultimate load-bearing capacity of steel–UHPC composite slabs under negative bending moments, aligns well with the test results.

Table 9. Theoretical calculation results of bending capacity under negative bending moments.

Specimen Number	$P_{u,Test}$ (kN)	$P_{u,Cal}$ (kN)	$P_{u,Test}/P_{u,Cal}$
N-SUCS-3.1%-150	116.3	112.4	1.03
N-SUCS-3.1%-200	112.6	112.4	1.00
N-SUCS-3.1%-250	110.5	112.4	0.98
N-SUCS-2.0%-200	86.8	90.85	0.96
N-SUCS-3.8%-200	128.2	125.3	1.02

5. Conclusions

A detailed numerical model of steel–UHPC composite slabs is established using ABAQUS software to conduct finite element analyses under positive and negative bending moments. The results are compared with experimental data to validate the accuracy of the finite element model. Based on the finite element analysis results, the stress and damage distribution of each component of the composite slabs under different loading conditions is analyzed. Finally, based on the experimental and finite element analyses, a calculation method for the load-bearing capacity of steel–UHPC composite slabs under positive and negative bending moments is proposed. The main conclusions are as follows:

(1) The load–deflection curves and failure modes obtained from the finite element analysis of the composite slabs show good agreement with the experimental results, indicating that the finite element model established in this study accurately reflects the mechanical behavior of the composite slabs under positive and negative bending moments.
(2) Under positive bending moment conditions, a shear connection degree of 100% in the pegs serves as the critical threshold that changes the mechanical behavior of the specimen. When the shear connection degree exceeds 100%, further increases do not lead to significant changes in the load-bearing capacity of the specimen, although a slight improvement in stiffness can be observed. Conversely, when the shear connection degree is less than 100%, the ultimate load-bearing capacity of the composite slab is governed by the shear capacity of the peg connections. At the ultimate load, significant plastic deformation occurs in the pegs within the shear span, and both the stiffness and load-bearing capacity of the composite slab decrease as the shear connection degree reduces.
(3) Under negative bending moment conditions, when the UHPC and steel plate are in full shear connection, further increasing the shear connection degree for composite slabs with the same reinforcement ratio results in a noticeable improvement in stiffness. Additionally, the crack development on the surface of the UHPC is effectively suppressed. However, the load-bearing capacity remains unchanged, as it is governed by the tensile strength of the UHPC and the reinforcement. For composite slab specimens with the same shear connection degree, increasing the reinforcement ratio of the UHPC effectively enhances both the load-bearing capacity and stiffness of the composite slab. This indicates that increasing the reinforcement ratio significantly improves the crack control capacity of the UHPC.
(4) Based on the experimental and finite element analysis results, the calculation formulas for the load-bearing capacity of steel–UHPC composite slabs under positive and negative bending moments are proposed. The difference between the calculated results and experimental results is less than 3%, which demonstrates the high accuracy of the calculation formulas proposed in this study.

Author Contributions: Conceptualization, X.G.; methodology, C.L.; software, X.G.; validation, B.Z. and X.Z.; formal analysis, C.L.; investigation, D.H., H.B. and X.Z.; resources, H.B.; data curation, B.Z. and X.Z.; writing—original draft preparation, X.G.; writing—review and editing, C.L.; visualization, D.H.; supervision, X.G.; project administration, X.G.; funding acquisition, X.G. All authors have read and agreed to the published version of the manuscript.

Funding: This work was supported by the National Natural Science Foundation of China [Grant No. 52208242], Key Research and Development Special Project of Henan Province (241111322600), the China Postdoctoral Science Foundation [Grant No. 2023M733356], Scientific and technological research and development projects of China Construction Seventh Engineering Division Corp Ltd. (047H2023TM047).

Institutional Review Board Statement: Not applicable.

Informed Consent Statement: Not applicable.

Data Availability Statement: All data generated or analyzed during this study are included in this article.

Conflicts of Interest: We declare that we have no financial and personal relationships with other people or organizations that can inappropriately influence our work. There is no professional or other personal interest of any nature or kind in any product, service, and/or company that could be construed as influencing the position presented in, or the review of, the manuscript entitled "A Numerical and Theoretical Investigation of the Flexural Behavior of Steel–Ultra-High-Performance Concrete composite slabs".

References

1. Amran, M.; Huang, S.S.; Onaizi, A.M.; Makul, N.; Abdelgader, H.S.; Ozbakkaloglu, T. Recent trends in ultra-high performance concrete (UHPC): Current status, challenges, and future prospects. *Constr. Build. Mater.* **2022**, *352*, 129029. [CrossRef]
2. Pan, B.Z.; Liu, F.; Zhuge, Y.; Zeng, J.J.; Liao, J.J. ECCs/UHPFRCCs with and without FRP reinforcement for structural strengthening/repairing: A state-of-the-art review. *Constr. Build. Mater.* **2022**, *316*, 125824. [CrossRef]
3. Zhang, Y.; Zhu, Y.P.; Qu, S.Q.; Kumar, A.; Shao, X.D. Improvement of flexural and tensile strength of layered-casting UHPC with aligned steel fibers. *Constr. Build. Mater.* **2020**, *251*, 118893. [CrossRef]
4. Bajaber, M.A.; Hakeem, I.Y. UHPC evolution, development, and utilization in construction: A review. *J. Mater. Res. Technol.* **2021**, *10*, 1058–1074. [CrossRef]
5. Shafieifar, M.; Farzad, M.; Azizinamini, A. Experimental and numerical study on mechanical properties of Ultra High Performance Concrete (UHPC). *Constr. Build. Mater.* **2017**, *156*, 402–411. [CrossRef]
6. Li, J.Q.; Wu, Z.M.; Shi, C.J.; Yuan, Q.; Zhang, Z.H. Durability of ultra-high performance concrete—A review. *Constr. Build. Mater.* **2020**, *255*, 119296. [CrossRef]
7. Dong, L.; Yang, Y.K.; Liu, Z.X.; Yang, T.; Xue, C.H.; Shao, R.Z.; Wu, C.Q. Effect of chloride ion migration behaviour on the microstructure and mechanical properties of ultra-high performance concrete: A review. *J. Build. Eng.* **2024**, *82*, 108233. [CrossRef]
8. Mousavinejad, S.H.G.; Sammak, M. Strength and chloride ion penetration resistance of ultra-high-performance fiber reinforced geopolymer concrete. *Structures* **2021**, *32*, 1420–1427. [CrossRef]
9. Zhongya, Z.; Kun, P.; Lihao, X.; Yang, Z.; Jun, Y.; Chuanbiao, W. The bond properties between UHPC and stone under different interface treatment methods. *Constr. Build. Mater.* **2023**, *365*, 130092. [CrossRef]
10. Li, Y.; Zhou, H.; Zhang, Z.; Yang, J.; Wang, X.; Wang, X.; Zou, Y. Macro-micro investigation on the coefficient of friction on the interface between steel and cast-in-place UHPC. *Eng. Struct.* **2024**, *318*, 118769. [CrossRef]
11. Gao, X.L.; Wu, K.; Guo, Y.Q.; Zhao, Y.C.; Guo, J.Y. Experimental and numerical study on flexural behaviors of damaged RC beams strengthened with UHPC layer using the bonding technology of post-installed reinforcing bar. *Constr. Build. Mater.* **2023**, *391*, 131835. [CrossRef]
12. Chen, C.; Cai, H.Y.; Cheng, L.J. Shear Strengthening of Corroded RC Beams Using UHPC-FRP Composites. *J. Bridge Eng.* **2021**, *26*, 04020111. [CrossRef]
13. Zeng, J.J.; Liang, Q.J.; Cai, W.J.; Liao, J.J.; Zhou, J.K.; Zhu, J.Y.; Zhang, L.H. Strengthening RC square columns with UHP-ECC section curvilinearization and FRP confinement: Concept and axial compression tests. *Eng. Struct.* **2023**, *280*, 115666. [CrossRef]
14. Sun, L.P.; Liu, Y.J.; Han, Q.; Wang, L. Conceptual design of lightweight assembled double-skinned UHPC composite pylons for large-span suspension bridges. *Structures* **2024**, *70*, 107725. [CrossRef]
15. Leng, J.C.; Yang, J.; Zhang, Z.Y.; Zou, Y.; Du, J.; Zhou, J.T. Mechanical behavior of a novel compact steel-UHPC joint for hybrid girder bridges: Experimental and numerical investigation. *J. Constr. Steel Res.* **2024**, *218*, 108742. [CrossRef]
16. Xiao, J.L.; Yang, T.Y.; Nie, X.; Li, B.Y.; Fan, J.S.; Shu, B.A. Experimental and numerical investigation on mechanical performance of continuous steel-UHPC composite slabs. *Eng. Struct.* **2022**, *270*, 114804. [CrossRef]
17. Gao, X.L.; Wang, M.L.; Guo, J.Y.; Li, H.X. Flexural behaviors of a novel precast hollow UHPC composite beam reinforced with inverted T-shaped steel: Experimental investigation and theoretical analysis. *J. Build. Eng.* **2024**, *86*, 108893. [CrossRef]
18. Gao, X.L.; Yin, C.L.; Guo, J.Y.; Chen, G.X.; Huang, H.R. Experimental and theoretical study on the shear behaviors of an innovative precast UHPC composite beams reinforced with inverted T-shaped steel. *J. Build. Eng.* **2024**, *96*, 110504. [CrossRef]
19. Tong, J.Z.; Chen, Y.L.; Li, Q.H.; Xu, S.L.; Zeng, T.; Gao, W. Experimental study on flexural performance of steel-UHTCC composite bridge decks considering different shear connection degrees. *Eng. Struct.* **2023**, *281*, 115738. [CrossRef]
20. Shao, X.D.; Cao, J.H.; Yi, D.T.; Chen, B.; Huang, Z.Y. Study on the basic performance of orthotropic anisotropic steel plate-thin layer RPC combination bridge decks. *China Highw. J.* **2012**, *25*, 40–45. (In Chinese) [CrossRef]
21. He, Z.G.; Lin, P.Z. Research on fatigue performance and optimal design of steel-UHPC composite slab. *Structures* **2022**, *43*, 682–695. [CrossRef]
22. Zhang, X.H.; Yang, X.Y.; Li, C.; Xu, F.; Wang, G.D. Friction affected fatigue behavior of steel-UHPC composite structures and the fatigue crack growth in studs. *Int. J. Fatigue* **2023**, *177*, 107949. [CrossRef]
23. Li, C.X.; He, L.F.; Pei, B.D.; Pan, R.S.; Shi, Y. Fatigue behavior and damage mechanism of steel plate-UHPC composite bridge deck subjected to negative bending moment. *Structures* **2024**, *65*, 106754. [CrossRef]
24. Xian, B.X.; Wang, G.D.; Ma, F.Y.; Fang, S.; Jiang, H.B.; Xiao, J.J. Shear performance of single embedded nut bolted shear connectors in precast steel-UHPC composite beams under combined tension-shear loads. *Case Stud. Constr. Mater.* **2024**, *21*, e03558. [CrossRef]
25. Zhang, Y.J.; Zhu, J.H.; Wu, L.L.; Lin, H.X.; Fang, Z.C.; Cen, Z.C.; Yang, Y.M.; Fang, S. Experimental and numerical analyses on the shear behavior of grouped single-embedded-nut high-strength bolts in steel-ultra-high-performance concrete composite slabs. *J. Build. Eng.* **2024**, *86*, 108829. [CrossRef]

26. Zhao, X.L.; Lin, G.; Zhou, Y.W.; Huang, Z.Y. Push-out test behavior and damage detection of steel-UHPC composite using PZT sensing. *Eng. Struct.* **2024**, *305*, 117725. [CrossRef]
27. Jia, J.F.; Ren, Z.D.; Bai, Y.L.; Li, J.; Li, B.; Sun, Y.G.; Zhang, Z.X.; Zhang, J.X. Tensile behavior of UHPC wet joints for precast bridge deck panels. *Eng. Struct.* **2023**, *282*, 115826. [CrossRef]
28. Ren, Z.D.; Jia, J.F.; Li, J.; Hu, M.H.; Li, B.; Han, Q.; Du, X.L. Shear behavior of precast bridge deck panels with UHPC wet joints. *Eng. Struct.* **2024**, *316*, 118569. [CrossRef]
29. Feng, Z.; Li, C.X.; Ke, L.; Yoo, D.Y. Experimental and numerical investigations on flexural performance of ultra-high-performance concrete (UHPC) beams with wet joints. *Structures* **2022**, *45*, 199–213. [CrossRef]
30. Hu, W.X.; Li, C.; Chen, B.C.; Liu, Y.J. Finite element analysis on shear behavior of headed studs in steel-UHPC composite slab. *Structures* **2023**, *52*, 464–475. [CrossRef]
31. Lu, K.W.; Xu, Q.Z.; Wang, M.D.; Yao, Y.M.; Wang, J.Q. Anchorage performance of bolt connection embedded in thin UHPC members. *Structures* **2021**, *34*, 1253–1260. [CrossRef]
32. Zhang, S.; Han, B.; Xie, H.B.; An, M.Z.; Lyu, S. Brittleness of Concrete under Different Curing Conditions. *Materials* **2021**, *14*, 7865. [CrossRef]
33. Luo, J.; Shao, X.D.; Fan, W.; Cao, J.H.; Deng, S.W. Flexural cracking behavior and crack width predictions of composite (steel plus UHPC) lightweight deck system. *Eng. Struct.* **2019**, *194*, 120–137. [CrossRef]
34. Wang, Y.; Cao, J.; Shao, X.; Shen, X. Flexural behavior and crack width prediction of UHPC-steel strip composite decks under sagging moments. *Eng. Struct.* **2023**, *293*, 116581. [CrossRef]
35. Zou, Y.; Zheng, K.D.; Zhou, Z.X.; Zhang, Z.Y.; Guo, J.C.; Jiang, J.L. Experimental study on flexural behavior of hollow steel-UHPC composite bridge deck. *Eng. Struct.* **2023**, *274*, 115087. [CrossRef]
36. Li, P.P.; Shao, X.D.; Cao, J.H.; Liu, Q.W.; Qiu, M.H.; Liu, G. Rational configuration and mechanical properties of the steel-UHPC composite deck system. *J. Constr. Steel Res.* **2024**, *221*, 108880. [CrossRef]
37. Taffese, W.Z.; Zhu, Y.P.; Chen, G.D. Explainable AI based slip prediction of steel-UHPC interface connected by shear studs. *Expert Syst. Appl.* **2025**, *259*, 125293. [CrossRef]
38. Xiao, J.L.; Zhou, M.; Nie, J.G.; Yang, T.Y.; Fan, J.S. Flexural behavior of steel-UHPC composite slabs with perfobond rib shear connectors. *Eng. Struct.* **2021**, *245*, 112912. [CrossRef]
39. Rabi, M. Bond prediction of stainless-steel reinforcement using artificial neural networks. *Proc. Inst. Civ. Eng. Constr. Mater.* **2023**, *177*, 87–97. [CrossRef]
40. Rabi, M.; Abarkan, I.; Shamass, R. Buckling resistance of hot-finished CHS beam-columns using FE modelling and machine learning. *Steel Constr.* **2023**, *17*, 93–103. [CrossRef]
41. Rabi, M.; Shamass, R.; Cashell, K.A. Description of the constitutive behaviour of stainless steel reinforcement. *Case Stud. Constr. Mater.* **2024**, *20*, e03013. [CrossRef]
42. Rabi, M.; Jweihan, Y.S.; Abarkan, I.; Ferreira, F.P.V.; Shamass, R.; Limbachiya, V.; Tsavdaridis, K.D.; Santos, L.F.P. Machine learning-driven web-post buckling resistance prediction for high-strength steel beams with elliptically-based web openings. *Results Eng.* **2024**, *21*, 101749. [CrossRef]
43. Abarkan, I.; Rabi, M.; Ferreira, V.P.F. Machine learning for optimal design of circular hollow section stainless steel stub columns: A comparative analysis with Eurocode 3 predictions. *Eng. Appl. Artif. Intell.* **2024**, *132*, 107952. [CrossRef]
44. Gao, X.L.; Wang, J.Y.; Bian, C.; Xiao, R.C.; Ma, B. Experimental investigation on the behaviour of UHPC-steel composite slabs under hogging moment. *Steel Compos. Struct.* **2022**, *42*, 765–777. [CrossRef]
45. Gao, X.L.; Wang, J.Y.; Yan, J.B. Experimental studies of headed stud shear connectors in UHPC Steel composite slabs. *Struct. Eng. Mech.* **2020**, *74*, 657–670. [CrossRef]
46. Liu, X.; Bradford, M.A.; Chen, Q.-J.; Ban, H. Finite element modelling of steel–concrete composite beams with high-strength friction-grip bolt shear connectors. *Finite Elem. Anal. Des.* **2016**, *108*, 54–65. [CrossRef]
47. Ataei, A.; Bradford, M.A.; Valipour, H.R. Finite element analysis of HSS semi-rigid composite joints with precast concrete slabs and demountable bolted shear connectors. *Finite Elem. Anal. Des.* **2016**, *122*, 16–38. [CrossRef]
48. Luo, Y.B.; Hoki, K.; Hayashi, K.; Nakashima, M. Behavior and Strength of Headed Stud-SFRCC Shear Connection. II: Strength Evaluation. *J. Struct. Eng.* **2016**, *142*, 04015113. [CrossRef]
49. Gao, X.L.; Wang, J.Y.; Guo, J.Y.; Liu, C. Axial tensile mechanical properties and constitutive relation model of ultra-high performance concrete under cyclic loading. *Acta Mater. Compos. Sin.* **2021**, *38*, 3925–3938. [CrossRef]
50. Cao, J.; Shao, X. Finite element analysis of headed studs embedded in thin UHPC. *J. Constr. Steel Res.* **2019**, *161*, 355–368. [CrossRef]
51. Kruszewski, D.; Zaghi, A.E.; Wille, K. Finite element study of headed shear studs embedded in ultra-high performance concrete. *Eng. Struct.* **2019**, *188*, 538–552. [CrossRef]
52. Nguyen, H.T.; Kim, S.E. Finite element modeling of push-out tests for large stud shear connectors. *J. Constr. Steel Res.* **2009**, *65*, 1909–1920. [CrossRef]

53. Kim, S.E.; Nguyen, H.T. Finite element modeling and analysis of a hybrid steel-PSC beam connection. *Eng. Struct.* **2010**, *32*, 2557–2569. [CrossRef]
54. Yan, J.B.; Li, Z.X.; Xie, J. Numerical and parametric studies on steel-elastic concrete composite structures. *J. Constr. Steel Res.* **2017**, *133*, 84–96. [CrossRef]
55. Nie, J.G.; Liu, M.; Ye, L. *Steel-Concrete Combined Structures*; Construction Industry Press: Beijing, China, 2005; pp. 68–72. ISBN 978-711-207-189-0. (In Chinese)
56. Li, W.G.; Shao, X.D.; Fang, H.; Zhang, Z. Experimental study on the bending performance of steel-UHPC composite plate. *Civ. Eng. J.* **2015**, *48*, 93–102. (In Chinese) [CrossRef]
57. MCS-EPFL. *Ultra-High Performance Fibre Reinforced Cement-Based Composites (UHPFRC): Construction Material, Dimensioning and Application*; Swiss Federal Institute of Technology: Zürich, Switzerland, 2016.

Disclaimer/Publisher's Note: The statements, opinions and data contained in all publications are solely those of the individual author(s) and contributor(s) and not of MDPI and/or the editor(s). MDPI and/or the editor(s) disclaim responsibility for any injury to people or property resulting from any ideas, methods, instructions or products referred to in the content.

buildings

MDPI

Article

Experimental Investigation of Infrared Detection of Debonding in Concrete-Filled Steel Tubes via Cooling-Based Excitation

Haonan Cai [1] and Chongsheng Cheng [1,2,*]

1 School of Civil Engineering, Chongqing Jiaotong University, Chongqing 400074, China; 611210080005@mails.cqjtu.edu.cn
2 State Key Laboratory of Mountain Bridge and Tunnel Engineering, Chongqing Jiaotong University, Chongqing 400074, China
* Correspondence: ccheng@cqjtu.edu.cn

Abstract: Debonding in concrete-filled steel tubes (CFSTs) is a common defect that often occurs during the construction phase of CFST structures, significantly reducing their load-bearing capacity. Current methods for detecting debonding in CFSTs using infrared thermography primarily rely on heat excitation. However, applying this method during the exothermic hydration phase presents considerable challenges. This paper proposes the innovative use of spray cooling as an excitation method during the exothermic hydration phase, providing quantitative insights into the heat conduction dynamics on steel plates for infrared debonding detection in CFSTs. The effects of atomization level, excitation distance, excitation duration, and water temperature in the tank on infrared debonding detection performance were examined. The timing of the maximum temperature difference under cooling excitation was analyzed, and the heat conduction characteristics on the surface of the steel plate during the cooling process were explored. A highly efficient and stable cooling excitation method, suitable for practical engineering detection, is proposed, providing a foundation for quantitative infrared debonding detection in CFSTs. This method does not require additional energy sources, features a simple excitation process, and results in a five-times increase in temperature difference in the debonded region after excitation.

Keywords: concrete-filled steel tube (CFST); infrared debonding detection; exothermic hydration phase; spray-cooling excitation

1. Introduction

CFSTs, which are hybrid structures consisting of high-strength steel encasing concrete, have found extensive applications in buildings and bridge engineering due to their exceptional mechanical performance and structural efficiency [1–4]. These structures offer significant benefits, including a high load-bearing capacity, ease of construction, and superior seismic performance. However, debonding between the steel tube and concrete in CFSTs is a common issue that undermines the synergy of the composite material, thereby diminishing its mechanical properties [5–11]. Yang et al. [9] reported that the mechanical performance of CFSTs decreases when the extent of debonding exceeds 0.0757%. When debonding occurs, the load-bearing capacity of CFST components can decrease by 10–32.1%, and the modulus can decline by 30–53% [5–8]. Furthermore, Lu et al. [10] observed that debonding can alter a bridge's natural frequency, affecting its dynamic performance. Thus, debonding in CFSTs is a critical concern for the safety and longevity of these structures.

https://doi.org/10.3390/buildings15030465

During the construction phase, debonding in CFSTs may result in large voids due to factors such as air bubble migration and concrete bleeding [12]. Han et al. [8] observed that air may not be expelled efficiently during the concrete filling process, leading to residual bubbles and subsequent void formation. Additionally, Chen et al. [13] identified concrete bleeding as a key factor in void development in CFSTs during arch bridge construction. While methods such as vibration and hammering can mitigate void formation during construction, they become ineffective once the concrete solidifies. Post-solidification repairs typically involve invasive procedures such as drilling and grouting [12]. Therefore, the timely and accurate detection of debonding in CFSTs during construction is crucial to prevent the need for such damaging repair methods and ensure structural integrity.

Infrared thermography stands as a nondestructive detection technique useful in identifying debonding in CFSTs. This method analyzes discrepancies in the temperature field across the CFST surface [14–16]. Its advantages over traditional methods, such as ultrasound [17], impact echo techniques [18] and acoustic emission [19], include noncontact operation and heightened efficiency. Due to the high detection efficiency, fast detection speed, and long detection distance of infrared defect detection technology, many researchers have recently conducted studies on infrared thermal imaging-based debonding detection in CFSTs [20,21]. A crucial step in enhancing the visibility of defects in infrared thermography is the application of manual active excitation during testing [22–27]. Predominantly, these excitation methods involve various heating techniques, including microwave heating [22], laser heating [23,24], and pulse heating [26,28,29]. These techniques have been extensively researched, focusing on aspects such as excitation distance, uniformity, and duration, as well as post-excitation visualization in the domain of defect detection in metal materials using infrared thermography. Deane et al. [25] employed a 250 W lamp as a thermal source, analyzing the resulting infrared images using techniques such as pulsed phase thermography transforms, the principal component technique, and cold image subtraction. Addressing the challenge of uneven illumination in infrared thermal wave nondestructive testing, Li et al. [26] developed an optimized optical model, ensuring uniform irradiance distribution on curved surfaces. Further advancing the field, Xu et al. [24] introduced a novel method that utilizes an optical excitation line laser as a heat source in infrared thermal imaging. This technique, specifically designed for debonding detection in fiber-reinforced polymer-reinforced concrete structures, overcomes the limitations inherent in traditional infrared methods, such as short heating distance, low thermal sensitivity, and high power requirements.

Most research on CFSTs has historically focused on testing during the operational phase, where exothermic hydration is not a factor [30,31]. Currently, research on conducting semi-real-time infrared tests during the hydration heat phase of CFSTs remains insufficient. During this phase, the hydration of concrete within the steel tube releases significant heat, which may compromise the reliability of traditional thermal excitation methods [30,31]. Li et al. [31] explored using infrared thermography to detect grouting defects in external prestressed tendon ducts during exothermic hydration. Their research, involving semi-real-time infrared tests conducted within 48 h after grouting, demonstrated the method's feasibility. Furthermore, Yang et al. [32] examined the effectiveness of infrared thermography in detecting CFST debonding under varying temperature conditions, simulating the exothermic hydration process of concrete at different water temperatures. Their findings indicated the successful identification of large defects, though smaller ones were overlooked. Additionally, they validated the use of water to simulate concrete's exothermic hydration through ANSYS simulations. Despite these advancements, no studies have yet incorporated external active excitation during the CFST hydration process. Cheng et al. [33] demonstrated that, when CFST is in the construction stage, using the concrete's hydration heat as an internal heat source can effectively enable infrared debonding detection. Factors such as debonding size, hydration heat temperature rise rate, and ambient temperature all influence detection efficiency. Cai et al. [34] found that during the heat absorption and release phases of debonding in CFSTs, the infrared images of the debonding region exhibit different characteristics. The infrared thermal contrast is linearly correlated with the interfacial heat flux of the CFST. When CFST is in the hydration heat phase, the heat from the CFST conducts outward. If heating excitation is applied at this time, the interfacial heat flux of the CFST will approach zero, inhibiting outward heat conduction and ultimately hindering infrared debonding detection.

Therefore, the active excitation of CFST in the hydration heat stage by cooling can enhance the outward conduction of CFST heat, thus improving the infrared detection capability. Cai et al. [35] proved the feasibility of using spray cooling as the excitation source to actively stimulate CFST in the hydration heat stage through model tests, and defined the relationship between the excitation strength and the temperature difference in the debonding area through finite element simulation. However, this process involves the spraying of mist or water to absorb and dissipate thermal energy, with the underlying heat transfer mechanisms being notably complex [36–38]. Pais et al. [39] identified four primary mechanisms in spray cooling: evaporation, forced convection, nucleation sites on heated surfaces, and secondary nucleation sites on droplet surfaces. Yan et al. [40] emphasized the superior cooling efficiency of spray cooling compared to other methods, with Pais et al. [39] demonstrating that its highest heat flux density can reach 1200 W/cm^2. However, the cooling efficiency is influenced by factors such as the atomization level of the water flow [41]. Omer and Ashgriz [41] conducted evaluations of various nozzle types, including dual fluid, spiral, hydraulic, ultrasonic, rotary, and electrostatic, examining their distinct atomization modes. In the context of infrared debonding detection in CFSTs, variables such as spraying distance, atomization level, water temperature, and excitation duration during spray cooling can impact both the cooling effect and the detection outcomes. The influence of these spray-cooling parameters on infrared debonding detection during the CFST's exothermic hydration phase remains unclear. Currently, no standardized, efficient, and convenient method for spray-cooling excitation exists. Furthermore, the quantified detection of CFST debonding under cooling excitation cannot be achieved with existing infrared techniques. Thus, a comprehensive exploration of the impact of spray-cooling parameters during the exothermic hydration phase is essential for developing a foundational understanding and methodology for quantitative infrared debonding detection in CFSTs. Accordingly, a stable, efficient, and convenient spray-cooling excitation method should be formulated.

This study aims to determine the effects of various spray-cooling parameters (i.e., atomization level, spraying distance, water temperature, and excitation duration) on infrared debonding detection during the exothermic hydration of CFSTs through indoor modeling experiments. Additionally, this study explores the mechanism of surface heat transfer during cooling excitation. By integrating the principles of spray cooling with the surface heat conduction mechanism of CFSTs, this research establishes a foundation for a quantitative approach to infrared debonding detection and proposes a highly efficient and stable spray-cooling method for engineering applications. This study involved 39 sets of indoor model experiments to determine the surface temperature field distribution of CFSTs under diverse cooling excitation parameters. The average temperature difference between debonded and non-debonding regions on steel plate surfaces during excitation was used as a metric to evaluate the effectiveness of infrared detection. Subsequently, this study analyzed the variation in surface temperature difference on the steel plate throughout the cooling excitation process, aiming to elucidate the impact of various parameters on infrared debonding detection in CFSTs.

2. Materials and Methods

2.1. Research Framework

For heating excitation, excitation duration and excitation distance are relevant parameters that need to be considered [22–27]. Similarly, for spray cooling, the temperature of the spray water and the atomization level affect the heat transfer of the steel tube, so these parameters also need to be taken into account. As shown in Figure 1, this study selects different excitation parameters (atomization level, spraying distance, tank water temperature, and excitation duration) to determine the cooling excitation conditions. Based on these conditions, spray-cooling excitation experiments are conducted, and infrared thermal imaging is employed to monitor the surface temperature of the steel plate, collecting temperature difference data throughout the excitation process, particularly the average temperature difference between debonded and non-debonded regions, which serves as a metric to evaluate the effectiveness of infrared detection. Subsequently, parametric analysis is performed to investigate the impact of excitation duration, spraying distance, and other factors on heat transfer efficiency and debonding detection accuracy. Finally, based on the analysis results, an optimal experimental method and strategy are proposed to optimize the spray-cooling excitation process, enhancing detection efficiency and stability, thereby providing theoretical support and technical guidance for the infrared debonding detection of CFSTs in practical engineering applications.

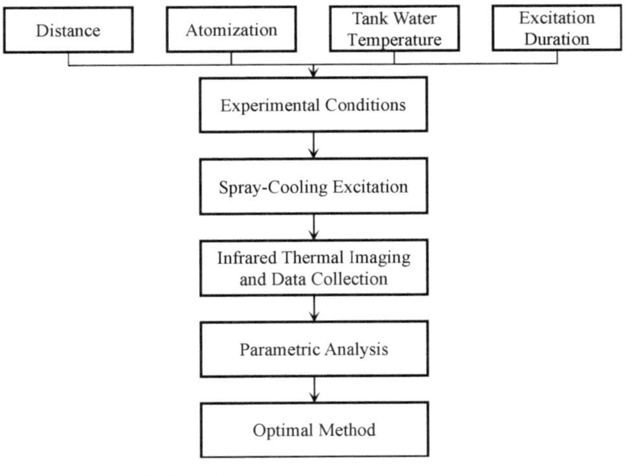

Figure 1. Research flowchart.

2.2. Spray-Cooling Excitation Method

2.2.1. Structure of the Spraying Device

This study utilizes an electric kettle equipped (Deli Electric Kettle, Deli Group Co., Ltd., Ningbo, China) with plain orifice nozzles to simulate the spray-cooling effect on the surface temperature field of a CFST during its exothermic hydration process. The specifications of the system are detailed in Table 1. Plain orifice nozzles, noted for their simplicity and widespread use, are employed in this setup (refer to Figure 2 for the structure) [41]. The temperature of the water in the tank is regulated by filling the tank with water at various temperatures.

Table 1. Specifications of spraying system.

Device Model	Tank Volume	Nozzle Type
DL581040	1000 mL	Plain orifice

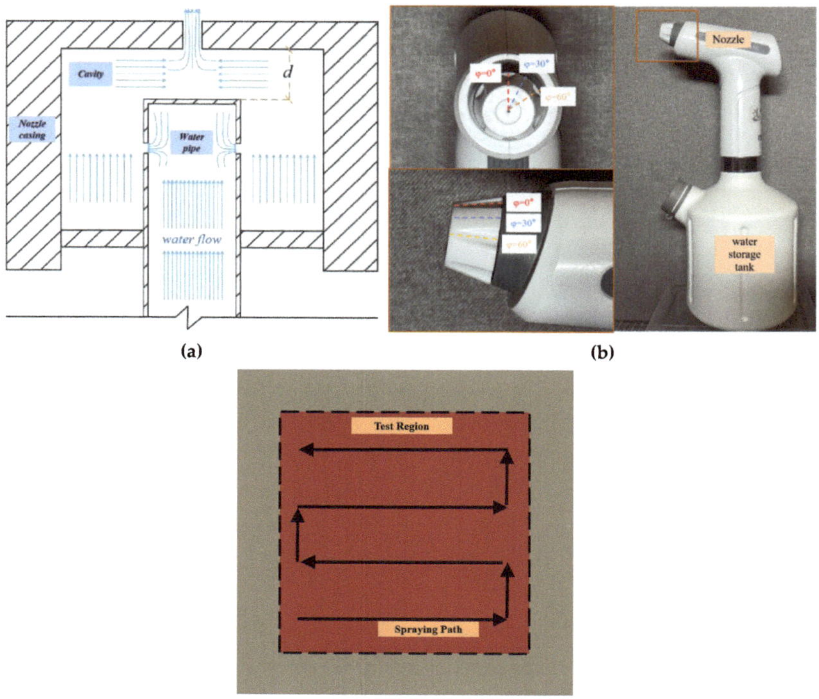

Figure 2. Spraying device details: (**a**) schematic layout of a plain orifice nozzle, (**b**) structure of the spraying device, and (**c**) design diagram of spraying path.

Figure 2a illustrates the nozzle structure utilized for cooling excitation in this study. The nozzle comprises an outer shell and an inner water pipe, which are screwed together to form a cavity. Water from the tank is pressurized by a pump into the nozzle pipe, passes through holes in the pipe, and is then sprayed out from apertures at the nozzle's top. The height of this cavity at the nozzle's apex (d) can be altered by adjusting the nozzle shell's angle (φ), as depicted in Figure 2b. An increase in angle φ results in a corresponding increase in height d of the cavity. According to Bernoulli's equation and the continuity equation [42], a decrease in d leads to an increase in the water's flow velocity at the nozzle outlet. Given that the outlet aperture is constant, changes in the internal cavity's water velocity affect the atomization degree of the water spray and the maximum spray distance. The Sauter mean diameter (SMD) is typically used to quantitatively assess the degree of liquid atomization [43,44]. Research indicates that as the liquid flow rate increases, the SMD of the droplets decreases, enhancing the atomization degree [45–47]. The excitation path, shown in Figure 2c, involves uniformly spraying water from the bottom to the top of the detection region at a constant speed, ensuring that the mist covers the entire test area evenly.

2.2.2. Control Parameters for Spray Excitation

As depicted in Figure 2, this study employs three different nozzle rotation angles ($\varphi = 0°$, $\varphi = 30°$, $\varphi = 60°$), corresponding to three distinct cavity heights: *dmin*, *dmid*, and *dmax* (with *dmax* > *dmid* > *dmin*). The morphology of water spray at these three cavity heights was captured using a camera (Sony ZVE-10, Sony Corporation, Japan, Tokyo), and the maximum spray distance was measured using a measuring tape (Delixi Electric Co., Ltd., Wenzhou, China).

Figure 3 shows the water spray morphology for the three nozzle rotation angles. A comparative analysis of Figure 3a–c reveals that the water forms a cone-shaped spray pattern, with the liquid ejected from all three nozzles in droplet form. Notably, the degree of atomization diminishes as the cavity height increases, a finding that aligns with previous research [46,47]. The spray distance achieved by the three nozzles extends with increasing cavity height, with the shortest distance observed for the $\varphi = 0°$ (*dmin*) nozzle (95 cm).

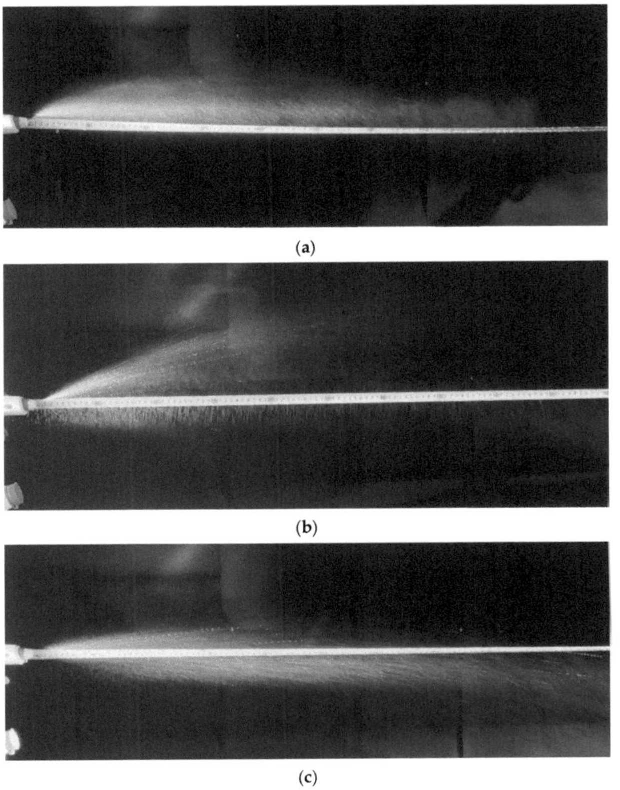

(**a**)

(**b**)

(**c**)

Figure 3. Water spray morphologies: (**a**) nozzle angle $\varphi = 0°$ and cavity height *dmin*, (**b**) nozzle angle $\varphi = 30°$ and cavity height *dmid*, and (**c**) nozzle angle $\varphi = 60°$ and cavity height *dmax*.

Table 2 outlines the relationship between nozzle rotation angles, cavity heights, and atomization levels. The atomization degree progressively declines as cavity height increases. In this study, three nozzle angles ($\varphi = 0°$, 30°, and 60°) were selected to generate three distinct cavity heights (*dmin*, *dmid*, and *dmax*), thereby defining three atomization levels: high, medium, and low. Given that the spray distance of droplets at the high atomization level is limited to 95 cm, the excitation distances used in subsequent tests in this study are set to be less than or equal to 95 cm.

Table 2. Nozzle rotation angles, cavity heights, and corresponding atomization levels.

Nozzle Rotation Angle φ	Cavity Height	Atomization Level
0°	*dmin*	High
30°	*dmid*	Medium
60°	*dmax*	Low

2.3. Experimental Study

2.3.1. Test Platform

As depicted in Figure 4a, this study designed a test platform to simulate the temperature field on the outer surface of a CFST during its exothermic hydration stage. The platform comprises a polyvinyl chloride (PVC) water tank (Hebei Feilong Plastic & Rubber Products Co., Ltd., Hebei, China), coated steel plates (Q235 steel, produced by Chongqing Iron and Steel (Group) Co., Ltd., Chongqing, China), a heating rod (SUSUN Co., Ltd., Zhejiang, China), temperature sensors (TP-100, Shanghai Anyi instruments Co., Ltd., Shanghai, China), prefabricated debonding samples (Kangtai Polymer Materials Co., Ltd., Zhuhai, China), and propellers (SUNSUN Co., Ltd., Zhejiang, China). In practical engineering, interface debonding occurs only at the contact between the concrete and the inner wall of the steel pipe. Therefore, the experiment focuses solely on simulating the temperature change at the concrete interface rather than the core area's temperature [32]. This is achieved through controlled-temperature hot water, a method validated by Yang et al. [32] for simulating concrete's heat generation. In this study, the exothermic hydration of concrete is similarly simulated using heated water. Given that CFSTs in practical engineering typically have diameters ranging from 1.4 to 1.6 m with low surface curvature, the influence of tube curvature on detection results is negligible at the close excitation distances used. For experimental convenience, rectangular steel plates are used to simulate CFSTs.

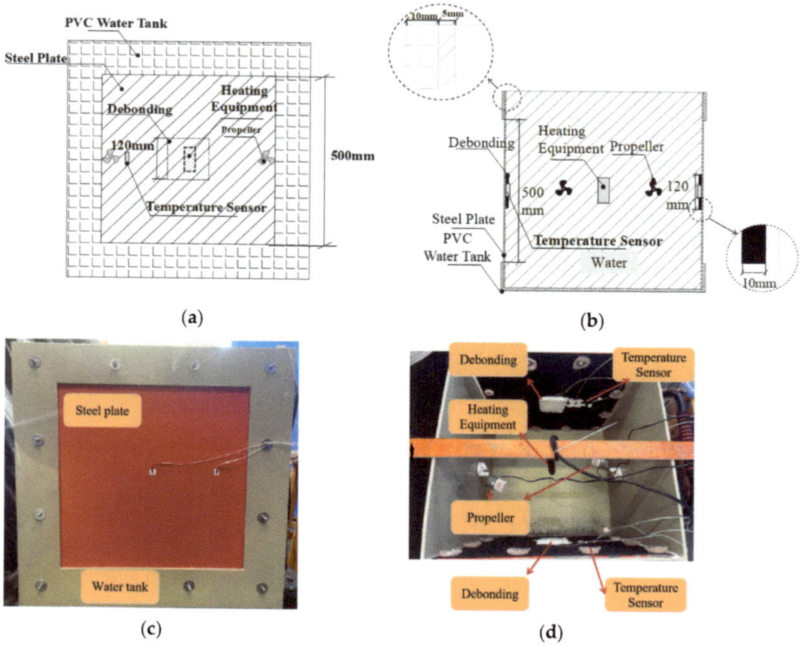

Figure 4. Test platform design and construction: (**a**) front view schematic, (**b**) side view schematic, (**c**) exterior front view, and (**d**) interior side view.

The PVC water tank, with internal dimensions of 700 mm × 700 mm × 700 mm (Figure 4b), has openings of 500 mm × 500 mm on the front and rear sides. Two steel plates (600 mm × 700 mm × 5 mm) are attached to the front and rear sides inside the tank to simulate CFST exteriors. Infrared cameras are positioned for clear views of the steel plate surfaces through these openings. The steel plates are coated with the same paint used on a CFST arch bridge in China.

Styrofoam, having thermal conductivity and specific heat capacity similar to air, is adhered to the steel plates inside the water tank to simulate debonding (Figure 4c). The prefabricated debonding voids, measuring 120 mm × 120 mm × 10 mm, are centrally located on the steel plates. To prevent water ingress into the voids, these areas are covered with cling wrap and sealed with glass adhesive. Heated water is poured into the tank to mimic concrete's exothermic hydration process. A heating rod, suspended in the tank, and temperature sensors, placed on the steel plates' inner sides, facilitate real-time temperature monitoring at the steel–water interfaces. The heating pump is activated to maintain the water temperature at a constant 55.5 °C ± 0.3 °C, reflecting the peak interface temperature during actual CFST exothermic hydration. Two propellers, mounted on the left and right sides of the tank, ensure a uniform water temperature distribution.

2.3.2. Experimental Design and Procedure

The experimental setup, as illustrated in Figure 4a,b, includes infrared cameras positioned on both the front and rear sides of the water tank. These cameras are responsible for detecting the surface temperatures of the two steel plates and capturing their temperature field data. During the experiments, the ambient conditions were controlled to maintain a stable room temperature of 29 °C with minimal airflow. Initially, water heated to 55.5 °C was poured into the tank until the steel plates were fully submerged. This water temperature was consistently maintained at 55.5 °C ± 0.3 °C using a heating pump and temperature sensors. After stabilizing the water temperature in the tank for 1 h, the cooling excitation test commenced. To optimize the experiment's duration, excitation was alternately conducted on the two steel plates. The infrared cameras recorded the temperature field data on the steel plate surfaces throughout the excitation process. The detection period encompassed the entirety of the tests under each excitation condition and continued for 1200 s after excitation. Enclosures were set up around the test platform to prevent ambient light from reflecting off the steel plate surfaces and interfering with the infrared camera readings.

The experiment investigated the impact of various water temperatures in the tank, excitation distances, excitation durations, and atomization levels on infrared debonding detection, employing the excitation method detailed in Section 2. The test conditions, as outlined in Figures 5 and 6, comprised two primary parts. The first part (Figure 6) examined the effects of different water temperatures inside the tank, excitation distances, and atomization levels on infrared detection, using a constant water volume of 500 mL for excitation. This part included excitation distances of 45, 60, 80, and 95 cm, along with water temperatures of 5.5, 10, and 28.5 °C (room temperature) and high and low atomization levels. Test conditions were labeled in the sequence of atomization level, water temperature, and excitation distance; for instance, H-5.5-45 indicates the high atomization level, a 5.5 °C water temperature, and a 45 cm excitation distance.

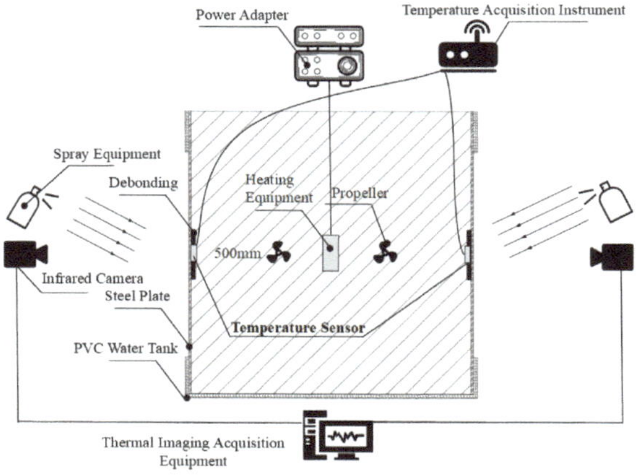

(a)

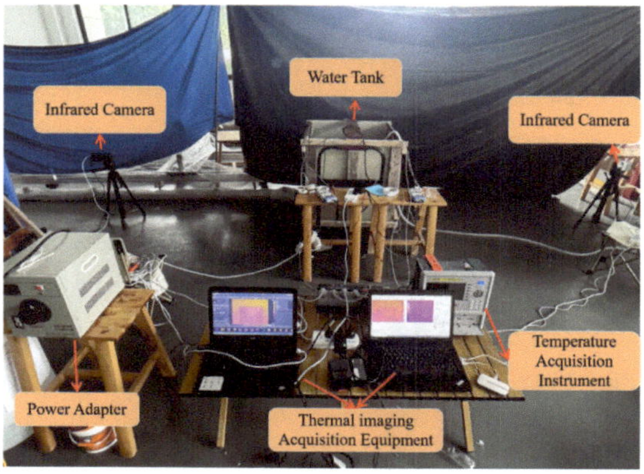

(b)

Figure 5. Experimental design and setup: (**a**) schematic and (**b**) photograph.

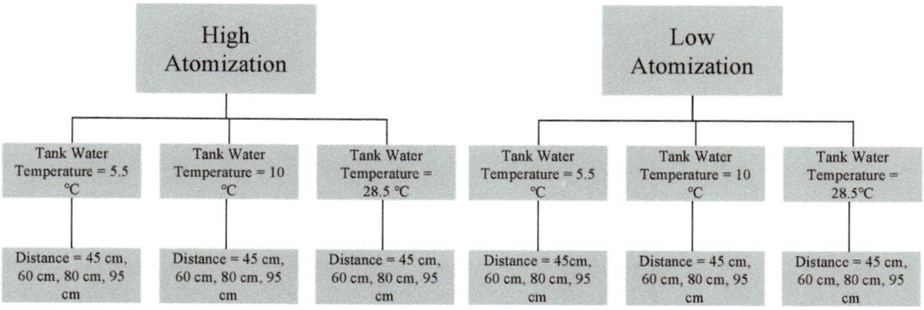

Figure 6. Set 1 of experimental conditions.

The second part of the experiment (Figure 7) focused on the effects of varying excitation durations and atomization levels on infrared detection. This part utilized excitation durations of 40, 80, 120, 200, and 350 s with high, medium, and low atomization levels while maintaining a constant water temperature of 28.5 °C (room temperature) and an excitation distance of 95 cm. Test conditions were labeled according to atomization level and excitation duration; for example, H-40s denotes the high atomization level with a 40 s excitation duration. To avoid interference between consecutive sets of conditions, a

1200 s interval was allotted following the completion of tests for each group. The water temperature inside the tank was verified using a thermometer (alcohol thermometer, accuracy ±0.1 °C, Shanghai Anyi Instruments Co., Ltd., Shanghai, China) before each cooling excitation to ensure compliance with the experimental requirements.

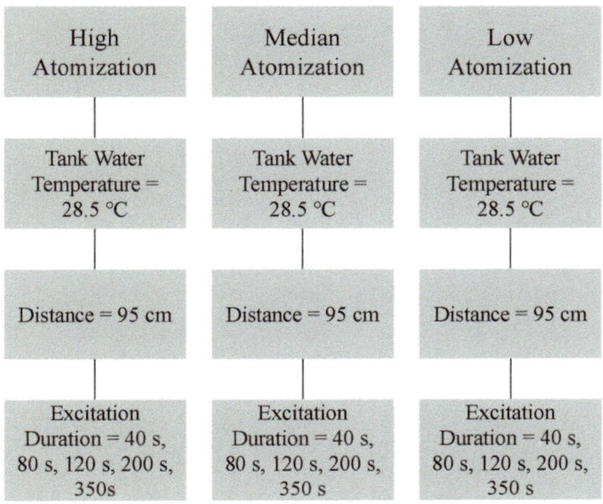

Figure 7. Set 2 of experimental conditions.

2.3.3. Infrared Thermal Imaging and Data Collection

To evaluate the effectiveness of infrared thermography in detecting debonding in CFSTs under cooling excitation, the steel plate area was monitored using long-wave thermal imaging cameras. Figure 8 shows the two infrared cameras used in the experiment: the FLIR A300 (Teledyne FLIR LLC, Goleta, CA, USA) and the MGS-F6 (Shanghai Magnity Technologies Co., Ltd., Shanghai, China). The specifications of these cameras are detailed in Table 3.

(**a**) (**b**)

Figure 8. Infrared imaging setup: (**a**) FLIR A300; (**b**) MGS-F6.

Table 3. Parameters of infrared thermal imaging cameras.

Camera Model	FLIR A300	MGS-F6
Detector type	Uncooled microbolometer	Uncooled microbolometer
NETD	<0.05 °C	<0.05 °C
Accuracy	±2 °C or ±2%	±0.7 °C or 0.7%
Resolution	320 × 240 pixels	640 × 480 pixels
Spectral range	7.5–13 μm	8–14 μm
Field of view	25° × 18.8°	25° × 19°

Both cameras are noncooled, long-wave infrared models. The FLIR A300 features a resolution of 320 × 240 pixels, a spectral range of 7.5–13 µm, and a noise equivalent temperature difference (NETD) below 0.05 °C (NETD is a metric used in infrared and thermal imaging systems to measure the device's sensitivity in detecting minor temperature variations). The MGS-F6 camera offers a higher resolution of 640 × 480 pixels, a spectral range of 8–14 µm, and an NETD below 0.05 °C. Both infrared cameras were positioned at the front and rear sides of the specimens. This setup ensured that the center of the infrared image aligned with the center of the steel plate, and the plane of the steel plate was parallel to the imaging plane of the camera. The image capture frequency for both cameras was set at one frame per second.

2.3.4. Data Analysis and Evaluation Indices

The temperature difference between the debonded and non-debonding regions serves as a critical index for evaluating the efficacy of infrared detection. Figure 9 illustrates the extraction of temperature fields from both the debonding region (T_D) and the non-debonding region (T_{non}) in thermal images. The relevant symbol abbreviations used in this study will be summarized in Appendix A.

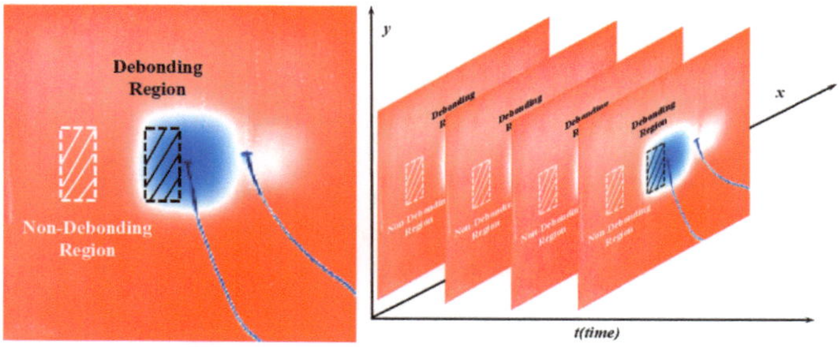

Figure 9. Temperature field extraction from debonded and non-debonding regions.

This study does not include an analysis of data from heat flux sensors affixed to the external surface of the steel plate (as shown in Figure 8). To mitigate the influence of these sensors, the left side of the steel plate's external surface was designated as the debonding region. The adjacent left part of the steel plate surface was selected as the non-debonding region. For a quantitative analysis of the impact of cooling excitation on detection effectiveness under different conditions, the following equation was employed to calculate the average temperature fields of both regions:

$$T_{Dmean} = \frac{\sum_{i=1}^{n} T_{Di}}{n} \qquad T_{nonmean} = \frac{\sum_{i=1}^{n} T_{noni}}{n} \tag{1}$$

where T_{Di} and T_{noni} represent the temperatures of pixel i in the debonded and non-debonding regions, respectively. The variable n denotes the number of pixels in the selected area of the thermal images. T_{Dmean} and $T_{nonmean}$ indicate the average temperatures of the debonded and non-debonding regions, respectively.

The average temperature difference between these regions, ΔT, is calculated using the absolute value of their difference, as follows [34]:

$$\Delta T = |T_{Dmean} - T_{nonmean}| \tag{2}$$

By extracting the temperature field data from each frame throughout the excitation process, as depicted in Figure 9, the average temperature difference for the entire excitation period can be obtained. An increase in the average temperature difference indicates improved resolution between the debonded and non-debonding regions. The value of ΔT under cooling excitation reflects the performance of infrared debonding detection. Furthermore, the maximum value of ΔT during the excitation period, maxΔT, is extracted as follows:

$$\max\Delta T = \max(\Delta T_1, \Delta T_2, \Delta T_3 \ldots \Delta T_n) \quad (3)$$

By analyzing the timing and magnitude of maxΔT, the optimal detection moment for a single condition and the degree of detection enhancement resulting from excitation can be effectively characterized. The time at which maxΔT occurs, denoted as $t_{\max\Delta T}$, is defined as the moment when the contrast between the debonding and non-debonding areas is at its peak (the optimal detection moment). An analysis of $t_{\max\Delta T}$ facilitates the determination of the best detection time.

3. Results and Discussion

3.1. Evolution of the Average Temperature Difference During Spray-Cooling Excitation

The temporal changes in the average temperature difference under cooling excitation reflect the overall trends observed in infrared detection. Figure 10 illustrates the evolution of the average temperature difference (ΔT) under (a) different excitation distances with fixed spray-cooling water volumes and (b) varying excitation durations. Due to space limitations, the figure selectively presents representative conditions from the experiment. However, the temperature difference exhibited similar patterns across other conditions.

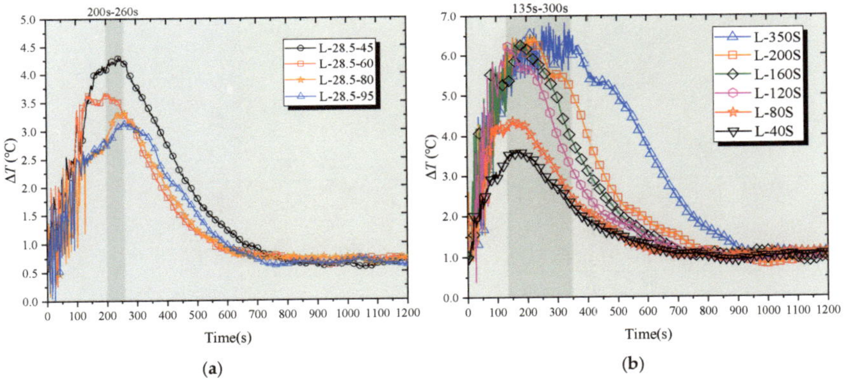

Figure 10. Temporal evolution of average temperature difference under cooling excitation: (**a**) for different excitation distances at constant water volume; (**b**) at various excitation durations.

In Figure 10a, the data demonstrate the relationship between excitation distance (45, 60, 80, and 95 cm) and the average temperature difference over time. For each condition shown, the water temperature in the tank was maintained at 28.5 °C, the total water volume of spray cooling was 500 mL, and the atomization level was selected as low. The figure indicates that the maximum average temperature difference is reached at a specific moment (approximately between 200 and 260 s), irrespective of the excitation distance. As the excitation distance increases (from 45 cm to 95 cm), the peak value of the average temperature difference progressively decreases. Approximately 750 s after the start of excitation, the value of the maximum average temperature difference for each condition reverts to the pre-cooling excitation level (around 0.75 °C).

Figure 10b presents results for conditions with a water temperature of 28.5 °C in the tank, an excitation distance of 95 cm, and a low atomization level for excitation durations ranging from 40 s to 350 s. These data show that the maximum average temperature difference increases with longer excitation durations. The average temperature difference for each condition returns to the pre-cooling excitation level approximately between 750 and 1000 s from the start of excitation. Given that experimental intervals of 1200 s (exceeding 900 s) were established between each group in this study, it can be concluded that the different experimental conditions are independent and do not influence each other.

3.2. Maximum Average Temperature Difference Relative to Excitation Distance and Water Temperature

Given that the average temperature under cooling excitation exhibits a rising and then declining trend, a maximum average temperature difference (maxΔT) value is present for each operating condition. This maxΔT value is indicative of the upper detection limit of infrared technology under cooling excitation. Both the spray distance and the water temperature within the tank significantly influence the maxΔT value. Consequently, Figure 11 plots the maxΔT values for three different tank water temperatures (5, 10, and 28.5 °C) against the excitation distance.

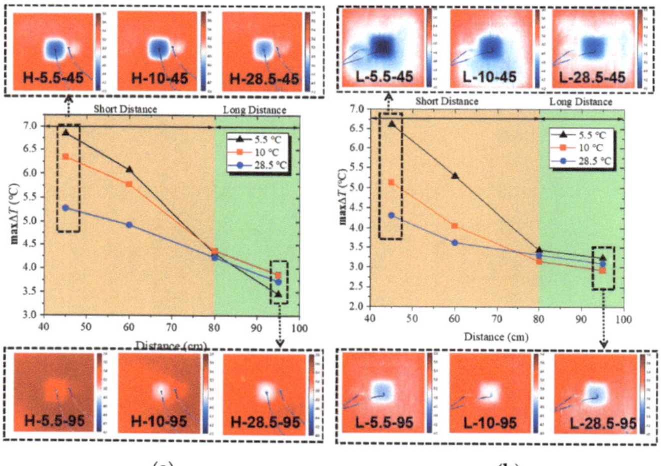

Figure 11. Maximum average temperature difference according to excitation distance at a (**a**) high atomization level and (**b**) low atomization level.

Figure 11 shows the maxΔT in relation to various spraying distances and atomization levels (high and low), along with infrared images captured at the moment when maxΔT values are observed. A noticeable low-temperature region, resembling the debonding area, is visible in the center of the infrared images, particularly for excitation distances of 45 and 95 cm. This feature aids in distinguishing the debonding. In the infrared image with a 45 cm excitation distance, the debonding region is more distinctly discernible due to a higher maxΔT compared to other excitation distances.

As shown in Figure 11a,b, when the water temperature in the tank is constant, the maxΔT decreases with an increase in excitation distance. Examining the blue curves (representing a water temperature of 28.5 °C) in Figure 11a,b, the decline in maxΔT is more pronounced at a high atomization level as the distance increases. Figure 11b demonstrates that the influence of distance on maxΔT is relatively minor when the spraying distance ranges between 80 and 95 cm. As shown in Table 4, the standard deviations of maxΔT at three different water temperatures are 0.06 and 0.17, respectively. At the high atomization level (Figure 11a), smaller droplet volumes and decreased momentum may render the droplets more susceptible to gravitational and air disturbances during flight. This leads to fewer droplets reaching the steel plate surface as distance increases, thus diminishing the excitation effect. In contrast, at the low atomization level (Figure 11b), the droplets are larger and possess greater initial kinetic energy, resulting in a less pronounced decrease in the number of droplets reaching the test surface with increasing distance. Furthermore, the decline in maxΔT with increasing excitation distance is less severe. The maxΔT remains relatively stable when the distance is between 80 and 95 cm. As shown in Table 5, the standard deviations of maxΔT at three different water temperatures are 0.12 and 0.13, respectively.

Table 4. Maximum average temperature difference according to excitation distance at a high atomization level.

Distance	5.5 °C	10 °C	28.5 °C	Standard Deviation
45 cm	6.85 °C	6.34 °C	5.28 °C	0.65 °C
60 cm	6.07 °C	5.77 °C	4.93 °C	0.48 °C
80 cm	4.31 °C	4.37 °C	4.23 °C	0.06 °C
95 cm	3.46 °C	3.87 °C	3.73 °C	0.17 °C

Table 5. Maximum average temperature difference according to excitation distance at a low atomization level.

Distance	5.5 °C	10 °C	28.5 °C	Standard Deviation
45 cm	6.62 °C	5.13 °C	4.32 °C	0.95
60 cm	5.29 °C	4.06 °C	3.63 °C	0.70
80 cm	3.44 °C	3.15 °C	3.31 °C	0.12
95 cm	3.24 °C	2.93 °C	3.1 °C	0.13

Figure 11a indicates that at distances below 80 cm (marked as the orange background) and with a constant excitation distance, a lower water temperature in the tank leads to an increased maximum average temperature difference. When the distance is 80 cm or greater (denoted as the green background), the maxΔT values for the three tested water temperatures (5.5, 10, and 28.5 °C) exhibit minimal variation (standard deviation of max$\Delta T \leq 0.17$), suggesting that the water temperature in the tank has a negligible effect on the maxΔT at these distances. A similar trend is observed in Figure 11b; for distances of 80 cm or more (green background), the maxΔT values across the three water temperatures are comparatively uniform (standard deviation of max$\Delta T \leq 0.13$), indicating a minimal impact of water temperature on maxΔT. This pattern can be attributed to the heat exchange between the droplets and the air during their flight. As droplets travel, they gradually warm up due to this exchange [44]. At shorter spraying distances (<80 cm), the flight duration of the droplets is limited, providing insufficient time for them to absorb significant ambient heat before reaching the steel plate. In such cases, the water temperature in the tank considerably influences the maxΔT. At longer distances (≥80 cm), the droplets experience more prolonged heating in the air, often reaching temperatures close to room temperature (28.5 °C) by the time they contact the steel plate. Under these conditions, the initial water temperature has almost no impact on the maxΔT.

In summary, under cooling excitation, the maxΔT between the debonded and non-debonding regions is significantly enhanced, ranging from 3 °C to 7 °C. Particularly at the low atomization level, when the spraying distance extends to 80–90 cm, the maxΔT

shows insensitivity to both the spraying distance and the water temperature in the tank. During this phase, the maxΔT (ranging from 3.1 °C to 3.4 °C) remains relatively stable. This temperature differential is substantial enough to effectively distinguish debonding areas.

3.3. Timing of the Maximum Average Temperature Difference Relative to the Excitation Duration

The timing of the maximum average temperature difference determines the optimal detection moment, and conducting the detection at this moment helps improve accuracy. Figure 12 and Table 6 illustrate the timing of the maximum average temperature difference ($t_{\max\Delta T}$) in relation to the excitation duration at various atomization levels. The duration of spray-cooling excitation affects the total heat exchange between the steel plate and water, consequently influencing the average temperature difference between the debonded and non-debonding regions. The dashed line in Figure 12 indicates that the occurrence of the maxΔT aligns with the duration of the excitation. However, the timing of maxΔT following spray-cooling excitation does not exhibit a direct linear correlation with the excitation duration. For excitation durations shorter than 120 s (marked as the green background in Figure 11) and at a consistent atomization level, the correlation between the timing of maxΔT and the excitation duration appears relatively weak. The moments of occurrence are clustered within a narrow range: from 135 s to 175 s at the low atomization level, from 95 s to 135 s at the medium level, and from 125 s to 165 s at the high atomization level. Conversely, when the excitation duration exceeds 120 s (indicated by the blue background) at the same atomization level, the timing of maxΔT tends to increase proportionally with the excitation duration. In this phase, the occurrence of maxΔT for both low and medium atomization levels closely approaches the conclusion of the excitation period.

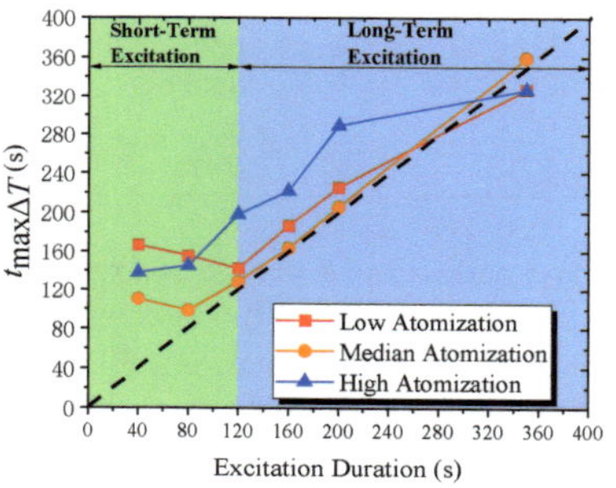

Figure 12. Temporal relation of average temperature difference to excitation duration.

Table 6. The timing of the maximum temperature difference under different excitation durations and atomization levels.

Excitation Duration (s)	Low Atomization	Median Atomization	High Atomization
40	166 s	110 s	138 s
80	155 s	98 s	145 s
120	142 s	128 s	198 s
160	186 s	163 s	222 s
200	225 s	205 s	289 s
350	327 s	359 s	327 s

Figure 13 illustrates the liquid morphologies on the steel plate surface at different excitation durations for the medium atomization level test group. In this study, the surface temperature of the steel plate in the specimen was maintained at 55.5 °C ± 0.3 °C, a temperature insufficient to induce boiling of the liquid droplets. The heat exchange between the liquid droplets and the steel plate primarily occurs through surface evaporation and forced convection resulting from the impact of heated surface droplets [31]. As shown in Figure 13, at an excitation duration of 40 s, the water on the steel plate surface predominantly exists as liquid droplets with limited mobility. With increasing excitation duration (from 40 s to 120 s), these droplets gradually coalesce and flow under the influence of gravity. Further extension of the excitation time (from 120 s to 350 s) leads to a noticeable downward flow of water on the steel plate surface. When the excitation duration is short (<120 s), the limited time and water volume prevent the formation of a fast-flowing water stream on the steel plate. Under these conditions, heat dissipation from the steel plate surface primarily occurs through evaporation. After excitation ceases, evaporation continues to contribute to heat transfer, further increasing the maximum average temperature difference. As the liquid gradually evaporates, the efficiency of heat transfer by evaporation decreases, leading to the peak in maxΔT. Consequently, a weak correlation is observed between the excitation duration and the timing of maxΔT. However, for longer excitation durations (>120 s), the continuous water flow on the steel plate surface effectively removes heat, primarily through forced convection. Once the excitation stops, this dominant convection rapidly diminishes, making it challenging to further expand the temperature difference between the debonded and non-debonding regions. This results in the temperature peak being reached sooner. Therefore, a more pronounced linear correlation is observed between the excitation duration and the timing of the maximum temperature difference.

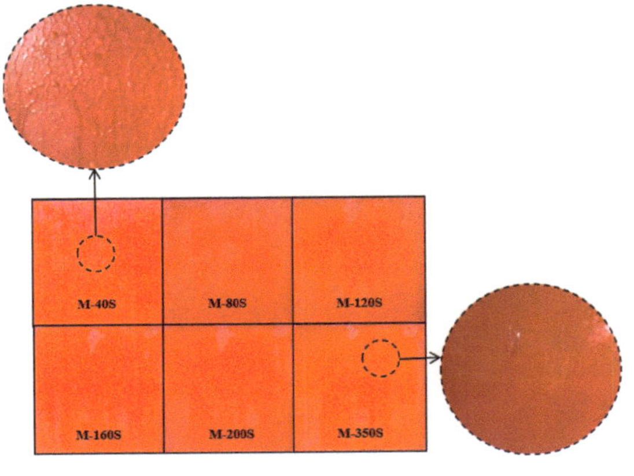

Figure 13. Liquid morphology on steel plate surface at medium atomization level over various excitation durations.

In summary, different excitation durations lead to variations in the water morphology on the steel plate, resulting in two distinct modes of heat exchange. This variation in heat exchange mechanisms may explain why the timing of the appearance of the maxΔT following spray-cooling excitation does not exhibit a strict linear correlation with the excitation duration.

3.4. Maximum Average Temperature Difference Relative to Excitation Duration

In practical detection, the larger the maxΔT, the more favorable it is for detection. Figure 14 and Table 7 present the maximum average temperature difference in relation to the excitation duration under various working conditions. When the excitation duration is short (<120 s, indicated as the green region), there is only a marginal difference in maxΔT

across the three atomization levels at the same excitation duration. This suggests a weak correlation between maxΔT and atomization level, implying that maxΔT is predominantly influenced by the excitation duration. Experimental results across the three atomization levels indicate a consistent increase in maxΔT corresponding to longer excitation durations. In cases where the excitation duration is extended (≥120 s, labeled as the blue region), maxΔT decreases with an increase in atomization level under the same excitation time. At a constant atomization level, maxΔT exhibits a gradual increase with longer excitation durations, tending to stabilize around a nearly constant value. Under long excitation durations (≥120 s), heat exchange on the steel plate surface is primarily dominated by forced convection. Due to the larger volume of liquid droplets at the low atomization level, the heat transfer on the steel plate surface is more intense compared to the high atomization level test group, leading to a more significant increase in maxΔT for the low atomization level group.

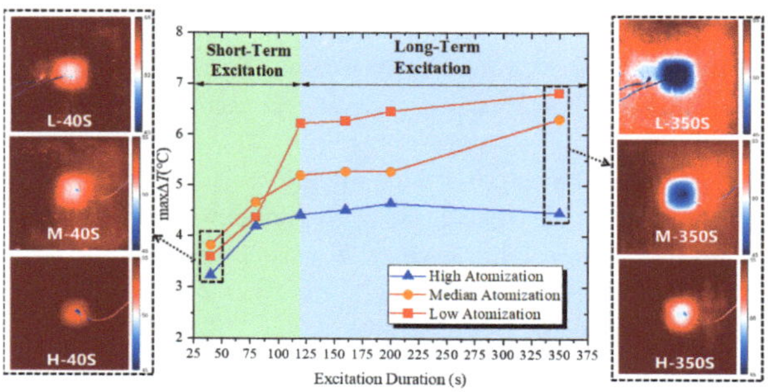

Figure 14. Maximum average temperature difference by excitation duration.

Table 7. Maximum average temperature difference by excitation duration.

Excitation Duration (s)	Low Atomization	Median Atomization	High Atomization
40	3.6 °C	3.82 °C	3.24 °C
80	4.37 °C	4.67 °C	4.2 °C
120	6.22 °C	5.2 °C	4.42 °C
160	6.27 °C	5.28 °C	4.52 °C
200	6.46 °C	5.28 °C	4.65 °C
350	6.81 °C	6.3 °C	4.46 °C

In summary, under consistent atomization levels and extended excitation durations, there is minimal variation in maxΔT relative to the excitation duration. Specifically, when long durations (≥120 s), low atomization levels, and extended distances (≥80 cm) are selected for spray-cooling excitation, the variation in maxΔT after excitation is minor (with a standard deviation of 0.23 °C). Under these conditions, maxΔT typically occurs close to the end of the excitation period, and its value shows limited correlation with the selection of various excitation parameters. Instead, the value of maxΔT may be more related to factors such as the debonding shape, CFST structure, material properties, and peak concrete temperature. Additionally, the range of maxΔT achieved under these conditions (between 6.22 and 6.81 °C), compared to the ΔT before cooling excitation (approximately 1 °C), increases by about five times, which is sufficient to detect the presence of debonding in typical scenarios. In practical detection, the detection personnel can use the above method to actively excite the CFST test area, without needing to control the water temperature during excitation. It is sufficient to let the water temperature approach ambient temperature.

After excitation, infrared images can be immediately collected to identify the debonding. Compared to traditional thermal excitation methods, such as microwave heating (900 W equipment power) [22] and flashlamp heating (single bulb power of 250 W) [25] the cooling excitation method proposed in this paper can be implemented while the object is in the exothermic phase, without requiring additional energy sources for excitation, making it a simpler and more suitable method for outdoor use.

4. Conclusions

The application of spray-cooling excitation significantly enhances the maximum average temperature difference between debonded and non-debonding regions, thereby improving the efficacy of infrared thermography for detecting debonding in CFSTs. Specifically, under the conditions of a low atomization level and extended spraying distance, the impact of both the spraying distance and the water temperature within the tank on the maximum average temperature difference is minimal.

When the spraying duration is short, the primary mode of heat removal from the steel plate surface is through surface evaporation. In such scenarios, the maximum average temperature difference occurs within a specific range and tends to increase with the duration of excitation. With prolonged excitation, the dominant mechanism for heat removal shifts to forced convective flow. Consequently, the variation in the maximum average temperature difference in response to the excitation duration becomes relatively minor.

We studied spray-cooling excitation with a long duration (≥120 s), low atomization level, and extended distance (≥80 cm). In this context, spray-cooling excitation significantly increases the maximum average temperature difference, and variations in the spray-cooling parameters have only a minor effect on this temperature difference. Additionally, the optimal moment for detection typically aligns with the end of the excitation period.

In practical testing, an excitation duration greater than 120 s, an excitation distance greater than 80 cm, and an excitation water temperature set to ambient temperature with a low atomization level can be selected. This method can increase the maximum average temperature difference by about five times compared to before excitation, without the need for additional manual control of the excitation water temperature, making it convenient for application in actual engineering. The use of water as the excitation source avoids environmental pollution and is cost-effective. Additionally, no extra energy supply equipment is required, which helps reduce project costs and improve project efficiency.

5. Limitations and Future Work

However, it is important to note that this study is limited to laboratory experiments and has not been applied in real-world engineering settings. The findings and methods presented here may not fully account for the complexities and variables present in actual engineering environments, and further validation in practical engineering projects is necessary to assess the method's performance and reliability under real-world conditions. Future research should consider adopting the method of spray-cooling excitation to minimize research variables, enabling a more focused and controlled investigation. This approach is expected to facilitate a comprehensive examination of how additional factors—such as CFST (concrete-filled steel tube) dimensions, debonding sizes, and the temperature of exothermic hydration—affect the maximum average temperature difference. A deeper understanding of these factors could lead to the more precise and quantitative detection of debonding, providing valuable insights for both theoretical and practical applications. Furthermore, exploring the development of highly mobile spray-cooling equipment would allow for the more efficient utilization of excitation water, enabling precise targeting and uniform cooling. This innovation would enhance detection efficiency by reducing water waste and energy consumption. Ultimately, the integration of such advanced equipment into engineering practices promises not only more efficient, stable, and convenient detection of debonding but also a more sustainable approach for large-scale industrial applications.

Author Contributions: H.C.: Conceptualization, Writing—Original Draft, Review and Editing, Data Curation, Formal Analysis, Validation. C.C.: Methodology, Writing—Review and Editing, Project Administration, Funding Acquisition. All authors have read and agreed to the published version of the manuscript.

Funding: This work was partially supported by the National Natural Science Foundation of China (No.52108267, 52478301), the Chongqing Natural Science Foundation of China (CSTB2022NSCQ-MSX1379), the Special Support Program of Chongqing Postdoctoral Research (2021XMT007), and the China Postdoctoral Science Foundation under Grant Number 2024T171100.

Data Availability Statement: The original contributions presented in this study are included in this article. Further inquiries can be directed to the corresponding authors.

Acknowledgments: The State Key Laboratory of Mountain Bridge and Tunnel Engineering is acknowledged for its invaluable support in the experimental phase of this research. The provision of advanced facilities and the conducive environment for scientific exploration offered by the laboratory were crucial in the successful execution of the experiments. Gratitude is extended for the unwavering commitment to fostering academic and scientific excellence, which has significantly contributed to the progression and completion of this study.

Conflicts of Interest: The authors declare no conflicts of interest.

Appendix A

T_{Dmean}: Average temperatures of the debonded regions;

T_{nonmean}: Average temperatures of the non-debonded regions;

ΔT: The average temperature difference between non-debonded regions and debonded regions;

$\max\Delta T$: The maximum value of ΔT during the excitation period;

$t_{\max\Delta T}$: The timing of the maximum average temperature difference.

References

1. Han, L.-H.; Li, W.; Bjorhovde, R. Developments and advanced applications of concrete-filled steel tubular (CFST) structures: Members. *J. Constr. Steel Res.* **2014**, *100*, 211–228. [CrossRef]
2. Chen, B.-C.; Wang, T.-L. Overview of Concrete Filled Steel Tube Arch Bridges in China. *Pract. Period. Struct. Des. Constr.* **2009**, *14*, 70–80. [CrossRef]
3. Zhu, Y.; Yang, H.; Yang, X.; Sun, F. Behavior of concrete-filled steel tubes subjected to axial impact loading. *J. Constr. Steel Res.* **2020**, *173*, 106245. [CrossRef]
4. Ou, Z.; Chen, B.; Hsieh, K.H.; Halling, M.W.; Barr, P.J. Experimental and Analytical Investigation of Concrete Filled Steel Tubular Columns. *J. Struct. Eng.* **2011**, *137*, 635–645. [CrossRef]
5. Liao, F.-Y.; Han, L.-H.; He, S.-H. Behavior of CFST short column and beam with initial concrete imperfection: Experiments. *J. Constr. Steel Res.* **2011**, *67*, 1922–1935. [CrossRef]
6. Xue, J.-Q.; Fiore, A.; Liu, Z.-H.; Briseghella, B.; Marano, G.C. Prediction of ultimate load capacities of CFST columns with debonding by EPR. *Thin-Walled Struct.* **2021**, *164*, 107912. [CrossRef]
7. Xue, J.-Q.; Briseghella, B.; Chen, B.-C. Effects of debonding on circular CFST stub columns. *J. Constr. Steel Res.* **2012**, *69*, 64–76. [CrossRef]
8. Han, L.-H.; Ye, Y.; Liao, F.-Y. Effects of Core Concrete Initial Imperfection on Performance of Eccentrically Loaded CFST Columns. *J. Struct. Eng.* **2016**, *142*, 04016132. [CrossRef]
9. Shi-Cong, Y.; Fu-Min, W.; Ping, Q.U. Brief Introduction to the Core Concrete's Empty Influence on the Mechanical Performance of Concrete Filled Steel Tube Components. *J. Chongqing Jiaotong Univ.* **2008**, *27*, 360–365.
10. Lu, Z.; Guo, C.; Li, G. Air void and ring gap effect on CFST arch bridges dynamic performance. *J. Constr. Steel Res.* **2021**, *177*, 106418. [CrossRef]

11. Zhang, Z.; Pang, K.; Xu, L.; Zou, Y.; Yang, J.; Wang, C. The bond properties between UHPC and stone under different interface treatment methods. *Constr. Build. Mater.* **2023**, *365*, 130092. [CrossRef]
12. Wen, Y.N.; Su, J.C. Overview of Concrete Filled Steel Tube Viod Countermeasures. *Appl. Mech. Mater.* **2013**, *256–259*, 1121–1124. [CrossRef]
13. Chen, X.; Liao, F.; Mohamed, H.S.; Wang, J.; Huang, Z.; Zhang, K. Behavior of CFRP strengthened concrete-filled steel tubes with spherical-cap gap under axial compression. *Structures* **2023**, *58*, 105339. [CrossRef]
14. Ciampa, F.; Mahmoodi, P.; Pinto, F.; Meo, M. Recent Advances in Active Infrared Thermography for Non-Destructive Testing of Aerospace Components. *Sensors* **2018**, *18*, 609. [CrossRef]
15. Clark, M.R.; McCann, D.M.; Forde, M.C. Application of infrared thermography to the non-destructive testing of concrete and masonry bridges. *NDT E Int.* **2003**, *36*, 265–275. [CrossRef]
16. Tang, Q.; Dai, J.; Bu, C.; Qi, L.; Li, D. Experimental study on debonding defects detection in thermal barrier coating structure using infrared lock-in thermographic technique. *Appl. Therm. Eng.* **2016**, *107*, 463–468. [CrossRef]
17. Iyer, S.; Sinha, S.K.; Tittmann, B.R.; Pedrick, M.K. Ultrasonic signal processing methods for detection of defects in concrete pipes. *Autom. Constr.* **2012**, *22*, 135–148. [CrossRef]
18. Hola, J.; Sadowski, L.; Schabowicz, K. Nondestructive identification of delaminations in concrete floor toppings with acoustic methods. *Autom. Constr.* **2011**, *20*, 799–807. [CrossRef]
19. Shen, W.; Bai, H.; Wang, F.; Li, C.; Du, F. Acoustic Emission characteristics and damage evolution of Concrete-Encased CFST columns under compressive load. *Eng. Fract. Mech.* **2024**, *311*, 110578. [CrossRef]
20. Shao, Z.; Zha, X. Research on detection of gap defect of CFST based on temperature method. *Structures* **2024**, *61*, 105977. [CrossRef]
21. Cheng, C.; Chen, D.; Shao, S.; Na, R.; Cai, H.; Zhou, H.; Wu, B. Revealing the Impact of Depth and Surface Property Variations on Infrared Detection of Delamination in Concrete Structures Under Natural Environmental Conditions. *Buildings* **2025**, *15*, 10. [CrossRef]
22. Cuccurullo, G.; Berardi, P.G.; Carfagna, R.; Pierro, V. IR temperature measurements in microwave heating. *Infrared Phys. Technol.* **2002**, *43*, 145–150. [CrossRef]
23. Fang, W.; Yang, X.; Wang, X.; Hu, G.; Tao, N.; Zhang, C. Sequential laser pulse thermal excitation method for the detection of defects in metallic materials. In Proceedings of the Infrared, Millimeter-Wave, and Terahertz Technologies VIII 2021, Nantong, China, 10–12 October 2021. [CrossRef]
24. Xu, Y.; Sohn, H. Nondestructive debonding detection of fiber reinforced plastics strengthened structure based on infrared thermal imaging with laser thermal excitation. In Proceedings of the Sensors and Smart Structures Technologies for Civil, Mechanical, and Aerospace Systems 2020, Online, 27 April–9 May 2020; pp. 128–144. [CrossRef]
25. Deane, S.; Avdelidis, N.P.; Ibarra-Castanedo, C.; Williamson, A.A.; Withers, S.; Zolotas, A.; Maldague, X.P.V.; Ahmadi, M.; Pant, S.; Genest, M.; et al. Development of a thermal excitation source used in an active thermographic UAV platform. *Quant. InfraRed Thermogr. J.* **2022**, *20*, 198–229. [CrossRef]
26. Li, D.; Yang, H.; Qin, Q. Optimal design of flash thermal excitation in infrared nondestructive testing system. In Proceedings of the 3rd IEEE International Conference on Electronic Information Technology and Computer Engineering, EITCE 2019, Xiamen, China, 18–20 October 2019; pp. 1771–1775. [CrossRef]
27. Zhang, Q.; Liu, J.; Gu, J.; Tian, Y. Study on coal-rock interface characteristics change law and recognition based on active thermal excitation. *Eur. J. Remote Sens.* **2022**, *55*, 35–45. [CrossRef]
28. Tang, Q.; Liu, J.; Dai, J.; Yu, Z. Theoretical and experimental study on thermal barrier coating (TBC) uneven thickness detection using pulsed infrared thermography technology. *Appl. Therm. Eng.* **2017**, *114*, 770–775. [CrossRef]
29. Bu, C.; Tang, Q.; Liu, Y.; Yu, F.; Mei, C.; Zhao, Y. Quantitative detection of thermal barrier coating thickness based on simulated annealing algorithm using pulsed infrared thermography technology. *Appl. Therm. Eng.* **2016**, *99*, 751–755. [CrossRef]
30. Sun, J.; Xie, J. Simulation analysis of the hydration heat of large diameter CFST arch and its effects on loading age. *Appl. Therm. Eng.* **2019**, *150*, 482–491. [CrossRef]
31. Li, S.; Han, S.; Wang, J.; Han, X.; Zheng, P.; Cui, C.; Gao, X.; Sun, S. Infrared thermography detection of grouting defects in external post-tensioned tendon ducts under construction hydration heat excitation. *NDT E Int.* **2023**, *134*, 102785. [CrossRef]
32. Yang, H.; Liu, W.; Wei, D.; Wang, G.; Shen, J. Study on the Detection of the Compactness of Concrete Filled Steel Tube with Model Temperature Difference. *China Concr. Cem. Prod.* **2018**, *5*, 78–81. (In Chinese)
33. Cheng, C.; Cheng, X.; Zhang, H.; Cai, H.; Zhou, J.; Na, R.; Wu, B. Experimental study on infrared detection of debonding in concrete-filled steel tubular structure under acceleratory period of hydration heat action. *Case Stud. Constr. Mater.* **2024**, *21*, e03928. [CrossRef]
34. Cai, H.; Cheng, C.; Wang, L.; Zhang, H.; Zhou, J.; Na, R.; Wu, B. Numerical and experimental study on the evolution of thermal contrast for infrared detection of debonding in concrete filled steel tubular structure. *Appl. Therm. Eng.* **2025**, *258*, 124743. [CrossRef]

35. Cai, H.; Cheng, C.; Na, R.; Zhang, H.; Zhou, J.; Jing, S.; Miao, C. Cooling-excited infrared thermography for enhancing the detection of concrete filled steel tube interfacial debonding at concrete hydration. *Case Stud. Constr. Mater.* **2024**, *20*, e02995. [CrossRef]
36. Rini, D.P.; Chen, R.-H.; Chow, L.C. Bubble behavior and nucleate boiling heat transfer in saturated FC-72 spray cooling. *J. Heat Transf.* **2002**, *124*, 63–72. [CrossRef]
37. Selvam, R.P.; Lin, L.; Ponnappan, R. Direct simulation of spray cooling: Effect of vapor bubble growth and liquid droplet impact on heat transfer. *Int. J. Heat Mass Transf.* **2006**, *49*, 4265–4278. [CrossRef]
38. Sellers, S.M. *Heat Transfer Resulting from the Evaporation of Liquid Droplets on a Horizontal Heated Surface*; Georgia Institute of Technology: Atlanta, GA, USA, 2000.
39. Pais, M.; Chow, L.; Mahefkey, E. Surface roughness and its effects on the heat transfer mechanism in spray cooling. *J. Heat Transf.* **1992**, *114*, 211–219. [CrossRef]
40. Zhibin, Y.; Rui, Z.; Fei, D.; Teck Neng, W.; Kok Chuan, T.; Kok Fah, C.; Poh Keong, C.; Yong Sheng, C. Spray Cooling. In *Two Phase Flow, Phase Change and Numerical Modeling*; Amimul, A., Ed.; IntechOpen: Rijeka, Croatia, 2011; Chapter 13.
41. Omer, K.; Ashgriz, N. Spray Nozzles. In *Handbook of Atomization and Sprays: Theory and Applications*; Ashgriz, N., Ed.; Springer: Boston, MA, USA, 2011; pp. 497–579.
42. Mikhailov, G.K. Chapter 9—Daniel Bernoulli, Hydrodynamica (1738). In *Landmark Writings in Western Mathematics 1640–1940*; Grattan-Guinness, I., Cooke, R., Corry, L., Crépel, P., Guicciardini, N., Eds.; Elsevier Science: Amsterdam, The Netherlands, 2005; pp. 131–142.
43. Hiroyasu, H.; Arai, M.; Tabata, M. Empirical Equations for the Sauter Mean Diameter of a Diesel Spray. *SAE Trans.* **1989**, *98*, 868–877.
44. Kowalczuk, P.B.; Drzymala, J. Physical meaning of the Sauter mean diameter of spherical particulate matter. *Part. Sci. Technol.* **2016**, *34*, 645–647. [CrossRef]
45. Tanasawa, Y. On the atomization of liquid jet issuing from a cylindrical nozzle. *Technol. Rep. Tohoku Univ.* **1955**, *19*, 135.
46. Harmon, D.B. Drop sizes from low speed jets. *J. Frankl. Inst.* **1955**, *259*, 519–522. [CrossRef]
47. Merrington, A.C.; Richardson, E.G. The break-up of liquid jets. *Proc. Phys. Soc.* **1947**, *59*, 1. [CrossRef]

Disclaimer/Publisher's Note: The statements, opinions and data contained in all publications are solely those of the individual author(s) and contributor(s) and not of MDPI and/or the editor(s). MDPI and/or the editor(s) disclaim responsibility for any injury to people or property resulting from any ideas, methods, instructions or products referred to in the content.

buildings

MDPI

Article

Flexural Response of UHPC Wet Joints Subjected to Vibration Load: Experimental and Theoretical Investigation

Bin Zhao [1], Jun Yang [2,3], Dingsong Qin [1], Yang Zou [2,3], Zhongya Zhang [2,3], Kaijie Zhang [2,3] and Jingchen Leng [2,3,*]

1 China Railway First Group Co., Ltd., Xi'an 710054, China
2 State Key Laboratory of Mountain Bridge and Tunnel Engineering, Chongqing Jiaotong University, Chongqing 400074, China
3 School of Civil Engineering, Chongqing Jiaotong University, Chongqing 400074, China
* Correspondence: 611220080003@mails.cqjtu.edu.cn

Abstract: This study aims to investigate the flexural performance of ultra-high-performance concrete (UHPC) wet joints subjected to vibration load during the early curing period. The parameters investigated included vibration amplitude (1 mm, 3 mm, and 5 mm) and vibration stage (pouring—final setting, pouring—initial setting, and initial setting—final setting). A novel simulated vibration test set-up was developed to reproduce the actual vibration conditions of the joints. The actuator's reaction force time-history curves for the UHPC joint indicate that the reaction force is stable during the initial setting stage, and it increases linearly with time from the initial setting to the final setting, trending toward stability after 16 h of casting. Under the vibration of 3 Hz-5 mm, cracks measuring 14 cm × 0.2 mm emerge in the UHPC joint. It occurs during the stage from the initial setting to the final setting. The flexural performance of wet joint specimens after vibration was evaluated by the four-point flexural test, focusing on failure modes, load-deflection curves, and the interface opening. The results show that all specimens with joints exhibited bending failure, with cracks predominantly concentrated at the interfaces and the sides of the NC precast segment. The interfacial bond strength was reduced by vibrations of higher amplitude and frequency. Compared with the specimens without vibration, the flexural strength of specimens subjected to the vibration at 3 Hz-3 mm and 3 Hz-5 mm were decreased by 8% and 19%, respectively. However, as the amplitude and frequency decreased, the flexural strength of the specimens showed an increasing trend, as this type of vibration enhanced the compactness of the concrete. Additionally, the calculation model for the flexural strength of UHPC joints has been established, taking into account the impact of live-load vibration. The average ratio of theoretical calculation values to experimental values is 1.01, and the standard deviation is 0.04, the theoretical calculation value is relatively precise.

Keywords: ultra-high performance concrete; live-load vibration; joints; flexural strength

1. Introduction

Ultra-high-performance concrete (UHPC), a new type of cement-based composite material widely used in various building structures, has always attracted significant attention regarding its durability issues [1–5]. Particularly when facing extreme loads such as earthquakes, explosions, impacts, and overload, many serious problems have been exposed. Due to the limitations in the mixing, transportation, and curing of UHPC, wet joints are inevitable in cast-in-place UHPC structures. Similarly, for precast NC structures, there will inevitably be UHPC components or parts that require on-site casting. In the early stages of concrete joints, when live loads are applied to engineering structures, the deflection of the wet joint connecting the two parts may differ. This differential deflection could adversely affect the interfacial bond strength between precast elements and longitudinal joints before they attain the necessary tensile strength and plasticity. Wet joints are particularly susceptible to flexural failure under the action of live-load vibrations. The impact of live load vibrations on the flexural strength of the joint has not yet been clearly defined. Consequently, investigating the flexural performance of both the integral interfaces and wet joint interfaces of UHPC is of paramount importance.

In order to explore the influence of complex and harsh environments on wet joints and to find countermeasures for dealing with the durability problems of UHPC wet joints subjected to live-load vibrations, numerous studies have been conducted. It was found by Deng et al. [6] that a T-shaped girder wet joint for a lightweight steel-UHPC composite structure showed better crack resistance than the traditional I-shaped wet joint. Graybeal et al. [7] investigated the structural performance of cast-in-place UHPC connections by static and cyclic tests. Similar studies were performed by Haber et al. [8] and Varga et al. [9], and the results showed that UHPC connections have superior mechanical performance than those of NC. In response to the above phenomena, Graybeal et al. [10] designed and deployed the cast-in-place UHPC connections. In addition, Lee et al. [11] and Hash et al. [12] investigated the flexural behavior of prefabricated reinforced concrete specimens connected with UHPC. Their studies indicated that vibrations applied during the initial and final setting stages of the joint can influence the splitting tensile strength of the concrete. Zhang et al. [13] investigated the shear strength of UHPC-NC specimens; the results show that the effect of surface preparation on the shear strength of the UHPC–stone interface is most significant. It was found by Guan et al. [14] that vibrations that started before the initial setting increased the early compressive strength of NC. Additionally, vibrations applied from the initial to the final setting stage increased the concrete's early strength. However, all types of vibrations started after the final setting was observed to decrease its early compressive strength and reduce the long-term strength. The study of Zhang et al. [15] and Yang et al. [16] indicates that continuous live-load vibrations have a significant effect on the cracking load and ultimate load of composite materials, but have no significant impact on their deformation performance. A vibration table was used by Wu et al. [17] to simulate the vibration caused by the live load during the construction of joints, which studied the impact of vibration on the compressive and splitting tensile strength of the concrete. Huang et al. [18] conducted a vibration table test to study the seismic performance of two prestressed beams with UHPC connections; it found that UHPC connections can provide strong resilience even under high-intensity earthquake ground motions. Wang et al. [19] and Leng et al. [20] prepared UHPC by a vibration mixing method and pointed out that at the early stage of hydration, the position of steel fibers in the UHPC matrix can move freely, which significantly affects the flexural response. The result shows that with the increase in amplitude and frequency, the flexural response of UHPC specimens decreases sharply due to the distribution and orientation of steel fibers. Zhang et al. [15] conducted vibration tests on 324 newly cast thin plate specimens with different vibration frequencies and durations of vibration as parameters, to clarify the impact of traffic vibration on the flexural response of PVA-ECCs, followed by four-point bending tests. The results indicate that continuous traffic vibration has a significant adverse effect on the flexural response of the composite material.

In summary, existing research on the impact of live-load predominantly builds upon studies of NC, with a strong emphasis on using a miniature vibration table and being only applicable to standard-sized specimens. However, the investigation into the impact of live-load vibration on UHPC remained inadequately comprehensive. To clarify the effects of live-load vibration, a four-point bending test was carried out in this study, specifically focusing on the role of vibration stage and amplitude. The key indexes such as the failure mode, load-deformation relationship, and interface opening were analyzed. At the same time, it introduces a live-load vibration coefficient to correct the calculation model for the bending strength of UHPC joints, and the model's calculated values match the test results.

2. Experimental Program

2.1. Specimen Design and Test Parameters

To investigate the effects of different vibration conditions on structural bending strength and failure modes, this study designed and produced UHPC-NC composite specimens with dimensions of 2000 mm × 350 mm × 200 mm. The joint size is 300 mm × 350 mm × 200 mm, with rebars arranged in an interlaced pattern in the joint. The spacing of rebars is 72.5 mm, and the cover thickness is 30 mm. The precast section size is 850 mm × 350 mm × 200 mm, the spacing of longitudinal rebars is 145 mm, and the spacing of transverse rebars is 150 mm. The configuration of the joint specimen is shown in Figure 1.

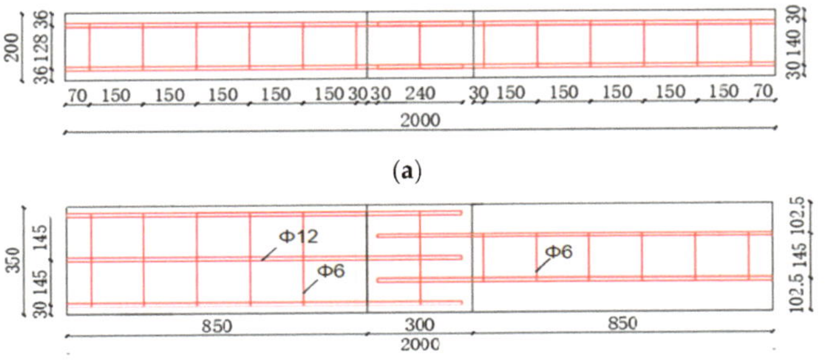

Figure 1. Configuration of the test specimens (unit: mm). (**a**) The elevation of the specimens. (**b**) The plan of the specimens.

The dynamic response parameters of engineering structures caused by live loads include frequency, amplitude, damping, stiffness [21], etc. In this study, the simulation of live-load vibration focuses on amplitude, while also incorporating the vibration stage as two parameters.

The grouping scheme of the test specimens is shown in Table 1. First, the vibration amplitude represents the differential deflection between the staged constructed structures. In this study, the vibration amplitude was designed following a real construction field. The maximum value of the first-order fundamental frequency and the maximum deflection of the background structure of this study are calculated by MIDAS/Civil (2022 v1.2) finite element analysis software and are 2.3 Hz and 5 mm, respectively. In summary, based on the suggestion by Ng et al. [22], the vibration amplitude varies from 1 mm to 5 mm.

Secondly, the impact of early live-load vibration on concrete joints is not yet clear. Consequently, an investigation into the impact of vibration exposure during the early stages of joint maturation was undertaken. To achieve this, the early maturation phase of the joint was delineated by the initial setting time and final setting time points, and enforced vibrations were applied at various developmental stages. The method of penetration resistance measured the setting time of the concrete, in accordance with the GBT50080-2016 [23].

Table 1. Specimen design parameter.

Test Parameters	Test Piece Number	Vibration Amplitude	Vibration Stage	Experimental Age
Not vibration	N-1	/	/	28 d
Vibration amplitude	F-1	1 mm	Pouring-final setting	28 d
	F-2	3 mm	Pouring-final setting	28 d
	F-3	5 mm	Pouring-final setting	28 d
Vibration stage	J-1	3 mm	Pouring-initial setting	28 d
	J-2	3 mm	Initial setting-final setting	28 d

"N" represents the specimen that is not vibrated, "J" represents the vibrated stage, and "F" represents the vibration amplitude. The specimen will be cured for 28 days for joint bending performance testing after the vibration.

The specimen fabrication process is shown in Figure 2. First of all, according to the design size of the specimen and the layout of the reinforcement, the integral formwork was designed. Subsequently, the prefabricated parts were cast synchronously with ordinary concrete. Immediately after pouring, the specimens were covered with wet burlap and a plastic sheet. The specimens were then cured at room temperature for 28 days. After the curing process, the surface at the junction of the wet joint and the pre-cast section was made rough to improve the adhesion of the joint. Finally, the joint UHPC for the live-load vibration test was cast.

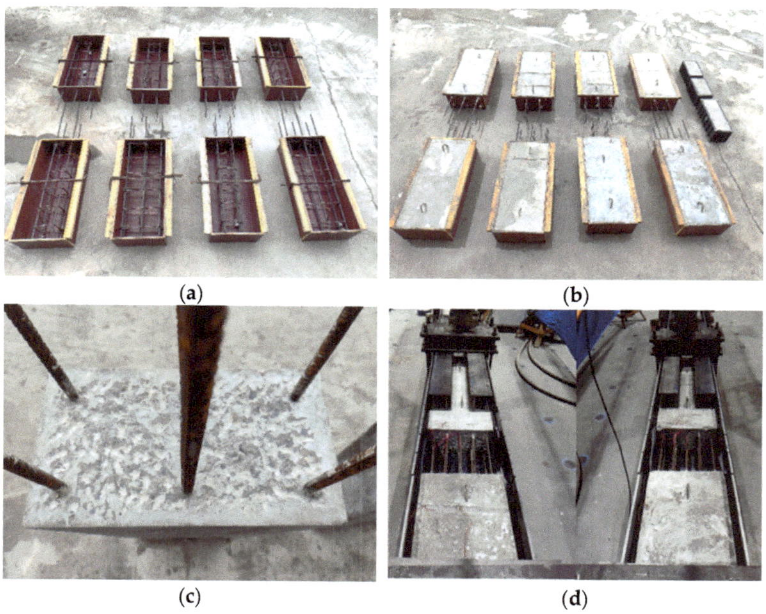

Figure 2. Fabrication process of the test specimens. (**a**) Manufacturing formwork and rebar. (**b**) Casting of Precast section concrete. (**c**) Roughing the interface. (**d**) Assembly of the steel framework.

2.2. Material Properties

The matrix ratio of UHPC in this study is shown in Table 2, the composition includes cement, silica fume, quartz sand, slag, mixed steel fibers, and coarse aggregate. The percentage by volume of mixed steel fibers is 2%, and the maximum diameter coarse bone particle size is 4 mm. The material of the precast sections is the grade of C60, and the cement is made of PO 42.5 normal portland cement, fine aggregate using river sand, coarse aggregate using crushed stone with a diameter of 4–18 mm, a water cement ratio of 0.3, and C60-grade concrete. The mixes used are shown in Table 2.

Table 2. Material mix proportion (Unit: kg/m^3).

Component	Cement	Silica Fume	Quartz Sand	Coarse Aggregate	Water	Mixed Steel Fiber
Mass ratio	771.2	192.8	848.3	231.4	173.5	170.1

The elastic modulus of UHPC material and C60 concrete is determined by Code for Design of Concrete Structures GB50010-2010 [24]. The mechanical properties of UHPC and C60 obtained from the test are shown in Table 3. The joint specimen is reinforced with HRB400, the diameter of the longitudinal rebar is 12 mm, and the diameter of the stirrup is 6 mm. The tensile strength of 6 mm rebar and 12 mm rebar is 421.3 MPa and 435.1 MPa, respectively, and their elastic modulus is 2.15 GPa.

Table 3. Mechanical properties of UHPC and C60.

Category	Cube Compressive Strength f_{cu}/MPa	Cube Tensile Strength f_{su}/MPa	Flexural Strength f_t/MPa	Elastic Modulus/GPa
UHPC	142.9	10.2	13.8	50.4
C60	61.8	/	6.5	36

2.3. *Simulated Vibrating Device and Test Setup*

2.3.1. Simulated Vibrating Device

To simulate the live-load vibration to the fresh casting UHPC wet joints, a simulated vibration device was designed. In this study, a simulated vibration device was designed to simulate the live-load vibration of newly poured UHPC wet joints. The device in this study consists of three fixture components: left, middle, and right U-shaped steel formwork, precast RC blocks, and the actuator. The prefabricated RC blocks in the U-shaped steel formwork are clamped with left, right, and central clamp. The middle fixture is connected to the actuator that connected to the reaction beam, achieving simultaneous simulation of the deformation vibration of two joints [25]. The layout of the simulated vibrating device is illustrated in Figure 3a,b.

2.3.2. Load Settings and Data Acquisition

Figure 4 shows the process of the joint live-load vibration test. After the calibration of the device, UHPC will be poured at the joint of the specimen. Once the joint has been poured, the actuator's amplitude and frequency will be aligned with the predetermined test parameters. Meanwhile, the vibration time will be recorded. When the vibration is completed, the joint specimen will be cured for 28 days before the bending strength tests.

The SIControl (V2.0) software is designed to facilitate the real-time acquisition and storage of critical experimental parameters, including frequency, amplitude, and actuator reaction force. Furthermore, it is capable of plotting the actuator reaction force as a function of time, thereby rendering the test procedure tangible and observable. The loading setup and data acquisition of the software are shown in Figure 4f. The study of the reaction forces exerted on the actuator during vibration and the surface of the joint after vibration found that the reaction forces on the actuator at the UHPC joint during vibration showed no significant changes at the initial setting stage. From the initial to final setting, the maximum and minimum reaction forces increased linearly with time, and the load growth tended to level off at 16 h after concreting. Under the live load vibration of 3 Hz-5 mm, the UHPC joint developed cracks measuring 14 cm in length and 0.2 mm in width. The cracks in the joint specimens under live load vibrations occurred during the initial to final setting stages of the joint material casting.

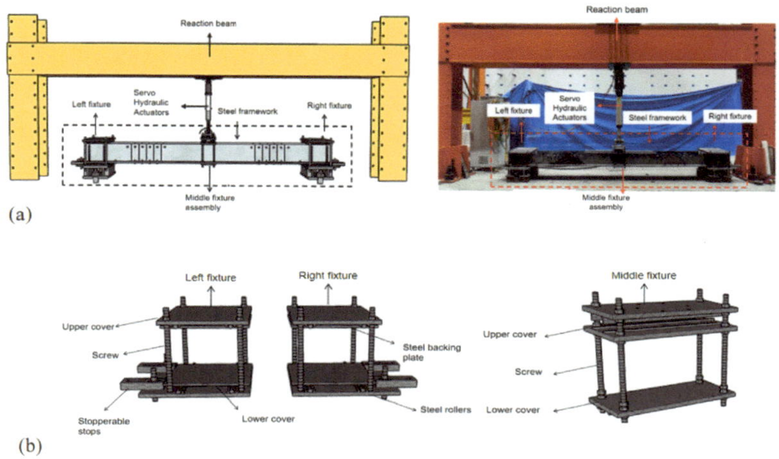

Figure 3. Schematic diagram of the (**a**) elevation view of the simulated vibrating device and (**b**) fixture component.

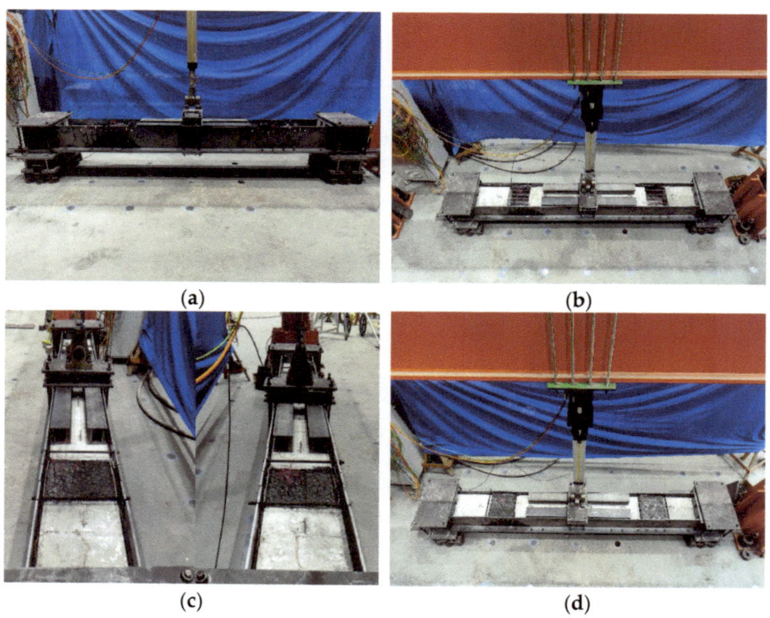

Figure 4. *Cont.*

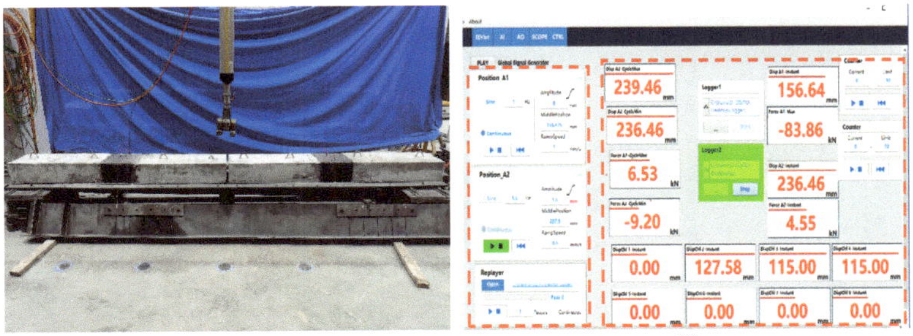

(e) (f)

Figure 4. Specimen production. (**a**) Assembling the test device. (**b**) Assembling the precast sections. (**c**) Pouring the wet joint. (**d**) Start live-load vibration. (**e**) Curing for 28 days. (**f**) SIControl software.

2.4. Four-Point Flexural Test and Instrumentation

2.4.1. Measuring Point Layout

The strain measurement points are mainly located in the rebars and concrete of the precast slab area near the joints, and the distribution of measurement points is shown in Figure 5. Linear variable displacement transducers (LVDTs) were used to monitor the interface opening and deflection of the specimen. The accuracy of the LVDT is 0.001 mm. Monitoring the strain of steel and concrete was to paste strain gauges on the surface of the specimen. Figure 6. shows the layout of the measuring points. In addition, the crack width of the specimen was recorded using a crack observation instrument with an accuracy of 0.02 mm.

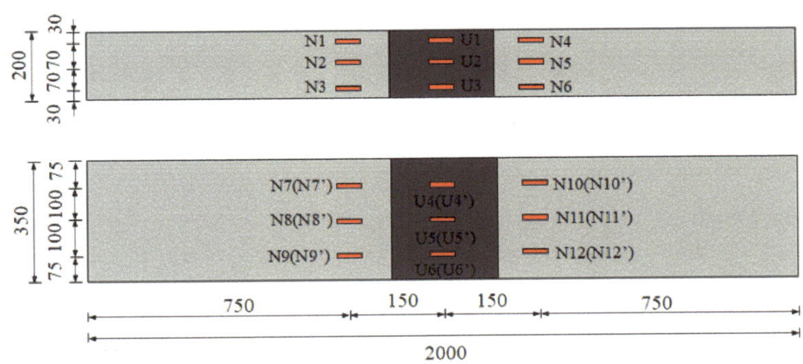

(**a**)

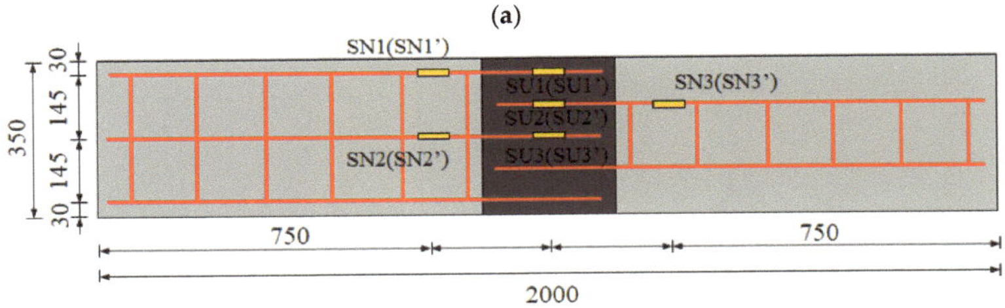

(**b**)

Figure 5. Strain measuring point arrangement. (**a**) Concrete measuring points. (**b**) Rebar measuring points.

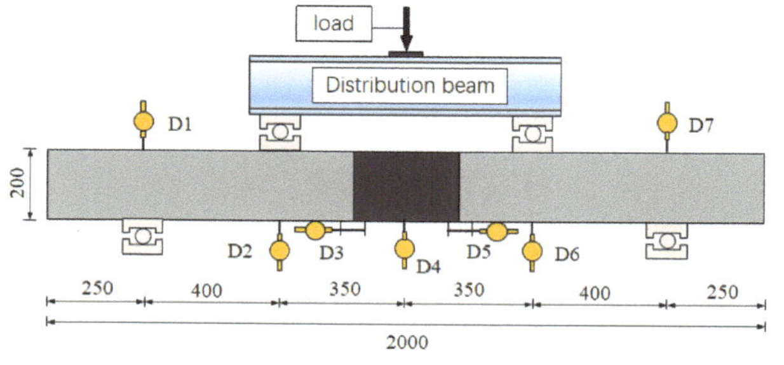

Figure 6. Displacement measuring point arrangement.

2.4.2. Loading Program

The loading device for the bending test of the joint specimen is shown in Figure 7. The specimens were loaded step by step with a hydraulic jack. In the initial stages, the specimen was preloaded to 10% of its ultimate load to eliminate the effects of minor gaps between the distribution beam and the specimen.

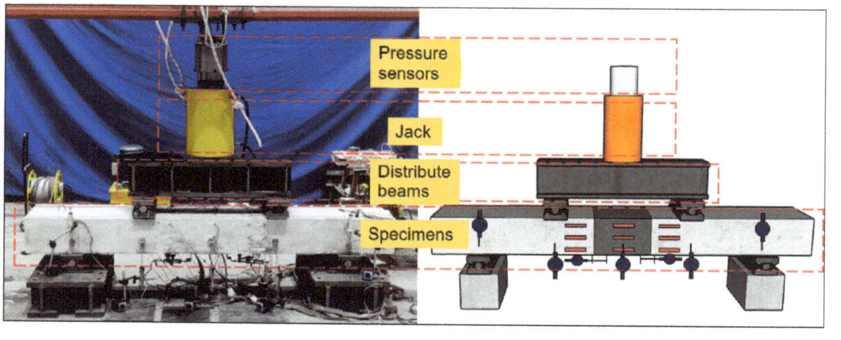

Figure 7. Joint-bending experiment.

During formal loading, force control was initially used, with each increment of load staged at 10% of the ultimate load, maintained for a duration of 2 min to facilitate the observation of the material response and the emergence of test phenomena. Subsequently, the loading intensity was escalated to 90% of the estimated ultimate load at a rate of 3 mm/min. Thereafter, a displacement control protocol was implemented, with the loading rate modulated to 0.05 mm/min. When the load on the specimen could no longer be increased, the specimen was considered to have failed. The test set-up is shown in Figure 8.

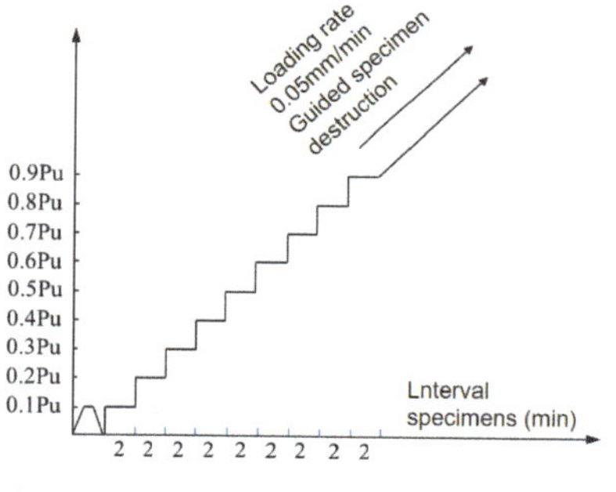

Figure 8. Experimental loading procedure diagram.

3. Test Results and Discussion

3.1. Failure Mode and Crack Pattern

Figure 9 shows the failure modes and crack patterns of each specimen. All specimens exhibited the same kind of failure, characterized by basically no cracks or failure phenomena in the joint; the main crack initiated at the bottom of the prefabricated section and the interface opening was apparent.

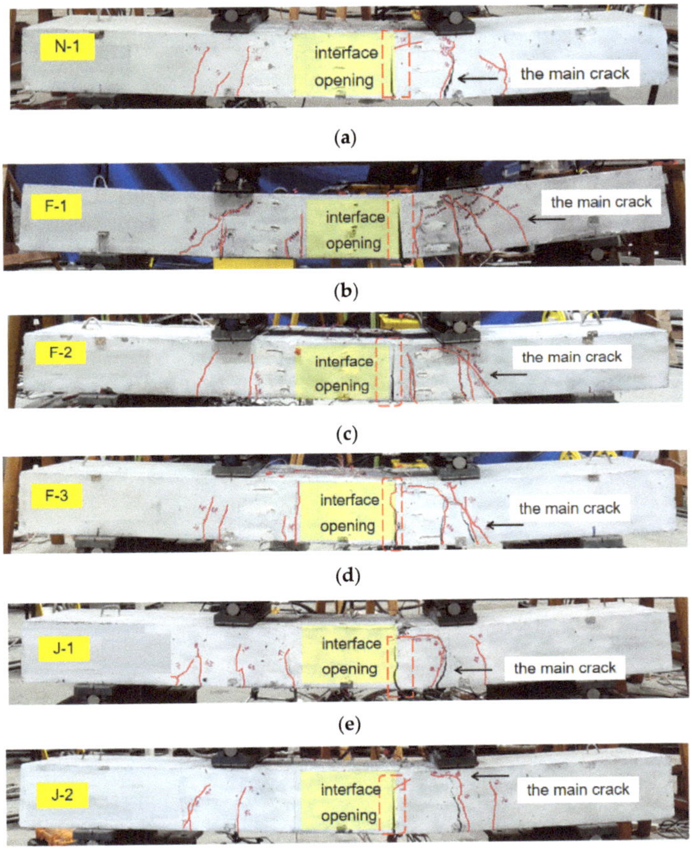

Figure 9. Failure model and crack pattern of the specimens. (**a**) Specimen N-1. (**b**) Specimen F-1. (**c**) Specimen F-2. (**d**) Specimen F-3. (**e**) Specimen J-1. (**f**) Specimen J-2.

During the elastic state, there were no visible cracks or interface openings. As the load increased, the first crack was observed at the bottom of the right prefabricated section. As depicted in Figure 9, the precast section below the left support of the distribution beam cracked, and the crack width was 0.11 mm and developed upward. However, the cracks in F-2 and F-3 appeared earlier than the F-1. This shows that the size of the vibration amplitude has a bad effect on the cracking load. In addition, the pure bending area in the middle of the prefabricated part will form the main crack as the load increases. Subsequently, the tensile reinforcement yielded, and the interface concrete was crushed, which is accompanied by the decrement of the rigidity.

3.2. Load-Deflection Curve Responses

3.2.1. Effects of the Vibration Properties

Table 4 summarizes the test results; the λ stands for the increase in ultimate load and the κ stands for the increase in peak displacement. To evaluate the impact of different vibration amplitudes, the load-deflection curves in the span during the test were plotted according to the test data. To facilitate the comparison, the average results for each group were selected for analysis in each group, as shown in Figure 10a. By analyzing Figure 10a, it can be found that the highest flexural strength of UHPC joint specimens under different vibration amplitudes is 3 Hz-1 mm (F-1), followed by 3 Hz-3 mm (F-2), and finally, 3 Hz-5 mm (F-3). The ultimate load of the F-1 specimen increased by 17% compared to the specimen that did not vibrate, while the ultimate load of the F-2 specimen decreased by 8% compared to the specimen that did not vibrate, and the ultimate load of the F-3 specimen decreased by 19% compared to the specimen that did not vibrate. This proves that low-amplitude live-load vibrations can be used to make the joints more compact. In addition, in order to improve the flexural strength of the joint specimens, the bond strength of the joint interface and the prefabricated segment can be increased. In contrast, high-amplitude live-load vibrations will destroy the bonding performance between the prefabricated segment and the joint interface, therefore reducing the flexural capacity of the joint specimen.

Table 4. Summary of critical results.

Specimen Set	Specimen Number	Cracking Loads P_{cr}(kN)	Ultimate Load P_u(kN)	λ	Peak Displacement δ_u(mm)	κ
N	N-1	47	143.2	1.00	24.9	1.00
F	F-1	55	168.1	1.17	30.1	1.21
	F-2	50	131.4	0.92	29.3	1.24
	F-3	47.5	116.7	0.81	30.4	1.22
J	J-1	50	148.1	1.03	22.8	0.92
	J-2	46	143.7	1.00	24.9	1.00

Note: When NC and UHPC develop cracks, the crack width is generally around 0.05 mm. Therefore, the load corresponding to a crack width of 0.05 mm is defined as the cracking load.

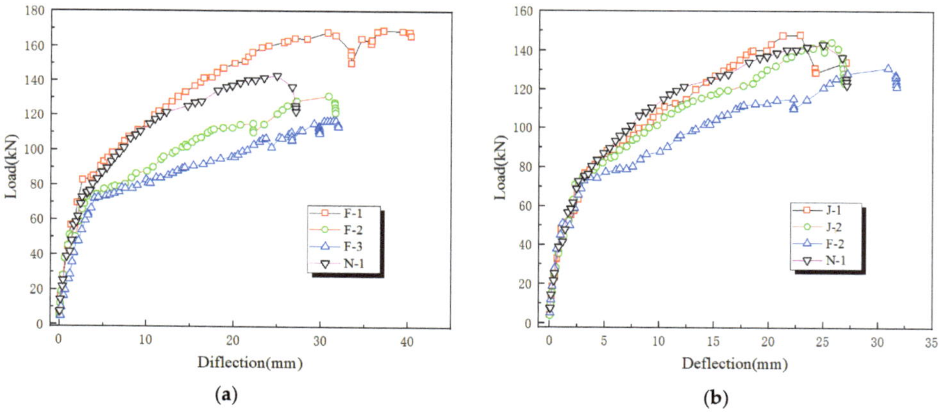

Figure 10. Load-midspan deflection curves. (**a**) Load-midspan deflection curves of the vibration amplitude specimens. (**b**) Load-midspan deflection curves of the vibration stage specimens.

The load-deflection curves at mid-span for different vibration stages of the specimens are shown in Figure 10b. By analyzing Figure 10b, it can be observed that the highest flexural strength of UHPC joint specimens under different vibration stages is pouring—initial setting (J-1) followed by initial setting—final setting (J-2), and finally pouring—final setting (F-2). The ultimate load of specimen J-1 increased by 3% compared to the specimen that did not vibrate, the ultimate load of specimen J-2 was equal to that of the specimen that did not vibrate, and the ultimate load of specimen F-2 decreased by 8% compared to the specimen that did not vibrated. This indicates that the timing of live-load vibration has no significant effect on the flexural strength of the specimens. The load-displacement curve trends for the vibration stage groups are essentially consistent, with no obvious discrepancy in the initial cracking load and the yield load.

3.2.2. Ductility and Stiffness

In general, the stiffness is used to measure the performance of the RC beam structure [26], and the bending stiffness K is used as the calculation target in this study. The displacement stiffness, as presented in Equation (1), was used to assess the stiffness of the specimens in this study.

$$K = \frac{P_y}{\delta_y} \tag{1}$$

where the P_y is the yield load, the δ_y is the yield displacement.

The ductility is used to measure the plastic deformation capacity of the specimen, the ductility of RC structure is generally defined as the ratio of the ultimate displacement δ_u to the yield displacement δ_y, and the calculation method for the ductility is shown in Equation (2).

$$\mu = \frac{\delta_u}{\delta_y} \tag{2}$$

The ductility and stiffness of the specimens obtained from Equations (1) and (2) are tabulated in Table 5.

Table 5. Summary of ductility and stiffness.

Specimen Number	P_y (kN)	δ_y (mm)	δ_u (mm)	K (kN/mm)	μ
N-1	75.98	3.79	27.34	20.05	7.21
F-1	83.01	3.81	41.59	21.78	10.91
F-2	77.52	4.69	31.48	16.41	6.71
F-3	75.13	4.82	33.44	15.56	6.93
J-1	75.74	3.60	27.90	20.83	7.75
J-2	78.45	3.87	33.43	20.15	8.64

By analyzing Figure 11, it can be seen that the ductility values of specimen F-1 has increased by 51.3% compared to the N-1 specimen; however, the vibration decreased the ductility of the specimens F-2 and F-3. This result implies that the low amplitude has a positive effect on the ductility, but the high amplitude has a negative effect and improves on the ductility of the specimen. Additionally, the ductility of J-series specimen increased compared to the specimen without vibration. The stiffness of specimen F-1 increased by 8.6% compared to the specimen that did not vibrate, while the specimen F-2 was decreased by 18.1% compared to the specimen that did not vibrate, and the specimen F-3 decreased by 22.3% compared to the specimen that did not vibrate. However, the vibration stage does not significantly improve based on the stiffness of the specimens.

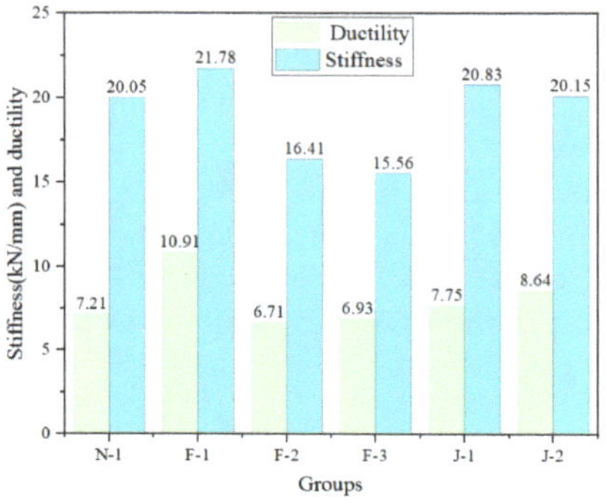

Figure 11. Ductility and stiffness.

3.3. Load-Crack Width Relationship

The maximum crack width of all regions at each load level was called the maximum crack width in the curve. During the loading process of specimens F-1 (3 Hz-1 mm), F-2 (3 Hz-3 mm), and F-3 (3 Hz-5 mm), cracks began to appear and propagate as the load increased. The F-1 specimen developed a crack on the surface of the bottom at 55 kN, and the specimen failed due to a through-crack at the right interface at 168.1 kN. The F-2 specimen developed a crack on the surface of the bottom at 50 kN and failed due to the crushing of the precast section and the interface opening at 131.4 kN. The F-3 specimen cracked on the bottom surface at 47.5 kN and failed due to diagonal cracking, crushing, and the interface opening at 116.7 kN. According to the failure modes, it was observed that the precast section below the right support of the distribution beam experienced rapid crack propagation, crushing of the precast section, and a sudden opening of the right interface, indicating that the vibration amplitude significantly affected crack development and specimen failure. No cracks were observed in the UHPC during loading, and the specimens failed in a ductile manner. The vibration did not obviously affect the cracking load of the specimens, and specimens with and without vibration showed similar behavior in the linear elastic range. The maximum crack width of the amplitude group specimen is shown in Figure 12a.

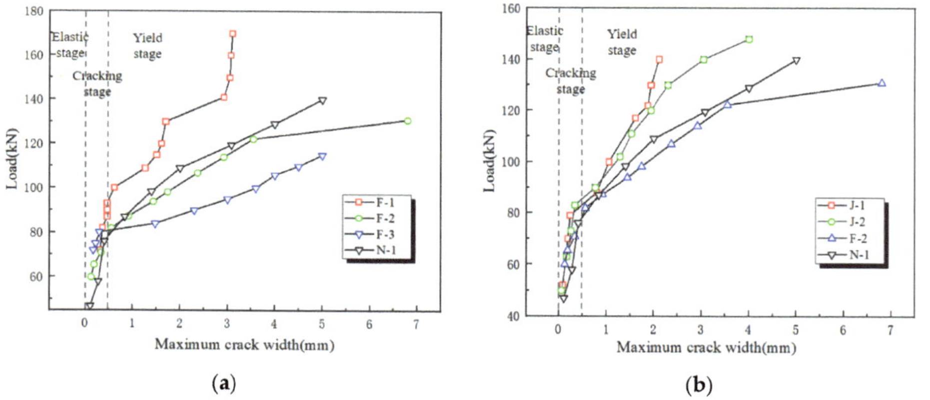

Figure 12. Load-maximum crack width curves of the test specimens. (**a**) Load-maximum crack width curves of the vibration amplitude specimens. (**b**) Load-maximum crack width curves of the vibration stage specimens.

The failure modes of specimens J-1, J-2, and F-2 indicate that in the specimens vibrated during the pouring to initial setting period, cracks first appeared at the bottom of the precast section below the right support of the distribution beam and gradually extended upward. At a load of 148.1 kN, the main crack expanded rapidly, leading to concrete crushing and interface opening, resulting in the ultimate failure of the specimen. Specimens that vibrated during the initial to final setting period exhibited similar failure patterns, with crack development and interface opening eventually leading to specimen failure. The specimens that vibrated during the pouring to final setting period had the same failure characteristics as the F-2 specimens, so further description is omitted. From the failure modes, all specimens showed rapid development of cracks in the precast section below the right support of the distribution beam, with concrete crushing above and a sudden opening of the right interface. No cracks were observed in the UHPC during loading, indicating a ductile failure of the specimens. The maximum crack width of vibration stage group specimens is shown in Figure 12b.

3.4. Load-Interface Opening

Figure 13a compares the load-interface opening curves of specimens under different vibration amplitudes. It indicates that specimen F-1 exhibited significant crack widths and interface opening widths at ultimate load, while the F-2 and F-3 vibration specimens showed faster interface opening speeds at the yield load, indicating that vibration significantly affects crack development and interface bonding. For the yield stage, an increase in vibration amplitude decreased the interface bonding strength, leading to an increased rate of interface opening. Consequently, the rotational angle of the interface increased, which decreased the post-yield flexural stiffness of the specimens. However, for specimens with lower vibration amplitudes, vibration increased the post-yield flexural stiffness. This is because low-amplitude and low-frequency vibrations increase the compactness of the UHPC and also enhance the interface bonding strength.

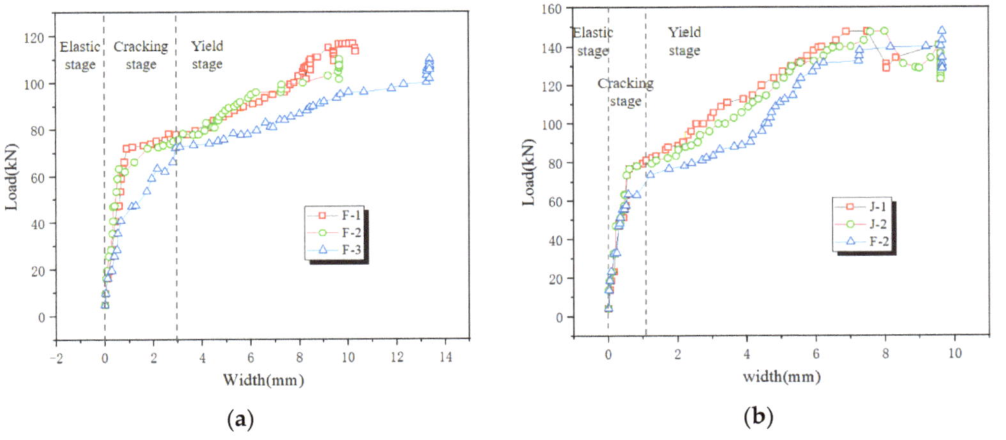

Figure 13. Load-interface opening curves. (**a**) Load-interface opening curves of the vibration amplitude specimens. (**b**) Load-interface opening curves of the vibration stage specimens.

Figure 13b compares the load-interface opening curves of specimens under different vibration stages. Interface opening width curves of specimens were shown under different vibration stages. The specimen vibrated during pouring to initial setting had a crack of left opening of 0.54 mm and a maximum crack width of 0.77 mm at the yield load, while the specimen that vibrated during pouring to final setting had a crack of the left opening of 0.40 mm and a maximum crack width of 0.56 mm at the yield load; the specimen that vibrated during the initial to final setting had a crack of left opening of 0.34 mm and a maximum crack width of 0.21 mm at the yield load. These results indicate that different vibration stages are related to the crack development speed and joint opening speed of UHPC wet joints, and that vibration significantly affects crack development and interface bonding.

However, for J-2, F-2, and F-3 specimens, the load become larger at the greatest crack width. This is because during the loading process of UHPC joint specimens, the steel fibers within it can effectively hinder further propagation of cracks. Once the cracks reach a certain extent, the bridging action of the fibers or changes in the material's microstructure may cause the cracks to cease expanding, but the specimen can still bear larger loads.

3.5. Load-Strain Curves

3.5.1. Load-Steel Strain Curves

To obtain the load-strain development curve of the specimen, the strain gauge should be laid on each longitudinal bar in the wet joint, so as to take the average strain value; the curve is as shown in Figure 14. Due to the consistent pattern of load-steel strain curves for joint specimens under different vibration stages and amplitude factors, the load-steel strain curves of the F-2 specimen from the vibration stage group (pouring-final setting) and the F-3 specimen from the vibration amplitude group (3 Hz-5 mm) are selected for analysis here.

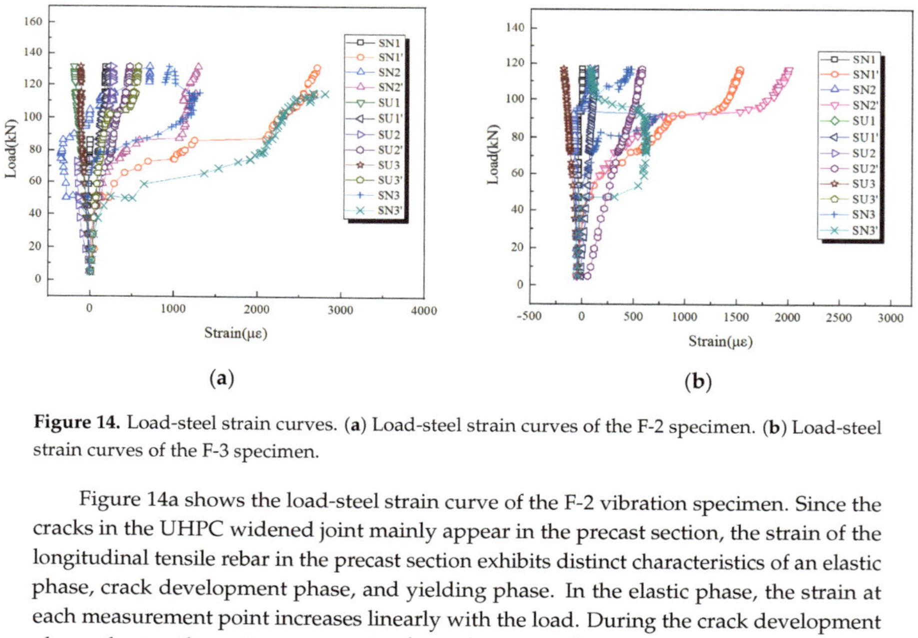

Figure 14. Load-steel strain curves. (**a**) Load-steel strain curves of the F-2 specimen. (**b**) Load-steel strain curves of the F-3 specimen.

Figure 14a shows the load-steel strain curve of the F-2 vibration specimen. Since the cracks in the UHPC widened joint mainly appear in the precast section, the strain of the longitudinal tensile rebar in the precast section exhibits distinct characteristics of an elastic phase, crack development phase, and yielding phase. In the elastic phase, the strain at each measurement point increases linearly with the load. During the crack development phase, the tensile strain at points SN1′, SN2′, and SN3′ is approximately twice that of the tensile strain at points SU1′, SU2′, and SU3′ in the joint, indicating that the rate of tensile strain increase in the longitudinal tensile rebar of the precast section is much greater than that in the joint. Under the ultimate load, the longitudinal tensile rebar inside the precast section yielded, while the longitudinal compressive rebar in the joint did not, due to the staggered arrangement of longitudinal rebar in the UHPC joint, with a rebar ratio approximately twice that of the precast sections on both sides of the joint, indicating that when the anchorage length in the UHPC joint is 28 cm, the rebar in the precast section can bear the load adequately, and the anchorage length is sufficient.

Figure 14b shows the load-steel strain curve for the F-3 vibration specimen. The curve of the F-3 specimen is similar to that of the F-2, and since the cracks in the UHPC-widened joint mainly appear in the precast section, the strain of the longitudinal tensile rebar in the precast section also exhibits distinct characteristics of an elastic phase, crack development phase, and yielding phase. After the specimen was cracked, the strain of the steel bar increased rapidly, and some concrete in the tensile zone failed. The slope of the load-strain curve did not change obviously when the strain exceeded the yield strain of the steel bar. Until the specimen yielded, the strain continued to increase, but the load remained basically unchanged. This phenomenon proves that the stress-strain relationship of steel bars in UHPC is different from that of bare bars.

3.5.2. Load-Concrete Strain Curve

Also, the load-concrete strain curves of the F-2 specimen from the vibration stage group (pouring-final setting) and the F-3 specimen from the vibration amplitude group (3 Hz-5 mm) are selected for analysis.

Figure 15 shows the load-concrete strain curves for the pouring-final setting vibration specimens. Figure 15a,b represents the load-concrete strain curves for the left and right precast sections, respectively. The compressive strain at the top of the right precast section is much greater than that at the top of the left precast section. The compressive strain at the top of the right precast section corresponding to the ultimate load has already exceeded the ultimate compressive strain, which matches the experimental phenomenon of the top crushing of the pure bending section of the right precast section when the joint specimen fails. The strain at the lateral measurement points of the precast section conforms to the plane assumption during the elastic stage of the specimen. The measurement points at the bottom of the precast section undergo tensile strain. As the concrete of the precast section reaches the ultimate tensile strain, the specimen cracks, and the strain gauges are subsequently damaged.

Figure 15c shows the load-concrete strain curve for the joint. Since the UHPC joint did not crack throughout the loading process, it can be seen that the strain of almost all measurement points increases linearly, and the lateral measurement points conform to the plane section assumption throughout the loading process.

Figure 16 shows the load-concrete strain curves for the F-3 vibration specimen. Figure 16a and Figure 16b represent the load-concrete strain curves for the left and right precast sections, respectively. The compressive strain at the top of the right precast section is significantly greater than that at the top of the left precast section. The compressive strain at the top of the right precast section corresponding to the ultimate load has already exceeded the ultimate compressive strain, which corresponds with the experimental observation of the top crushing in the pure bending region of the right precast section when the joint specimen fails. The strain at the lateral measurement points of the precast section conforms to the plane assumption during the elastic stage of the specimen. The measurement points at the bottom of the precast section experience tensile strain, and as the concrete of the precast section reaches its ultimate tensile strain, the specimen begins to crack, leading to the subsequent damage of the strain gauges.

Figure 16c shows the load-concrete strain curve for the joint. Since the UHPC joint did not crack throughout the entire loading process, it can be observed that the strain at almost all measurement points changes linearly, and the lateral measurement points conform to the plane assumption throughout the loading process.

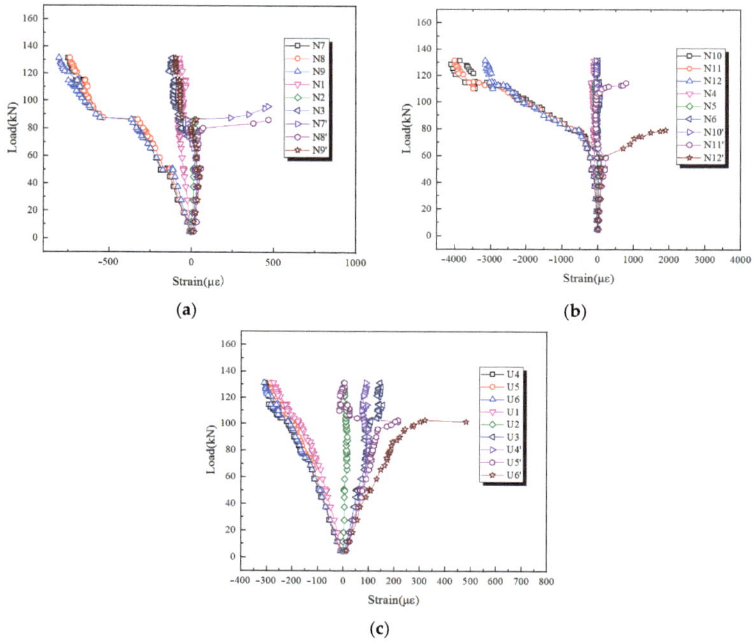

Figure 15. Load-concrete strain curves of the F-2 specimen. (**a**) Load-concrete strain curve of the left precast section. (**b**) Load-concrete strain curve of the right precast section. (**c**) Load-concrete strain curve of the joint.

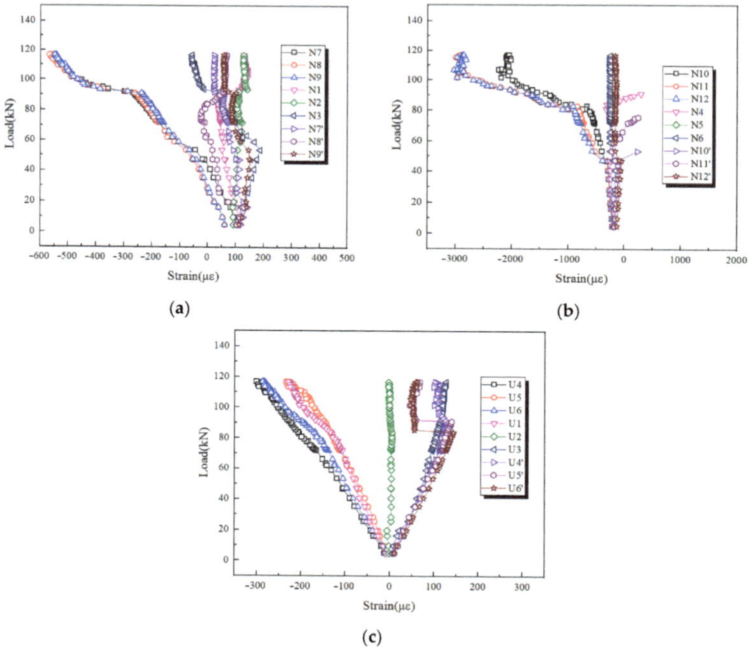

Figure 16. Load-concrete strain curve of the F-3 specimen. (**a**) Load-concrete strain curve of the left precast section. (**b**) Load-concrete strain curve of the right precast section. (**c**) Load-concrete strain curve of the joint.

4. Prediction of the Flexural Strength Considering Vibration

The flexural strength of UHPC wet joints was calculated by the strut-and-tie model. The basic assumptions of the strut-and-tie model configuration are as follows:

(1) In the strut-and-tie model, the tensile members within the structure are subjected to axial tensile forces, while the compression members are subjected to axial compressive forces. It is posited that the nodes within the framework do not sustain bending moment loads and are in compliance with the equilibrium equations of forces;

(2) The compression members are composed of concrete from a certain area, the shape of which is not uniform, and the random distribution of coarse and fine aggregates in concrete leads to a decrease in the strength of the compression member concrete. Therefore, the strength of the compression members cannot be simply taken as the compressive strength of concrete; instead, a strength influence factor should be considered for correction. According to the ACI318-2011 code [27], the calculation method for the effective strength of compression member concrete is shown in Equation (3).

$$f_{cu} = 0.85\beta_n f_c' \tag{3}$$

where f_{cu} is the effective strength of compression member concrete and β_n is the strength influence factor. When all members in the node area are concrete compression members, the value is taken as 1.00; when there is one tensile member in the node area, the value is taken as 0.75; when there are two tensile members, the value is taken as 0.6. f_c is the design compressive strength of concrete, obtained from the basic mechanical performance tests of concrete materials.

Anchored rebar joints primarily facilitate load transfer through the cohesive interaction between the rebar and the surrounding concrete matrix. Based on the "circuitous force flow" transfer mechanism of anchored rebar joints [28], the principle of circuitous force transfer in the post-cast wet joint composed of anchored rebar, transverse rebar, and UHPC joint can be idealized as the calculation diagram of the strut-and-tie model, which is displayed in Figure 17. In the wet joint, circuitous force are transferred between the anchored rebar, transverse rebar, and joint. The anchored rebar and transverse rebar act as tension members, represented by solid lines in the diagram; the joint between the anchored rebars acts as a compression member, also represented by solid lines in the figure; and the compression member and tension members are in a state of force equilibrium at node C. In the diagram, B represents the concrete compression member, and T represents the steel tension member.

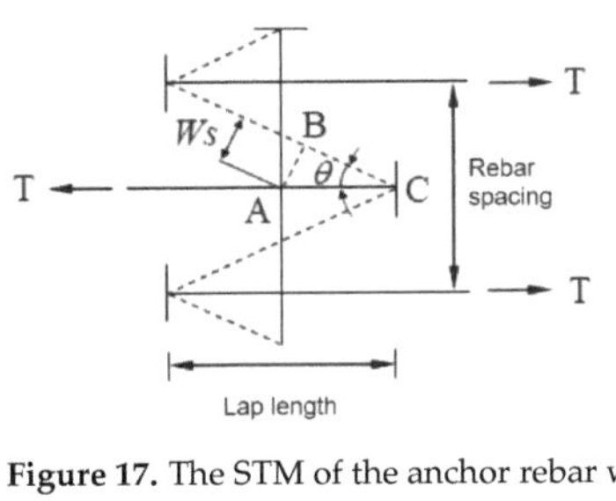

Figure 17. The STM of the anchor rebar wet joint.

On the basis of the wet joint strut-and-tie model (STM) of the anchor rebar shown in Figure 17, the final bearing capacity of the STM of joint is controlled by the yield of the anchor rebar tension rod AC, the crushing of the diagonal concrete compression rod BC, and the yield of the transverse rebar tension rod AB, which means it depends on the compressive strength of the joint and the yield strength of the rebar. The ideal failure mode for wet joints of anchor rebars is that after the anchor rebar tension rod AC yields, the concrete compression rod BC is crushed or the transverse rebar tension rod AB yields, resulting in ductile failure of the joint. On the contrary, if the yield of the anchoring steel rod occurs after the concrete compression rod is crushed or the transverse steel rod yields, the anchoring rebar cannot exert its bearing capacity, and the failure of the joint is brittle failure rather than ductile failure.

As shown in Figure 17, the internal force expressions for the concrete compression rod AC, the anchoring steel rod AB, and the transverse steel rod BC within the wet joint STM are derived from the force equilibrium condition at node A, which are calculated according to Equations (4)–(6).

$$F_s = \frac{T}{2\cos\theta} \tag{4}$$

$$F_h = T \tag{5}$$

$$F_l = \frac{T}{2}\tan\theta \tag{6}$$

where T is the tensile force of anchoring rebar tie rod; F_s is the pressure of concrete compression rod; F_h is the tensile strength of anchoring rebars and tie rods; F_l is the tensile force of transverse rebar tie rod; θ is the angle between two opposite anchoring rebars at adjacent positions, $\cos\theta = \frac{2l}{\sqrt{4l^2+s^2}}$; l is the overlap length of two opposite anchoring rebars at adjacent positions; and s is the spacing between two adjacent anchoring rebars in the same direction.

The final bearing capacity of the wet joint of anchored rebars is controlled by the ultimate compressive strength of the concrete compression rod and the ultimate tensile bearing capacity of the rebar tension rod. The ultimate compressive strength of the concrete compression rod is calculated according to Equation (7).

$$F_s = f_{cu}A_{strut} = 0.85\beta_n f_c' A_{strut} \tag{7}$$

where F_s is the ultimate compressive strength of concrete compression rod; A_{strut} is the cross-sectional area of the concrete compression bar, $A_{strut} = DW_s = \frac{Dl\sin\theta}{2}$; D is the height of the concrete compression rod taken as the diameter of the anchoring plate at the end of the anchoring rebar; and W_s is the width of the concrete compression bar.

The ultimate compressive strength of the concrete compression rod and the ultimate tensile bearing capacity of the anchored steel rod can be used to determine the STM ultimate bearing capacity of the wet joint of the anchored steel rod. They are calculated according to Equations (8)–(10).

$$T_{us} = 2F_s\cos\theta = 2F_s\frac{2l}{\sqrt{4l^2+s^2}} = \frac{3.4f_c' A_{strut} l}{\sqrt{4l^2+s^2}} \tag{8}$$

$$T_{uh} = F_h = f_{yh}A_h \tag{9}$$

$$T_{ul} = \frac{2f_{lh}A_l}{\tan\theta} = \frac{4f_{lh}A_l l}{s} \tag{10}$$

where T_{us} is ultimate tensile bearing capacity of concrete compression bar; T_{uh} is the ultimate tensile bearing capacity of anchored rebars and tie rods; T_{ul} is the ultimate tensile bearing capacity of transverse rebar tie rods; f_{yh} is the yield strength of anchored rebars; f_{lh} is the yield strength of transverse rebars; A_h is the cross-sectional area of anchored rebars; and A_l is the cross-sectional area of transverse rebars.

The ultimate tensile bearing capacity of the entire joint specimen is related to the number of anchoring rebars on one side of the joint and is calculated according to Equation (11).

$$T_u = N \times \min\left(\frac{2.72f_c'A_{strut}l}{\sqrt{4l^2+s^2}}, f_{yh}A_h, \frac{4f_{lh}A_l l}{s}\right) \tag{11}$$

In this study, the post poured wet joint bears a pure bending load, and the rebars and concrete in the joint are well bonded. The bonding between the precast section of the joint specimen and the post poured wet joint is reliable. The calculation method of the bending capacity of a single reinforced rectangular section bending member can be referred to calculate the bending capacity of the wet joint. The calculation diagram of the bending capacity of the wet joint is shown in the following Figure 18.

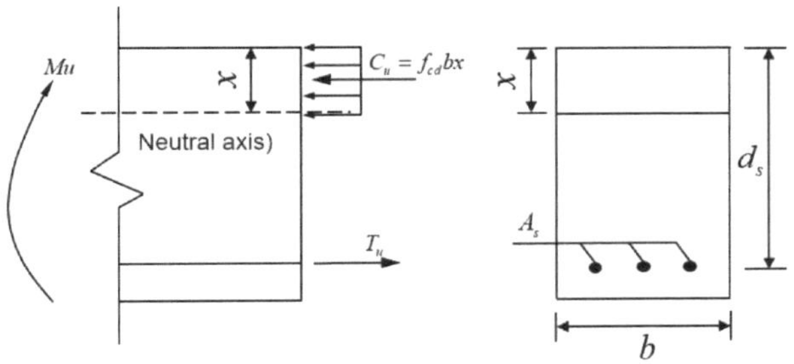

Figure 18. Calculation diagram of the wet joint.

The position x of the neutral axis of concrete and the bending capacity of the joint M_u can be obtained based on two equilibrium conditions: the sum of circuitous force in the horizontal direction on the section is equal to zero and the sum of moments at the point of resultant force in the compressed area is equal to zero. They are calculated according to Equations (12) and (13).

$$x = \frac{C_u}{0.85f_c'b} = \frac{T_u}{0.85f_c'b} \tag{12}$$

$$M_u = T_u\xi\left(d_z - \frac{x}{2}\right) = T_u\xi\left(d_z - \frac{T_u}{1.7f_c'b}\right) \tag{13}$$

where C_u is the resultant force of concrete in the compression zone of the section; T_u is the joint force of rebars in the tensile zone of the cross-section; b is joint width; d_s is the effective height of the cross-section; and ξ is the impact coefficient of live-load vibration, taken as 1 when considering no live-load vibration. When considering live-load vibrations $\xi = k\alpha + k_1$, where α is the amplitude of the live-load vibration, k and k_1 are constants, k is the intercept, and k_1 is the influence coefficient of the live-load vibration amplitude. By substituting the experimental data into Formula (13), it can be calculated that the k is 1.208 and the k_1 is -0.125; therefore, the $\xi = -0.125\alpha + 1.208$.

In summary, the final bending capacity of UHPC joint specimens was calculated by substituting each group of specimens into Formula (13). The comparison results with the experimental values are shown in Table 6.

As shown in Table 4, the ratio of the theoretical calculation value to the experimental value of the ultimate bending capacity of each specimen is between 0.96 and 1.05, with an average value of 0.99 and a standard deviation of 0.035. The accuracy of the theoretical calculation value is high, which can provide theoretical support for the calculation of joint flexural capacity under live-load vibration.

Table 6. The summary of calculated values and experimental values.

Test-Piece	Mc	Mtes	Mc/Mtes
W-1	28.17	28.64	0.98
F-1	32.59	33.62	0.97
F-2	27.53	26.28	1.05
F-3	22.46	23.34	0.96

Note: Mc is the calculated value, Mtes is the experimental value, both in kN·m.

5. Conclusions

The following conclusions were obtained:

- After the specimens were destroyed, UHPC joints barely showed any cracks but the initial and main cracks were observed within the pure bending zone of the NC. These cracks were particularly evident at the top of the precast section, where the concrete was crushed. It indicated that UHPC wet joints have excellent crack resistance;
- The vibrations of higher amplitude and frequency will reduce the interfacial bond strength of the wet joints. Compared with the specimens without vibration, the post-yield flexural stiffness and flexural strength of the specimens reduced by 24.39%. Additionally, the flexural strength of specimens subjected to the vibration at 3 Hz-3 mm and 3 Hz-5 mm were decreased by 8% and 19%, respectively;
- Vibration with lower amplitude or frequency was found to increase the flexural strength of joint specimens. As the amplitude and frequency of vibration decrease, the live-load vibration shows an increasing trend in the flexural strength of the specimens. This is due to the fact that such vibration helps to enhance the compactness of the concrete;
- Low amplitude vibration has a positive effect on ductility. However, compared to specimens without vibration, the stiffness of the 3 Hz-1 mm specimen increased by 8.6%, whereas the stiffness of the 3 Hz-3 mm and 3 Hz-5 mm specimens decreased by 18.1% and 22.3%, respectively;
- A calculation model for the flexural strength of UHPC joints was established, considering the impact of live-load vibrations. A live-load vibration coefficient was introduced to refine the calculation formula. The average ratio of theoretical calculation values to experimental values is 1.01, with a standard deviation of 0.04, indicating a high level of accuracy.

Author Contributions: Conceptualization and supervision, B.Z.; methodology and resources, Z.Z.; software and supervision, J.L.; validation and formal analysis, D.Q.; investigation and writing—original draft, Y.Z.; data curation and project administration, J.Y.; project administration and date curation, K.Z. All authors have read and agreed to the published version of the manuscript.

Funding: This research was funded by [the Science and Technology Project of China Railway Group Co., Ltd.] [Grant No. 2022-ZD-11], [the Natural Science Foundation of Chongqing, China] [Grant No. CSTB2023NSCQ-443BHX0229], [the Research and Innovation Program for Graduate Students in Chongqing] [Grant Nos. CYB23245, CYS240448], and [the Science and Technology Research Program of Chongqing Municipal Education Commission] [Grants No. KJQN202300723, KJQN202400701].

Data Availability Statement: The data presented in this study are available on request from the corresponding author due to privacy and legal reasons.

Acknowledgments: All individuals included in this section have consented to the acknowledgement.

Conflicts of Interest: The authors declare that they have no known competing financial interests or personal relationships that could have appeared to influence the work reported in this study.

References

1. Li, C.; Feng, Z.; Ke, L.; Pan, R.; Nie, J. Experimental study on shear performance of cast-in-place ultra-high performance concrete structures. *Materials* **2019**, *12*, 3254. [CrossRef] [PubMed]
2. Jang, H.-O.; Lee, H.-S.; Cho, K.; Kim, J. Experimental study on shear performance of plain construction joints integrated with ultra-high performance concrete (UHPC). *Constr. Build. Mater.* **2017**, *152*, 16–23. [CrossRef]
3. Semendary, A.A.; Svecova, D. Bond characterization and interfacial coefficients at precast UHPC and cast-in-place UHPC interface. *J. Mater. Civ. Eng.* **2021**, *33*, 04021143. [CrossRef]
4. He, H. Research on Key Technologies of Widening Existing Concrete Box Girder Bridges with Steel-Concrete Composite Structures. Master's Thesis, Southeast University, Nanjing, China, 2019.
5. Zhou, J.; Leng, J.; Yang, J.; Zhang, Z.; Du, J.; Zou, Y. Experimental investigation on shear behavior of damaged and acid rain-corroded RC T-beams strengthened with ultra-high-performance concrete. *Eng. Struct.* **2025**, *327*, 119618. [CrossRef]
6. Deng, S.; Shao, X.; Yan, B.; Wang, Y.; Li, H. On Flexural Performance of Girder Wet Joint for Lightweight Steel-UHPC Composite Bridge. *Appl. Sci.* **2020**, *10*, 1335. [CrossRef]
7. Graybeal, B.A. *Behavior of Field-Cast Ultra-High Performance Concrete Bridge Deck Connections Under Cyclic and Static Structural Loading*; FHWA, Report No. FHWAHRT-11-023; FHWA: Washington, DC, USA, 2010.
8. Haber, Z.B.; Graybeal, B.A. Lap-spliced rebar connections with UHPC closures. *J. Bridge Eng.* **2018**, *23*, 04018028. [CrossRef]
9. Varga, I.D.L.; Haber, Z.B.; Graybeal, B.A. Enhancing shrinkage properties and bond performance of prefabricated bridge deck connection grouts: Material and component testing. *J. Mater. Civ. Eng.* **2018**, *30*, 04018053. [CrossRef]
10. Graybeal, B.A. *Design and Construction of Field-Cast UHPC Connections*; FHWA, Report No. FHWA-HRT-14-084; FHWA: Washington, DC, USA, 2014.
11. Lee, J.K.; Lee, S.H. Flexural behavior of ultra-high-performance fiber-reinforced concrete moment connection for precast concrete decks. *ACI Struct. J.* **2015**, *112*, 451–462.
12. Harsh, S.; Darwin, D. *Traffic-Induced Vibrations and Bridge Deck Repairs*; American Concrete Institute: Indianapolis, IN, USA, 1986; pp. 36–42.
13. Zhang, Z.; Pang, K.; Xu, L.; Zou, Y.; Yang, J.; Wang, C. The bond properties between UHPC and stone under different interface treatment methods. *Constr. Build. Mater.* **2023**, *365*, 130092. [CrossRef]
14. Guan, Y.; Wei, J. Analysis of the Influence of Bridge Vibration Caused by Vehicle Load on the Compressive Strength of Concrete. *J. Highw. Transp. Res. Dev.* **2015**, *11*, 244–246.
15. Zhang, X.; Liu, S.; Yan, C.; Wang, X.; Wang, H. Effects of Traffic Vibrations on the Flexural Properties of Newly Placed PVA-ECC Bridge Repairs. *Materials* **2019**, *12*, 3337. [CrossRef]
16. Yang, J.; Leng, J.; Zhou, J.; Chen, R.; Yu, K.; Jiang, Z.; Zou, Y.; Zhang, Z.; Du, J. Study on the impact of vehicle-induced vibration on the flexural behavior of UHPC joints in widened bridges. *Urban Lifeline* **2024**, *2*, 20. [CrossRef]
17. Wu, Z.; Qi, L.; Ye, J.; Sun, J. The influence of vibration deformation vibration on the performance of connecting concrete between new and old bridges. *J. Highw. Transp. Res. Dev.* **2017**, *13*, 279–282.
18. Huang, C.; Song, J.; Zhang, N.; Lee, G.C. Seismic performance of precast prestressed concrete bridge girders using field-cast ultrahigh-performance concrete connections. *J. Bridge Eng.* **2019**, *24*, 04019046. [CrossRef]
19. Wang, S.; Yu, L.; Yang, F.; Xu, L.; Wu, K.; De Schutter, G.; Yang, Z. Effect of steel fiber distribution on the mechanical properties of UHPC caused by vehicle-bridge coupling vibration. *Compos. Part B Eng.* **2022**, *245*, 110201. [CrossRef]
20. Leng, J.; Yang, J.; Zhang, Z.; Du, J.; Zou, Y.; Zhou, J. Effect of vehicle-induced vibration on the strength, nano-mechanical properties, and microstructural characteristics of ultra-high-performance concrete during hardening process. *Cem. Concr. Compos.* **2024**, *148*, 105487. [CrossRef]
21. Pan, H.; Zhao, Q.; Fu, J. Research progress on disturbance performance of early aged concrete. *Bull. Chin. Ceram. Soc.* **2017**, *36*, 64–70.
22. Ng, P.L.; Kwan, A. Effects of traffic vibration on curing concrete stitch: Part II—Cracking, debonding and strength reduction. *Eng. Struct.* **2007**, *29*, 2881–2892. [CrossRef]
23. *GBT50080-2016*; Standard for Test Method of Performance on Ordinary Fresh Concrete. National Standards of the People's Republic of China: Beijing, China, 2016.
24. *GB 50010-2010*; Code for Design of Concrete Structures. Ministry of Housing and Urban Rural Development of the People's Republic of China: Beijing, China, 2015.
25. Chongqing Jiaotong University. Test Device for Simulating Concrete Vibration of Joints of Widened Bridges. China Patent 202211280827.3, 19 October 2022.
26. Wu, G.; Jiang, J.; Wu, Z.; Tian, Y.; Zhang, M. Experimental study on prestressed high-strength steel wire rope bending reinforcement of reinforced concrete beams. *J. Civ. Eng.* **2007**, *12*, 17–27.

27. *ACI Committee 318*; Building Code Requirement for Structural Concrete (ACI 318-08) and Commentary (ACI 318R-08). American Concrete Institute: Indianapolis, IN, USA, 2008.
28. Li, L.; Jiang, Z. Flexural Behavior and Strut-and-tie Model of Joints with headed bar details Connecting Precast Members. *Perspect. Sci.* **2016**, *7*, 253–260. [CrossRef]

Disclaimer/Publisher's Note: The statements, opinions and data contained in all publications are solely those of the individual author(s) and contributor(s) and not of MDPI and/or the editor(s). MDPI and/or the editor(s) disclaim responsibility for any injury to people or property resulting from any ideas, methods, instructions or products referred to in the content.

buildings

MDPI

Article

Flexural Performance of RC Beams Strengthened with High-Strength Steel Wire Mesh and UHPC

Chao Zhu [1], Chunlin Du [1], Yanxin Qi [2,3], Zhimei Jiang [2,3,*], Zhongya Zhang [2,3], Jun Yang [2,3], Yinbin Li [4] and Jun Cheng [2,3]

1 Chongqing Design Group Co., Ltd., Chongqing 400015, China
2 State Key Laboratory of Mountain Bridge and Tunnel Engineering, Chongqing Jiaotong University, Chongqing 400074, China; zhangzhongya@cqjtu.edu.cn (Z.Z.)
3 School of Civil Engineering, Chongqing Jiaotong University, Chongqing 400074, China
4 Guizhou Provincial Transportation Planning Survey and Design Institute Co., Ltd., Guiyang 550081, China
* Correspondence: jiangzhimei@mails.cqjtu.edu.cn

Abstract: This study proposes a strengthening technique comprising a combination of high-strength steel wire mesh and ultra-high performance concrete (UHPC) to address the challenge of the insufficient bearing capacity of existing structures. The tensile performance of high-strength wire mesh and the crack resistance of UHPC were comprehensively considered in this technique. To evaluate the influence of the steel fiber volume ratio and the high-strength steel mesh strengthening ratio on the axial tensile performance, uniaxial tensile tests were carried out on two sets of dumbbell-shaped specimens. A constitutive model of the wire mesh UHPC that matched the experimental results was established. The finite element analysis of RC beams strengthened with high-strength wire mesh and UHPC was carried out, based on this constitutive model. The experimental results indicated the following: (a) The crack resistance and ultimate strength of the specimen reinforced with the high-strength steel wire mesh were effectively enhanced, with enhancement ratios of 97.8% and 124.8%, respectively. (b) The embedded interactions between the steel wire mesh and UHPC were simulated by considering the material nonlinearity. The finite element modeling of RC beams strengthened with wire mesh UHPC was achieved. (c) Positive correlations were observed between the thickness of the UHPC layer, the steel fiber volume ratio, and the high-strength wire mesh layer with the flexural capacity of the strengthened beams. The cracking and ultimate moments were maximally enhanced by 96.2% and 99.4%, respectively.

Keywords: strengthening; UHPC; high-strength steel wire mesh; RC beam; flexural capacity

1. Introduction

Reinforced concrete (RC) structures were damaged by loads, environmental erosion, and other factors during service, which inevitably led to structural performance degradation. The normal utilization of the structure could be affected, posing potential safety hazards [1]. UHPC, as a new type of cementitious composite material [2], offers significant advantages in terms of durability, elastic modulus, and tensile and compressive strength [3,4]. The high density and low porosity of UHPC also provided it with a higher anti-deterioration ability [5,6], which makes it an ideal choice for strengthening existing concrete structures [7,8].

The application of UHPC in RC beam strengthening achieved both the light weight and high efficiency of the strengthened structure, i.e., the ultimate tensile strength far exceeded that of ordinary concrete (NC), making the strengthened structure thinner and lighter [9]. Scholars have carried out considerable research on the flexural performance of RC beams strengthened by employing the characteristics of UHPC [10]. Wei J et al. [11] conducted tests on the impact resistance of RC with UHPC-jacketed RC beams and concluded that the strengthening effect of UHPC was excellent. Zhang et al. [12] observed that the pre-damage of RC beams was negatively correlated with the cracking and flexural capacity of the UHPC layer. Ke, L. [13] and Nadir, W. [14] found that UHPC strengthening could inhibit the generation and development of cracks. Meanwhile, the bearing capacity and durability of the components could be improved. Dhafer Mirdan et al. [15] demonstrated that the application of a UHPC layer on the tensile side of RC beams significantly enhanced their flexural stiffness (by 20–132%) and effectively delayed the initiation of cracks. Yoo, D. Y. et al. [16] firmly bonded the UHPFRC strengthening layer to the RC beam. The experiment showed that even with a relatively thin UHPFRC layer, the flexural stiffness and load-carrying capacity of the component can be effectively improved. In addition, other scholars [17] conducted numerical simulation experiments using finite element software to research the flexural performance of RC beams strengthened with UHPC. All the experiments indicated that the flexural capacity of the strengthened RC beams was significantly enhanced. Yu et al. [18] established a multi-scale analysis method to simulate the mechanical behavior of the NC–UHPC interface. The results indicated that the interface roughness and the strength of the concrete substrate have a positive effect on the mechanical properties of the interface. Krishna et al. [19] determined the tensile and compressive behavior of UHPC. A numerical model was established to simulate the behavior of UHPC.

The strengthening effect of high-strength steel wire mesh on the mortar matrix has been demonstrated. The strengthening layer composed of high-strength steel wire mesh and mortar can effectively improve the extreme bearing capacity and stiffness of the RC beam. Paschalis et al. [20] used a steel mesh cement wrapping to increase the load-bearing capacity of columns. Compared to NC columns, the bearing capacity of fiber-reinforced steel mesh cement mortar restraint columns was increased by 51%. Shafieifar et al. [21] effectively improved the deformability and ductility of concrete specimens by adding new composite materials composed of high-strength steel rings, wire mesh, and modified high-strength mortar. Zhao et al. [22] investigated the influence of mesh type, wrapping method, and loading on the performance of RC square columns. The results indicated that the strength of columns reinforced with high-strength steel wire mesh mortar was significantly improved. Moreover, UHPC itself exhibits extremely high durability. Through the synergistic effect with high-strength steel wire mesh, it can not only improve the bearing capacity and overall stiffness of the structure but also delay the aging and damage process caused by the external environment. This greatly improves the durability and long-term service performance of the structure and provides a reliable guarantee for structural strengthening in complex environments [23]. Especially under high loads or extreme conditions, it can effectively reduce the safety risks caused by structural failure. In practical engineering, the propagation speed and depth of structural cracks directly impact its durability and safety. By increasing the strengthening ratio of high-strength steel wire mesh, crack propagation can be effectively delayed, ensuring that the structure maintains stable mechanical performance over a long period.

In summary, studies on RC structures strengthened with high-strength steel wire mesh and UHPC are scarce. The failure modes of RC beams strengthened with thin layers of high-strength steel mesh UHPC, as well as the impact of parameters such as strengthening layer thickness and steel fiber volume on their flexural capacity, remain unknown. Corresponding calculation methods are still lacking. In this study, the axial tensile mechanical behavior characteristics of high-strength steel mesh UHPC materials were explored through theoretical analysis, model tests, and finite element numerical analysis. The influence of various parameter changes on the ultimate bearing capacity and deflection of the strengthened beams was analyzed. The flexural mechanical performance of RC beams strengthened with high-strength steel wire mesh UHPC was investigated.

2. Experimental Program

2.1. Design of Axial Tension Specimen

A dumbbell-shaped uniaxial tensile specimen was designed with a cross-sectional width and thickness of 40 mm and 30 mm, respectively. The inclination angle of 75.26° on the side of the end was set to facilitate the use of the fixture. The high-strength steel wire mesh is installed in the middle of the specimens. The detailed dimensions are shown in Figure 1.

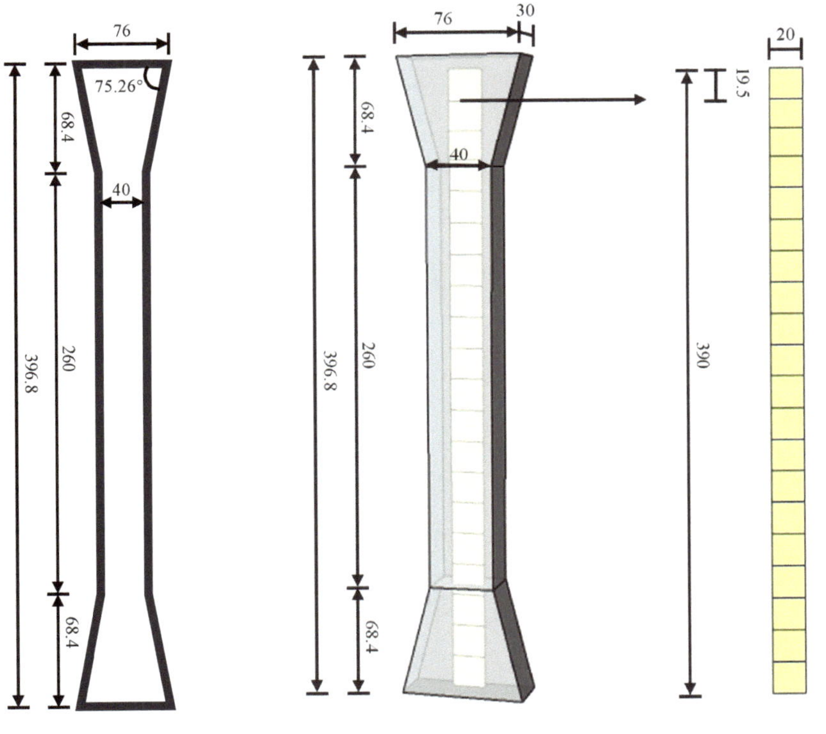

Figure 1. Dimensions of axial tension specimen (unit: mm).

A total of 21 dumbbell-shaped uniaxial tensile specimens in seven categories were designed and divided into two groups (see Table 1). The experimental parameters included the thickness of the UHPC layer, the steel fiber volume, and the strengthening ratio of the high-strength steel wire mesh. The detailed specimen configuration is shown in Table 1. In the first group (SU-1~3), the thickness of the specimen was 30 mm. There was no high-strength steel wire mesh layer, and only the steel fiber volume was different. The second group (SU-4~7) included four combinations of different steel fiber volumes and strengthening ratios of the high-strength steel mesh.

Table 1. Design table of uniaxial tensile specimens.

Group	Number	Thickness/mm	Steel Fiber Volume v_f/%	Strengthening Ratio of High-Strength Steel Wire Mesh ρ_w/%	Number of Specimens
1	SU-1	30	1	/	3
	SU-2	30	2	/	3
	SU-3	30	3	/	3
2	SU-4	30	2	0.52% (one layer)	3
	SU-5	30	2	1.04% (two layers)	3
	SU-6	30	3	0.52% (one layer)	3
	SU-7	30	3	1.04% (two layers)	3

Note: ρ_w is equal to the steel wire mesh cross-sectional area/specimen cross-sectional area × 100%.

2.2. *Test Material*

2.2.1. UHPC

The mix proportions of UHPC are shown in Table 2. Referring to GB/T 31387-2015 [24], three types of UHPC compressive strength standard cube specimens, with dimensions of 100 mm × 100 mm × 100 mm and different volume fractions of steel fiber volume, were produced. After curing for 28 days, the compressive strengths of the UHPC were measured. The test results are shown in Table 3.

Table 2. The mix proportions of UHPC (km/m^3).

Ingredient	Cement	Silica Fume	Quartz Sand	Quartz Powder	Water Reducer	Water
Unit weight	1	0.25	1.1	0.3	0.019	0.18

Table 3. Compressive strength of UHPC.

Steel Fiber Volume	Compressive Strength/MPa			Average/MPa
	Specimen 1	Specimen 2	Specimen 3	
1%	135.7	131.3	125.9	131.0
2%	137.9	141.3	143.5	140.9
3%	147.4	146.2	149.9	147.8

2.2.2. High-Strength Steel Wire Mesh

The high-strength steel wire mesh utilized in this study was fabricated by welding cold-drawn low-carbon steel wires with a diameter of 2 mm, forming a mesh with a 15 mm opening size. Six 20 cm long steel wire specimens were removed from the high-strength steel wire mesh and tested for their uniaxial tensile properties, according to GB 1499.2-2018 [25]. The results are shown in Table 4.

Table 4. Uniaxial tensile performance index of high-strength steel wire mesh.

Indicators	Yield Strength (MPa)	Ultimate Strength (MPa)	Elastic Modulus (GPa)
Average	490	721	204

2.3. *Loading Scheme*

The experiment was loaded using an MTS mechanical universal testing machine. To avoid cracking at the variable cross-section of the specimen, the steel sheets were bonded with epoxy resin adhesive to enhance the strength and stiffness at the variable cross-section. Before testing, two shallow marks were made 25 mm from the mid-span on both sides of the specimen. The blades of the extensometer were then clamped into these shallow marks, and the extensometer was fixed onto the specimen using a stretched rubber band. The test was performed under displacement-controlled loading. Initially, a pre-load of 0.5 kN was applied to securely clamp the specimen at both ends, followed by continuous loading at a rate of 0.05 mm/min. The loading procedure and measurement setup are illustrated in Figure 2.

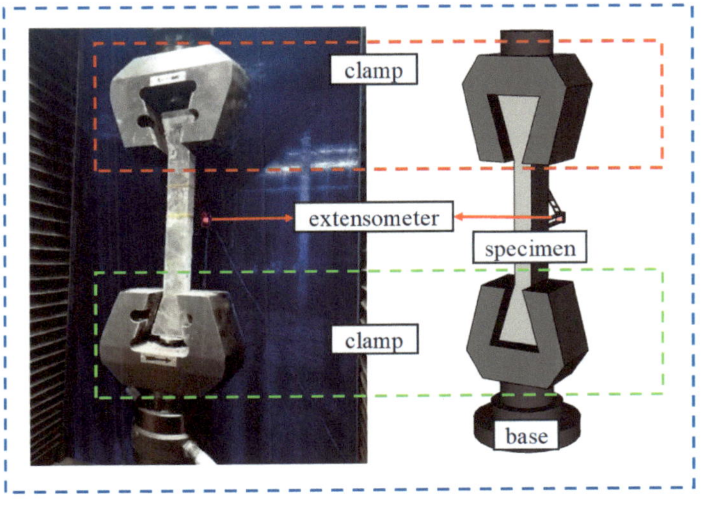

Figure 2. Test setup.

3. Analysis and Discussion of Results

3.1. Failure Mode

The SU-1~SU-3 specimens were not stiffened with high-strength steel wire mesh. The only difference among these groups was in the volume fraction of the steel fibers, denoted as v_f. The specimens were damaged by over-extension of the primary crack, with significant steel fiber pullout observed following damage. The SU-1 group specimens were mainly damaged by excessive extension of the primary crack. Due to the low steel fiber dosage, the specimens could not effectively limit the crack extension after cracking, which led to the damage of the specimens under lower strains. The specimens of group SU-2 were able to bridge the steel fibers at the primary crack to limit the extension of the crack during the loading process, but they were still damaged because of the extension of the primary crack. The specimens of group SU-3 included a higher steel fiber dosage. When the specimens were subjected to tensile loading, they were able to effectively resist the formation and extension of cracks at the initial stage. With the increase in load, although the primary cracks started to appear, the numerous steel fibers played an effective bridging role between the cracked surfaces and prevented the further extension of the cracks. This bridging effect enabled the specimens to maintain a high level of structural integrity before reaching the ultimate load capacity. Notably, the primary cracks were not always located at the mid-span position, as shown in Figure 3.

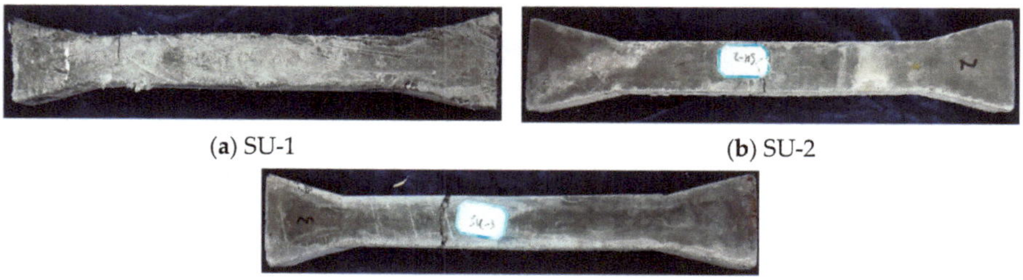

(**a**) SU-1 (**b**) SU-2 (**c**) SU-3

Figure 3. Failure modes of pure UHPC specimens.

Compared to the pure UHPC specimens, the specimens strengthened with high-strength steel wire mesh exhibited significantly lower propagation rates of the primary cracks during the strain-softening stage. The relative slip between the high-strength wire mesh and UHPC matrix was not observed. However, the phenomenon of the early pull-off of the high-tensile steel wire mesh occurred in groups SU-5 and SU-7 during the strain-softening stage. This resulted in a stress–strain curve characterized by a stepwise rather than a smooth decrease in the softening phase. The failure modes of the specimens are illustrated in Figure 4.

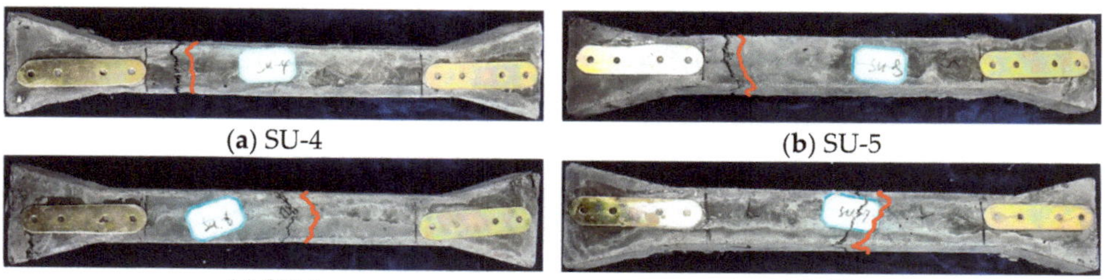

(**a**) SU-4 (**b**) SU-5 (**c**) SU-6 (**d**) SU-7

Figure 4. Failure modes of UHPC specimens with high-strength steel wire mesh.

3.2. Stress–Strain Curve

The stress–strain curves of the pure UHPC specimens are shown in Figure 5. The fiber-reinforced concrete can be classified into strain-hardening and strain-softening materials in terms of their axial tensile constitutive relationship [26]. It can be seen from the stress–strain curves that UHPC materials in this test were strain-hardening materials; however, the degree of hardening of the materials varied with different steel fiber volumes.

The stress–strain curves in the SU-1 group (Figure 5a) increased linearly during the initial loading stage. A continuous increase in strain, with almost no change in stress, was observed after the elastic limit was reached, exhibiting an inconspicuous strain-hardening behavior. This phenomenon is attributed to the fact that the steel fiber volume is only 1%, and the limited number of steel fibers cannot effectively restrain the crack development after the specimen cracks. The curves entered the softening stage once the ultimate stress was reached. However, some specimens displayed noticeable elastic unloading behavior due to the primary crack occurring outside the gauge measurement range. The significant strain-hardening behavior was observed in the SU-2 group (Figure 5b). After reaching the elastic limit, the slope of the curve decreased significantly; however, the stress still increased with increasing strain until the ultimate stress was reached. Notably, the stress–strain curves of specimens SU-2(1) and SU-2(2) are very similar, demonstrating good consistency.

Besides, the stress–strain curve for specimen SU-2(3) continued to rise after the elastic limit but exhibited elastic unloading once the ultimate strength was reached. This behavior is attributed to the primary crack of specimen SU-2(3) being located outside the gauge measurement range, despite the presence of micro-cracks within the gauge distance of the indenter. A similar phenomenon was observed in specimens SU-3(1) and SU-3(3).

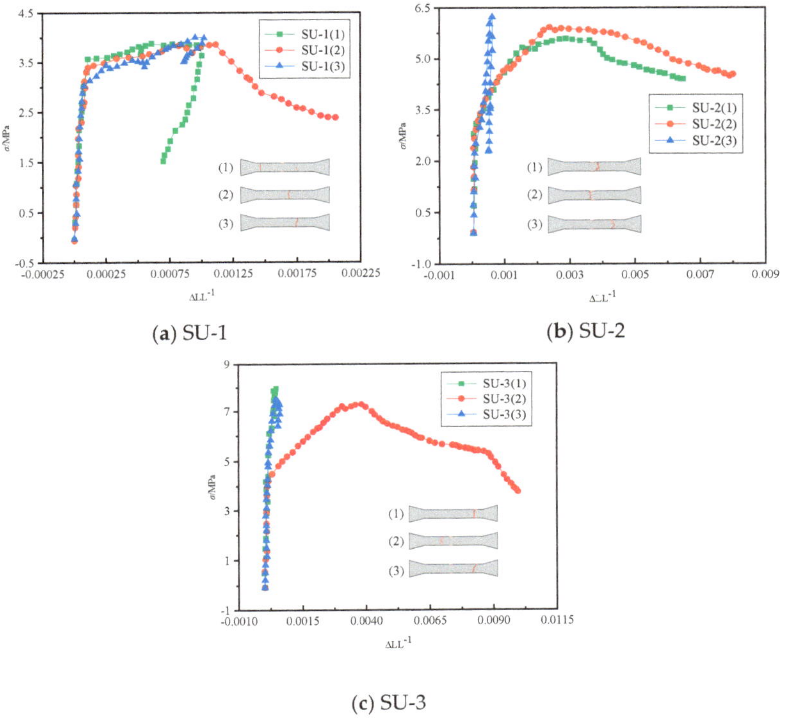

Figure 5. Stress–strain curves of pure UHPC specimens with different steel fiber strengthening ratios: (**a**) v_f = 1%; (**b**) v_f = 2%; (**c**) v_f = 3%.

To summarize, not all portions of UHPC entered the strain-hardening stage after reaching the elastic limit under axial tensile force. The uncracked portion remained linearly elastic, even as the stress reached the strain-hardening stage. The appearance of microcracks, however, caused plastic deformation throughout the material, resulting in the slope of the stress–strain curve decreasing in the strain-hardening stage. After UHPC reaches its ultimate tensile stress, the further propagation of the main crack causes the material to exhibit strain-softening behavior. The gradual decrease in stresses carried by the material, however, resulted in the retraction of other micro-cracks, with elastic unloading behavior in other parts (except for the primary cracks).

The stress–strain curves of the second group of specimens are shown in Figure 6. The SU-5 specimen group contained only two specimens with test data because of improper demolding. The specimens with the primary crack located outside the electronic indenter scale exhibited similar characteristics to those of the SU-1~SU-3 specimens. Specifically, the tested section continued to rise, with an approximately elastic slope, after reaching the material's elastic limit and entered the elastic unloading stage after reaching the ultimate strength. Comparing Figure 5a with Figure 6a reveals that the stiffening effect of the high-strength steel mesh effectively enhances the cracking stress and initial cracking stress of the specimen, while significantly increasing the ultimate strain. This results in a more gradual strengthening segment in the stress–strain curve of the specimen, making it closer to the stress–strain curve of low-strain-hardening materials. It can be observed that the enhancement of the axial tensile performance of the specimens by the increase in the high-strength wire mesh layer was relatively limited. However, the slope of the strain-hardening stage of the SU-5 specimen was significantly increased compared to that of the SU-4 specimen. This indicated that while increasing the enhancement ratio of high-tensile steel wire mesh could not substantially improve the axial tensile performance, it effectively improved its axial stiffness during the strain-hardening stage.

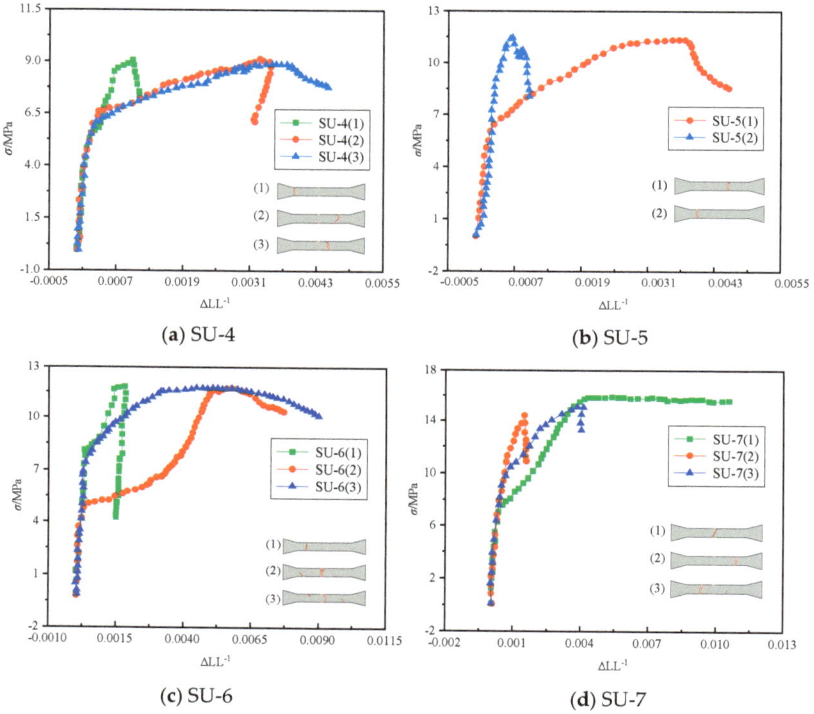

Figure 6. Stress–strain curves of UHPC specimens with different strengthening ratios of steel fiber and high-strength steel wire mesh: (**a**) $v_f = 2\%$, $\rho_w = 0.52\%$; (**b**) $v_f = 2\%$, $\rho_w = 1.04\%$; (**c**) $v_f = 3\%$, $\rho_w = 0.52\%$; (**d**) $v_f = 3\%$, $\rho_w = 1.04\%$.

The strain-hardening stage of the SU-6 and SU-7 specimens existed in two stages with different slopes, both of which included horizontal segments before they reached their ultimate strengths. Combined with the observation of the experimental process, it is observed that the better synergistic effect of both the steel fibers and the high-strength steel wire mesh was produced in the SU-6 and SU-7 specimens with the higher steel fiber content. It is noteworthy that the high-tensile steel wire mesh itself consisted of bright round steel wires. During specimen cracking, there was no effective bonding between the steel wire surface and the UHPC matrix at the cracks, which led to partial bond slip. The stress–strain curves of some specimens thus showed a smoother strain-hardening stage. However, the interlocking force of the high-strength steel mesh grid can effectively prevent further slippage of the steel wires, leading to a redistribution of internal forces between the steel fibers and the steel wires at the crack, resulting in a synergistic load-bearing effect and the increased axial stiffness of the specimen. Compared to specimen SU-6, specimen SU-7 exhibits stronger synergy between the high-strength steel mesh and the steel fibers after cracking due to the higher strengthening ratio of the high-strength steel mesh, making the slippage of steel wires less noticeable.

3.3. Characteristic Parameters

The relevant characteristic parameters in Figure 5 are summarized in Table 5 and Figure 7. A positive correlation between steel fiber volume and initial cracking stress–strain, ultimate stress, and strain of pure UHPC specimens was observed. The average ultimate stress and strain of the SU-2 specimens increased by 52.1% and 147.3%, respectively, compared with those of the SU-1 specimens. The average ultimate stress and strain of the SU-3 specimens increased by 28.5% and 43.6%, respectively, compared with those of the SU-2 specimens. The average cracking stress and strain of the SU-3 specimens increased by 45.9% and 36.0%, respectively, compared with those of the SU-2 specimens. However, the cracking stress and strain of the SU-1 specimens were higher than those of the SU-2 specimens. This is attributed to the fact that the SU-1 specimens already showed microcracks before reaching the strain-hardening stage. Meanwhile, the primary crack of the SU-2 specimen was not within the marked range, and its cracking load was not captured in a timely manner.

Table 5. Characteristic parameters of pure UHPC.

Number	Cracking Stress/MPa	Cracking Strain/10^{-6}	Ultimate Stress/MPa	Ultimate Strain/10^{-6}
	3.59	106.1	3.89	1009.3
SU-1	3.37	99.7	3.83	1082.6
	3.11	85.3	4.04	1041.1
Average	3.36	97.0	3.92	1044.3
	3.18	82.6	5.65	2685.1
SU-2	3.31	92.3	5.98	2479.8
	3.03	83.7	6.26	/
Average	3.17	86.2	5.96	2582.5
	4.59	109.7	8.05	/
SU-3	4.35	112.4	7.29	3707.3
	4.95	129.7	7.65	/
Average	4.63	117.3	7.66	3707.3

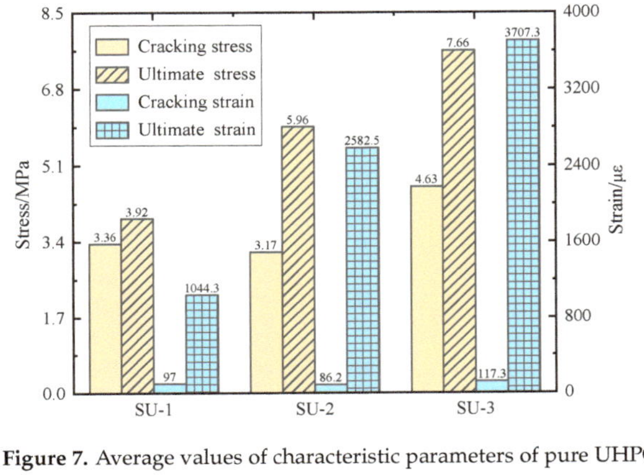

Figure 7. Average values of characteristic parameters of pure UHPC.

The relevant characteristic parameters in Figure 6 are summarized in Table 6 and Figure 8. It can be seen that the steel fiber volume significantly influenced the initial cracking stress of the specimens. The average initial cracking stress of the SU-6 and SU-7 specimens, with 3% steel fiber volume, increased by 32.4% and 22.0%, respectively, compared with that of the SU-4 and SU-5 specimens. The corresponding initial cracking stresses also increased significantly. The initial cracking stress of the SU-6 and SU-7 specimens with different enhancement ratios of high-strength wire mesh differed by only 13.3%. The ultimate stresses of the specimens were positively correlated with both the steel fiber admixture and the high-strength wire mesh enhancement ratio as the ultimate strength was reached. In particular, the average ultimate strength of the SU-7 group reached 15.15 MPa, which was more than twice that of the pure UHPC specimens.

Table 6. Characteristic parameters of UHPC with high-strength steel wire mesh.

Number	Cracking Stress/MPa	Cracking Strain/10^{-6}	Ultimate Stress/MPa	Ultimate Strain/10^{-6}
SU-4	4.92	142.3	9.11	/
	4.73	131.7	9.01	3397.1
	5.64	155.5	8.93	3726.5
Average	5.10	143.2	9.02	3561.8
SU-5	6.45	174.7	11.47	3724.7
	6.09	275.6	11.71	/
	/	/	/	/
Average	6.27	225.2	11.59	3724.7
SU-6	8.19	257.4	11.96	/
	4.77	133.6	11.73	5727.5
	7.30	231.5	11.91	5466.3
Average	6.75	135.5	11.87	5596.9
SU-7	7.07	189.1	15.85	8334.7
	7.74	246.3	14.45	/
	8.13	202.7	15.14	/
Average	7.65	212.7	15.15	8334.7

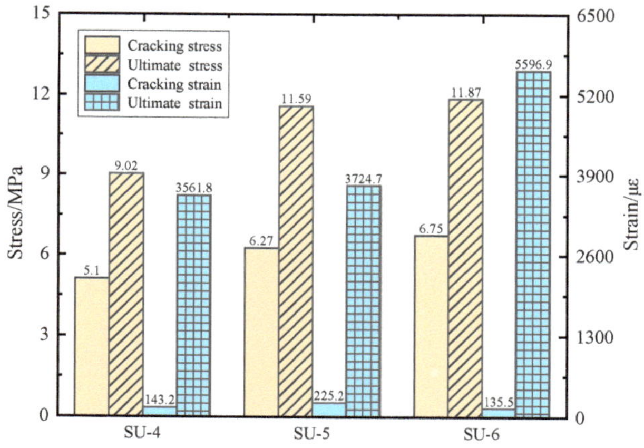

Figure 8. Average values of characteristic parameters of UHPC with high-strength steel wire mesh.

4. Strain-Hardened Tensile Constitutive Model

Based on the analysis of the test results presented in Section 3, the UHPC used in this test exhibited three stages as the stress level increased during axial tension.

(I) Elastic stage: The specimen was uncracked.
(II) Strain-hardening stage: The primary crack developed. The steel fibers at the crack, however, were still effective in bridging the forces, with a further rise in stress and a rapid increase in strain.
(III) Strain-softening stage: The primary crack extended rapidly once the ultimate strength was reached, and the stress decreased.

These characteristics of the stress–strain curve were close to those of the constitutive relationship of typical UHPC strain-hardening proposed by Wille [26], as shown in Figure 9.

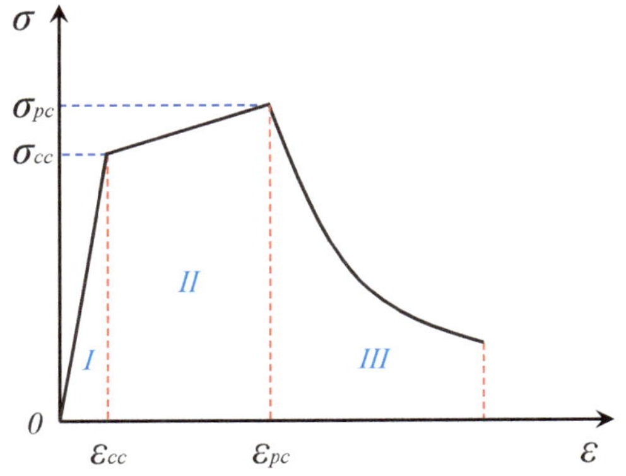

Figure 9. Constitutive model of UHPC tensile strain-hardening.

In the chart, the stress and strain of high-strength steel mesh UHPC at the limit of the elastic stage are denoted into σ_{cc} and ε_{cc}, respectively. The stress and strain at the limit of the variable strain-hardening stage are denoted as σ_{pc} and ε_{pc}, respectively.

In order to simplify the analysis and capture the primary behavior of the material prior to reaching ultimate strength, a two-stage UHPC axial tensile constitutive model containing an elastic segment and strain-hardening stage model was established, according to the methodology described in the literature [26].

As shown in Figure 9, the parameters that determine the two-stage axial tensile constitutive model for UHPC included the stress and strain at cracking, as well as the ultimate stress and strain. In stage I, the UHPC has not yet cracked, and the steel fibers and UHPC matrix work synergistically within the material. Therefore, the cross-section can achieve load equilibrium, as follows:

$$E_{cc}\varepsilon_{cc}A_{cc} = E_u\varepsilon_u A_u + E_s\varepsilon_s A_s \tag{1}$$

$$A_s = v_f \times A_{cc} \tag{2}$$

where E_{cc} and A_{cc} are the modulus of elasticity and the cross-sectional area, respectively, when UHPC reaches the elastic limit. E_u, ε_u, and ε_u are the modulus of elasticity, the strain, and the cross-sectional area of the UHPC matrix, respectively. E_s, ε_s, and A_s are the elastic modulus, the strain, and the cross-sectional equivalent area of the steel fibers, respectively.

Given that when uncracked, $\varepsilon_{cc} = \varepsilon_u = \varepsilon_s$, $A_{cc} = A_u + A_s$, we can obtain the following:

$$E_{cc} = (1 - v_f)E_u + v_f E_s \tag{3}$$

The elastic modulus of the steel fibers is 210 GPa, while the elastic modulus of the UHPC is approximately equal to its elastic modulus under compression [5]. Its compressive elastic modulus can be directly adopted. However, since no test on the elastic modulus of UHPC had been previously conducted, an approximate formula for calculating the elastic modulus, proposed in the literature [5], was used here, as follows:

$$E_u = k_0 \times f_{cc}^{1/3} \tag{4}$$

where k_0 is the conversion factor; f_{cc} is the compressive strength of UHPC.

The stage I constitutive model of UHPC can be determined only by determining any of the parameters σ_{cc} and ε_{cc}. The existing literature showed that the elastic limit strength (σ_{cc}) of UHPC under axial tension was related to the tensile strength (σ_{ct}), the volume fraction of the steel fiber (v_f), and the slenderness ratio of the steel fiber (l/d_f). The relationship is expressed as follows:

$$\sigma_{cc} = \sigma_{ct} + \alpha v_f \frac{l}{d_f} \tag{5}$$

where α is the effect coefficient of the fiber. The elastic ultimate strength (σ_{ct}) is the arithmetic average of the cracking strength of UHPC when v_f = 0%.

Since the parameters σ_{cc} and ε_{cc} had been confirmed, the stage II model of the UHPC strain-hardening constitutive model can be determined by further determining the parameters σ_{pc} and ε_{pc}. During the strain-hardening stage, UHPC cracks appeared and entered the multiple cracking stage. After the UHPC cracked under tension, the uncracked portion did not completely enter the plastic stage, and its stress–strain curve continued to rise in an approximately elastic state. The extension of microcracks is the main reason for the decrease in the slope of the curve. As the steel fiber volume increased, the width of all microcracks in the tested section would increase, while the crack extension rate would decrease. If only the influence of the parameter v_f was considered, the stress–strain slope of stage II and its linear fitting functions σ_{pc} and v_f can be established based on the test results. The stage II constitutive model of UHPC could be established.

In summary, the following constitutive relationship model can be established:

$$\begin{cases} \sigma = \left[E_u + (E_s - E_u)v_f\right]\varepsilon & 0 \le \varepsilon < \varepsilon_{cc} \\ \sigma = \sigma_{ct} + \frac{\alpha v_f l}{d_f} + \beta v_f(\varepsilon - \varepsilon_{cc}) & \varepsilon_{cc} \le \varepsilon \le \varepsilon_{pc} \end{cases} \tag{6}$$

where σ is the UHPC axial tensile stress; ε is the UHPC axial tensile strain; β is the strain-hardening fitting factor. The ε_{cc} and ε_{pc} are calculated using the following formula:

$$\varepsilon_{cc} = \frac{\sigma_{ct} + \alpha v_f \frac{l}{d_f}}{E_u + (E_s - E_u)v_f} \tag{7}$$

$$\varepsilon_{pc} = \varepsilon_{cc} + \gamma v_f \frac{l}{d_f} \tag{8}$$

where γ is the fitting factor of the UHPC ultimate stress.

According to the results of this test, the parameters in the model are as follows:

Bringing the parameters in Table 7 into Equation (6), the constitutive relationship model was calculated as follows:

$$\begin{cases} \sigma = \left[6.95 \times 10^3\left(1 - v_f\right)f_{cc}^{1/3} + 2.1 \times 10^5 v_f\right]\varepsilon & 0 \le \varepsilon < \varepsilon_{cc} \\ \sigma = 3.01 + 50.7v_f + 3.2 \times 10^4 v_f(\varepsilon - \varepsilon_{cc}) & \varepsilon_{cc} \le \varepsilon \le \varepsilon_{pc} \end{cases} \tag{9}$$

Table 7. Constitutive model parameters.

$K_0/10^3$	σ_{ct}	α	$\beta/10^4$	$\gamma/10^{-4}$
6.95	3.01	0.39	3.20	9.0

By incorporating the steel fiber volume fraction v_f into Equation (9), the constitutive relationship model of UHPC under different steel fiber volumes could be calculated. These calculated results were compared to the experimental results in Figure 10.

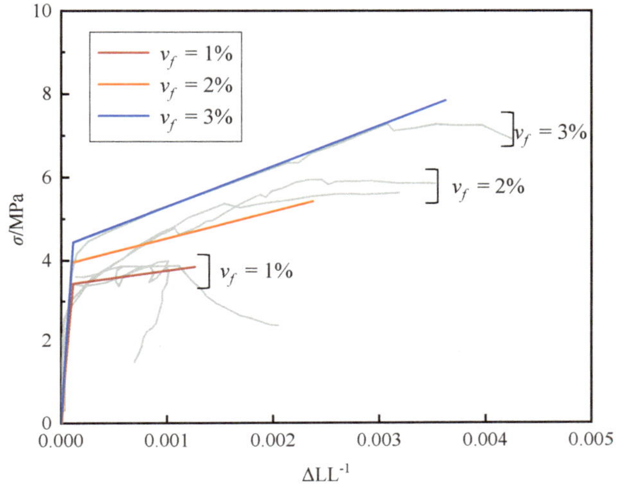

Figure 10. Comparison of the constitutive model with the experimental results.

A good simulation effect of the constitutive relationship model, with v_f = 1% and v_f = 3%, could be observed in Figure 10. However, since the cracking stress of the SU-2 group was generally lower, even lower than that of the SU-1 group, the simulation effect of the constitutive relationship model with v_f = 2% was relatively inferior to that of the other two groups.

5. Finite Element Analysis

5.1. Simulation of High-Strength Steel Wire Mesh UHPC Axial Tension Specimens

5.1.1. Analytical Methods

In this section, the finite element software ABAQUS/Standard 2022 was employed to replicate the experiments conducted in Section 3 to obtain an effective simulation method for high-strength steel wire mesh UHPC. Based on this, the finite element simulation of the RC beam strengthened with a thin layer of high-strength steel mesh UHPC was carried out. The relationship between the flexural performance of the strengthened beams and the thickness of the high-strength steel wire mesh UHPC layer, steel fiber volume, and high-strength steel wire mesh enhancement ratio parameters were discussed. Additionally, by simulating the nonlinear bond–slip effect of the UHPC–NC interface, the influence of interface roughening and implantation rebar on the bond stability of the UHPC–NC interface was analyzed. Finally, a reference for the finite element analysis of RC beams strengthened with high-strength steel wire mesh UHPC was provided.

5.1.2. Elements

A total of four materials, i.e., NC, UHPC, high-strength steel mesh, and reinforcement, were used in the finite element simulation in this study. The simulation elements used in each part are shown in Table 8.

Table 8. Elements used in each section of the simulation.

NC	UHPC	Reinforcement	Interface Reinforcement	High-Strength Steel Wire Mesh	UHPC–RC Bonding
C3D8R	C3D8R	T3D2	B31	B31	COH3D8

5.1.3. Interaction

Embedded region constraints were used to simulate the constraints between the reinforcement and the concrete. The existence of slip between the high-strength steel mesh and UHPC has been experimentally proven in this paper. However, the interlocking force of the mesh could effectively mitigate further slippage. To capture this behavior, truss elements were employed in the transverse direction of the high-strength steel wire mesh to simulate its interaction with the UHPC. The embedded region constraint was applied between the selected unit and the UHPC. The specific effects are shown in Figure 11.

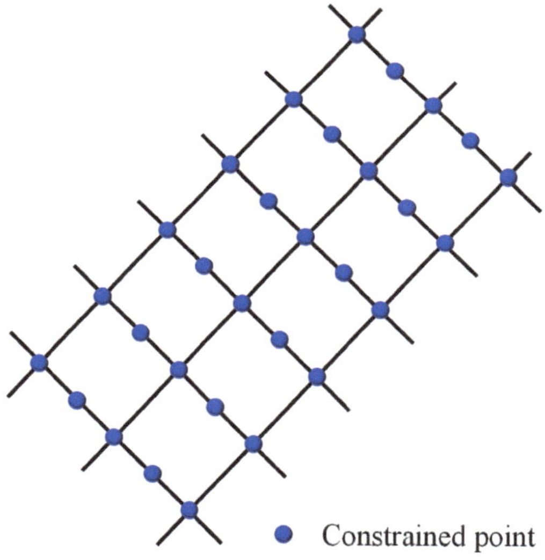

Figure 11. Schematic diagram of high-strength steel wire mesh–UHPC constraints.

Beam elements were adopted to simulate the interface reinforcement, which was connected to UHPC and NC through embedding constraints. The UHPC–NC interface bonding effect was simulated by setting the cohesive element with zero thickness.

The interface debonding effect of the UHPC–NC was determined by the UHPC–NC bilinear bond–slip model fitted by Yang et al. [27], based on experimental results. The tensile strength of the NC was applied as the strength limit in the bond–slip model. The resulting model is shown in Figure 12.

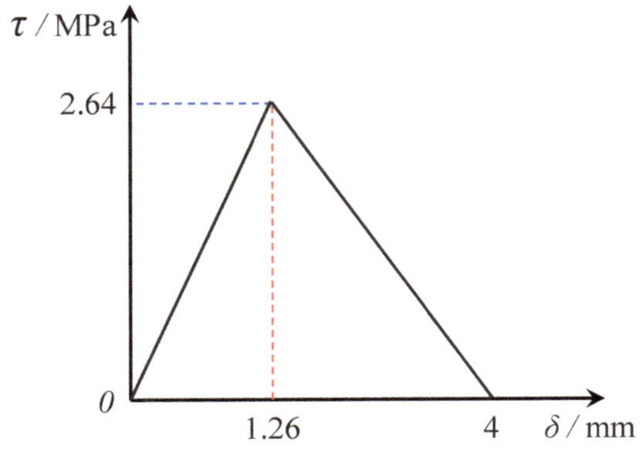

Figure 12. UHPC–NC interface bond–slip relationship.

5.1.4. Material Constitutive Model

The elastic–plastic constitutive model was adopted for high-strength steel wire mesh and steel reinforcement, based on the relevant provisions in Appendix C of GB 50010-2010 and related specifications [28–31]. An obvious yield stage and strengthening stage of the steel reinforcement used in the strengthened test were observed during the axial tension test. However, the high-strength steel wire mesh only displayed a strengthening stage, and its yield stage had not been observed. The constitutive relationship is set as shown in Figure 13.

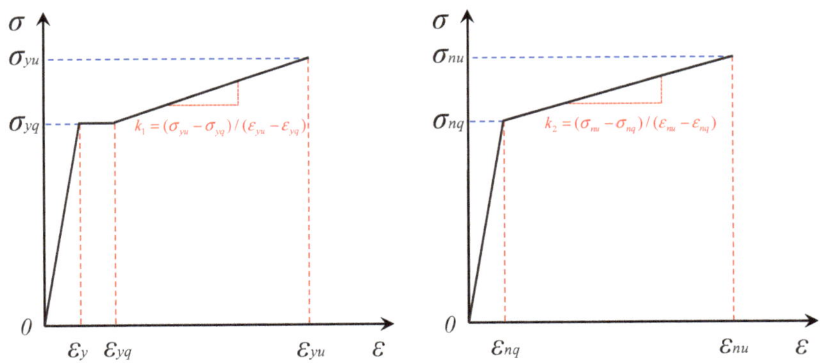

Figure 13. The constitutive relationship between steel bars and high-strength steel wire mesh.

In the figure, k_1 and k_2 are the slopes of the strengthening stage in the tensile constitutive relationship between the steel bar and the steel wire, respectively. The yield strength of the reinforcement with diameters of 8 mm and 10 mm, determined by axial tensile tests, were 421.1 MPa and 434.0 MPa, respectively.

The stress–strain relationship of concrete was set according to the relevant provisions of Appendix C in GB 50010-2010, and the standard value of concrete compressive strength was used as its ultimate strength.

The ascending branch of the axial tensile stress–strain relationship for UHPC was determined using the constitutive model in Section 4. Based on the test results, the softening branch of the UHPC tensile stress–strain relationship was established, as shown in Figure 14 (the ultimate strain was the measured average value of the test).

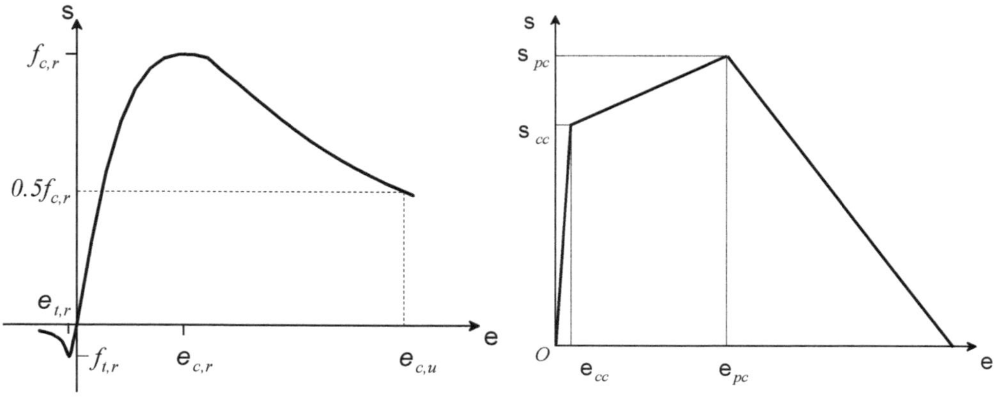

Figure 14. Stress–strain relationship of NC and UHPC.

The parameter values of NC and UHPC were all based on the results of the team's previous tests [27,32]. The viscosity parameter, which mainly affected the convergence performance of the model, was uniformly set to 0.0005. In addition, the d_k of NC was set according to the relevant research results in Reference [27], as shown in Equation (10). The damage to the UHPC material was calculated according to Equation (11) [27].

$$d_k = \frac{(1-\eta_k)\varepsilon_k^{\sim in}E_0}{\sigma_k+(1-\eta_k)\varepsilon_k^{\sim in}E_0} \tag{10}$$

$$d_k = A_0 e^{-\varepsilon_{norm}^{in}/t_0} + B_0 \tag{11}$$

where the subscript $k = c, t$ represents axial compression and axial tension, respectively. $\varepsilon_k^{\sim in}$ is inelastic strain; E_0 is the initial elastic modulus of the material; η_k is 0.6 under compression and 0.9 when under tension. ε_{norm}^{in} is the normalized inelastic strain; t_0 represents the rate of change of the damage factor with the increase in inelastic strain, which here is 0.6. A_0 and B_0 are related parameters that can be determined using Equation (12).

$$\begin{cases} A_0 = \frac{1}{e^{-1/t_0}-1} \\ B_0 = -\frac{1}{e^{-1/t_0}-1} \end{cases} \tag{12}$$

5.1.5. Finite Element Verification

To confirm that the mechanical performance of the high-strength steel mesh UHPC composite material matched the test results, the finite element model of the SU-4~SU-7 groups was established based on the relevant parameters specified in Section 4. The element sizes of the high-strength steel mesh and UHPC in the model were set to 2 mm and 5 mm, respectively. A vertical fixed constraint was applied to one end surface, and a displacement was applied to the other end. The details of the established model are shown in Figure 15.

The internal forces of the UHPC unit and the steel wire in the mid-span section of the specimen were separately integrated. The two internal forces were added and divided by the cross-sectional area to obtain the cross-sectional equivalent stress. The cross-sectional equivalent stress was combined with the strain of the UHPC unit at the mid-span to derive the axial tensile stress–strain relationship of the SU-4 to SU-7 specimens, which was then compared with the test results, as shown in Figure 16.

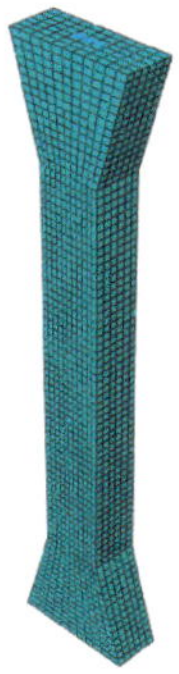

Figure 15. Finite element model meshing.

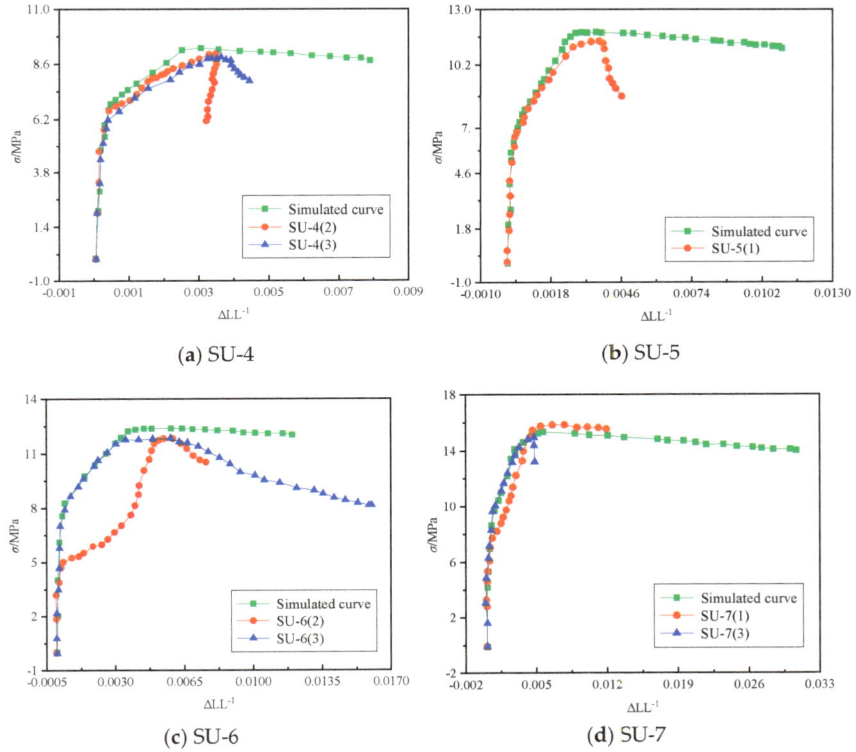

(**a**) SU-4 (**b**) SU-5

(**c**) SU-6 (**d**) SU-7

Figure 16. Simulation results of SU-4~7 specimens.

It can be seen that the calculation results for each finite element model were consistent with the test results before the stress–strain curve reached the descending stage. The simulation results of the SU-6 specimen differ greatly from the curve of the SU-6(2) specimen, primarily because the relative slip between the high-strength steel wire mesh and UHPC was not simulated in the model. However, the ultimate strength and ultimate tensile strain remained relatively close.

It is noteworthy that the model poorly matched the simulation results for the descending section (except for SU-7). Preliminary analysis showed that the deformation between the high-strength steel wire mesh unit and the UHPC unit in the model was relatively well-coordinated. However, the main crack of the specimen expanded rapidly after reaching the ultimate strength in the test. The strains of the UHPC and the steel wire at the main crack were extremely inconsistent. The stress at the crack rapidly redistributed from UHPC to the steel wire, resulting in a sharp reduction in load-bearing capacity. Consequently, the descending branch in the model was relatively gradual, while the test curve was steeper. In summary, the finite element simulation results could effectively simulate the test curve.

5.2. Finite Element Simulation of RC Beams Strengthened with High-Strength Steel Wire Mesh UHPC

When the simulation results for the high-strength steel wire mesh UHPC were in good agreement with the test results, the further finite element simulation analysis of its strengthened RC beam was carried out. The effects of various parameters on the flexural mechanical properties of RC beams were explored. To ensure more accurate and reliable simulation results, the experimentally verified constitutive model and the UHPC–NC interface bond–slip relationship were used. However, both exhibited strong nonlinear behavior characteristics and required more computing resources. Therefore, a small RC beam was selected as the strengthened object for the parametric analysis of RC beams strengthened with high-strength steel mesh UHPC, optimizing computational efficiency while maintaining model accuracy.

5.2.1. Structural Features

A rectangular cross-sectional RC beam was designed, with a length of 1500 mm and a cross-sectional size of 150 mm × 200 mm. The beams were made of C50-grade concrete, as defined in GB 50010-2010. The detailed dimensions of the designed RC beam are shown in Figure 17.

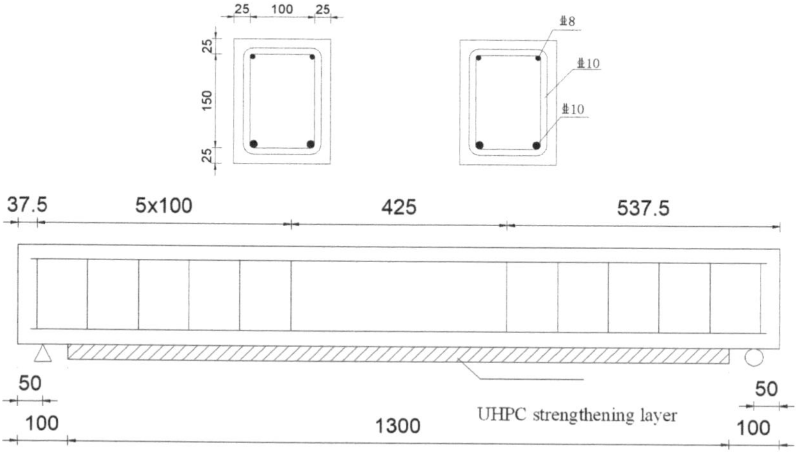

Figure 17. The size and reinforcement of the strengthened beam (unit: mm).

The effects of the strengthening method, high-strength steel wire mesh enhancement ratio (ρ_w), strengthening layer thickness (t), and steel fiber volume (v_f) on the flexural performance of RC beams strengthened with high-strength steel wire mesh UHPC were studied in this section. A total of 11 RC beam models were established, and the variable settings are shown in Table 9.

Table 9. Strengthened specimen design.

Number	Strengthened Layer Parameters			
	Strengthening Methods	Thickness t/mm	Steel Fiber Volume Fraction v_f/%	Enhancement Ratio of High-Strength Steel Mesh ρ_w
UHS-0	/	30	/	/
UHS-1	/	30	2%	/
UHS-2	High-Strength Steel Mesh	30	2%	0.70% (1 layer)
UHS-3	High-Strength Steel Mesh	30	2%	1.40% (2 layer)
UHS-4	High-Strength Steel Mesh	30	2%	2.09% (3 layer)
UHS-5	High-Strength Steel Mesh	30	0%	1.40% (2 layer)
UHS-6	High-Strength Steel Mesh	30	1%	1.40% (2 layer)
UHS-7	High Strength Steel Mesh	30	2%	1.40% (2 layer)
UHS-8	High-Strength Steel Mesh	30	3%	1.40% (2 layer)
UHS-9	High-Strength Steel Mesh	20	2%	1.05% (1 layer)
UHS-10	High-Strength Steel Mesh	10	2%	2.09% (1 layer)

The finite element model of the specimen was established based on the relevant conclusions in Section 4. The distance between the two supports was 1400 mm, and the pure bending section in the mid-span was 400 mm. The established geometric model and the boundary conditions are shown in Figure 18.

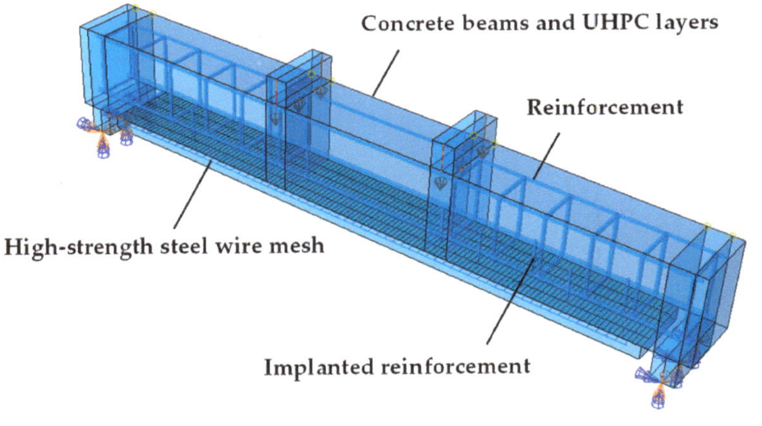

Figure 18. Boundary conditions and load settings.

To achieve high solution accuracy and convergence, while minimizing computation time, the final mesh sizes for the model were determined as follows: 30 mm in the horizontal direction, 15 mm in the vertical direction, and 10 mm along the beam. The mesh size of the UHPC thin layer was 10 mm. The mesh size for the longitudinal reinforcement, vertical reinforcement, and stirrups was 25 mm. The mesh size of the interface reinforcement was 10 mm. The mesh size of the high-strength steel wire mesh was 10 mm.

After the mesh sizes were determined, the finite element modes of 11 strengthened beams were established. Using the UHS-3 specimen model as an example, the various parts of the model are shown in Figure 19.

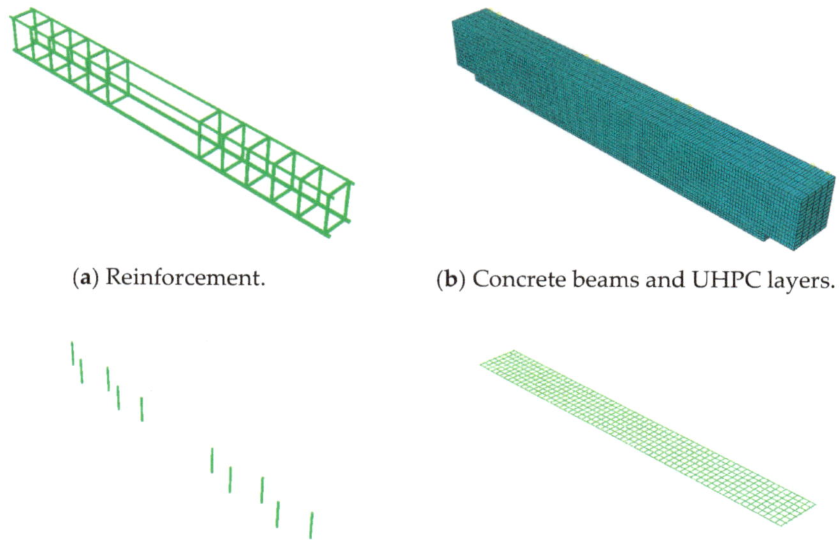

(a) Reinforcement. (b) Concrete beams and UHPC layers.

(c) Implanted reinforcement. (d) High-strength steel wire mesh.

Figure 19. Finite element model of each part.

5.2.2. Failure Modes

Specimens UHS-0, UHS-1, UHS-2, UHS-4, UHS-5, UHS-9, and UHS-10 all experienced longitudinal reinforcement yield failure. Except for UHS-4, the failure of the other specimens was attributed to an insufficient strengthening layer and inadequate tensile side stiffness of the section. By observing the loading process of UHS-4, it was found that the strengthening layer showed a high enhancement ratio of high-strength steel wire mesh, which gave the strengthening layer a high axial stiffness. The extension of the crack was effectively suppressed, and the specimen also displayed a sufficient cross-sectional compression area when the longitudinal reinforcement yielded.

UHS-3 and UHS-6 to UHS-8 were all crushed by the concrete. Because the relatively weak strengthening layer could not effectively limit the extension of cracks, UHS-3, UHS-6, and UHS-7 failed due to the small height of the cross-section compression zone under the limited state. Notably, the excessive strength of the strengthening layer caused UHS-8 to become an over-reinforced beam. The concrete in the compression zone was crushed before the reinforcement layer and the longitudinal reinforcement softened or yielded.

The stress cloud diagram of the UHS-0 to UHS-10 specimens under the ultimate state is shown in Figure 20. It can be seen that each specimen exhibited a localized stress concentration at the top loading point. Meanwhile, the stress peak value generally exceeded the standard value of C50 concrete compressive strength of 32.4 MPa. Except for UHS-3 and UHS-6 to UHS-8, the concrete stress at the top of the mid-span surpassed the standard value of C50 concrete compressive strength, which aligns with the observed failure mode.

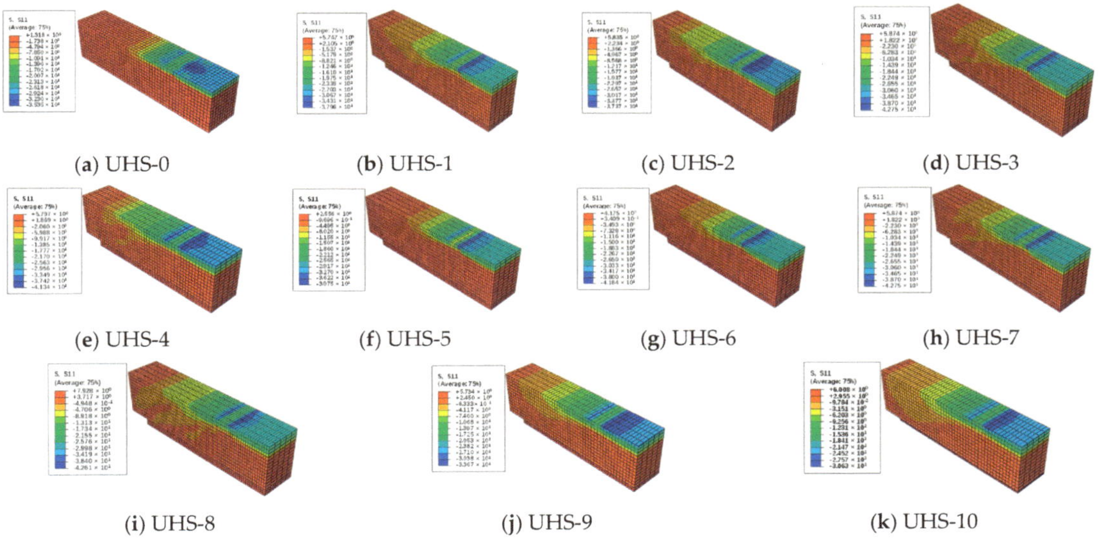

Figure 20. Stress cloud diagram under UHS-0~10.

5.2.3. Load-Displacement Curve

The load-deflection curves of the UHS-0 to UHS-10 specimens are shown in Figure 21. It can be seen that the vertical stiffness of the specimens was at its maximum before cracking occurred. As the load continued to increase, cracks formed in the concrete in the tensile zone of the specimen. Meanwhile, the vertical stiffness of the specimen decreased significantly. However, the stress on the strengthening layer and the longitudinal reinforcement increased as the internal force of the specimen was redistributed. The vertical stiffness of the specimen regained a certain degree of increase. Then, the deflection of the specimen increased rapidly until the longitudinal reinforcement yielded, the reinforcement layer softened, or the concrete in the compressive zone crushed. At this point, the specimen reached its ultimate load.

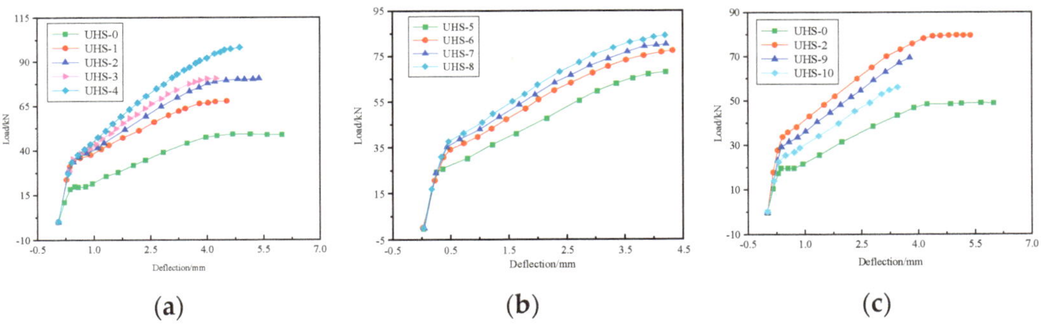

Figure 21. The load-deflection curve varies with ρ_w (**a**), v_f (**b**), and t (**c**).

The variation trend of the load-deflection curve with ρ_w in the strengthening layer is shown in Figure 21a. Before the concrete beam cracked, the vertical stiffness of the strengthened specimen was significantly higher than that of UHS-0, and the linear deviation point of the curve was also higher. After cracking, the influence of ρ_w on the vertical stiffness of the specimen was more obvious, generally increasing as ρ_w increased.

As ρ_w increased, the ultimate bearing capacity of the specimen effectively improved, but its deflection scarcely increased. In contrast, the deflection of the specimen increased significantly with the increase in the thickness of the strengthening layer. The load and deflection of the specimen in the cracking, wire yielding, and ultimate load stages for the extracted model are shown in Table 10.

Table 10. Load deflection at cracking, wire yielding, and load limit.

Number	Crack		Yield of Steel Wire		Load Limit		Failure Mode
	Load/kN	Deflection/mm	Load/kN	Deflection/mm	Load/kN	Deflection/mm	
UHS-0	15.9	0.23	/	/	49.2	5.28	A
UHS-1	25.4	0.25	/	/	68.0	4.52	A
UHS-2	27.9	0.27	65.6	2.81	80.1	5.35	A
UHS-3	28.7	0.27	75.1	3.23	81.1	4.18	B
UHS-4	28.3	0.27	86.8	3.46	98.1	4.82	A
UHS-5	23.2	0.23	59.4	2.96	69.6	4.31	A
UHS-6	25.4	0.25	69.9	3.07	78.3	4.33	B
UHS-7	27.9	0.28	75.1	3.23	81.1	4.18	B
UHS-8	31.2	0.30	80.5	3.39	85.2	4.26	B
UHS-9	22.9	0.24	60.1	3.02	70.2	3.80	A
UHS-10	19.7	0.25	54.2	3.11	57.2	3.46	A

Note: A represents the yield failure of the longitudinal bar; B represents the crushing failure of concrete in the compression zone.

Among all the parameters, t and v_f had the most significant impact on the cracking load of the specimen. When other parameters were kept constant, the cracking load of the UHS-2, with a reinforcement layer thickness of 30 mm, increased by 41.6% compared to that of the UHS-10, which had a reinforcement layer thickness of 10 mm. The cracking load of the UHS-8, with a steel fiber volume of 3%, increased by 34.5% compared to that of the UHS-5 without steel fiber.

When the steel wire yielded, the UHS-4 specimen with v_f = 2.09% carried the largest load, reaching 86.8 kN. Once the limit was reached, the ultimate load of the specimens strengthened with a high-strength steel wire mesh layer was greatly improved. Among them, the thickness of the UHPC contributed the most significant improvement in bearing capacity, and the ultimate load of the UHS-4 specimen was 98.1 kN. Notably, the effect of the high-strength steel wire mesh enhancement ratio on the ultimate load of the specimen was more complex. When v_f increased from 0.70% to 1.40%, the ultimate load of the specimen scarcely increased. However, when v_f increased from 1.40% to 2.09%, the ultimate load increased by 21.0%. This indicates that when using a thin layer of high-strength steel mesh UHPC for strengthening, if the stiffness of the strengthening layer is weak and cannot effectively restrain the crack expansion in the beam, only a sufficiently high v_f can effectively increase the ultimate load of the strengthened beam.

5.2.4. Interfacial Damage and Peeling

The cohesive element was used to simulate the bond–slip behavior of the UHPC–NC interface in the model. Therefore, the debonding and damage of the UHPC–NC interface could be obtained by monitoring the degradation rate of the cohesive element stiffness (SDEG) and the stress along the beam direction.

The stress and stiffness degradation of the UHPC–NC interface of the UHS-2 specimen at the time the longitudinal reinforcement yielded is given in Figure 22. It can be seen that the units on both sides of the UHPC–NC interface had almost completely degraded, and the stiffness degradation rate had reached 0.93. Both sides of the UHPC–NC interface had separated, and the stress peak on both sides of the UHPC–NC interface had shifted toward the mid-span. Some units of the UHPC–NC interface had reached the stress peak, which was tentatively attributed to concrete cracking in the mid-span.

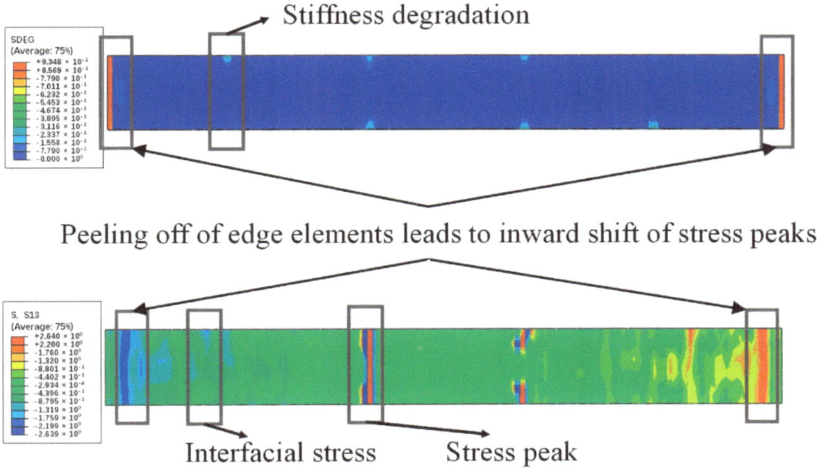

Figure 22. Stiffness degradation (**top**) and interfacial stress (**bottom**) of rebar yield.

The relationship between the stress at the UHPC–NC interface and the position of the interface shear key of the UHS-2 specimen is shown in Figure 23. The stress peaks on both sides of the interface remained outside the anchoring range of the shear bonds, and their tendency to peel was effectively restrained by the interface shear bonds. The anchoring effect of the outermost shear key ensured that its stress never reached the peak stress of the interface.

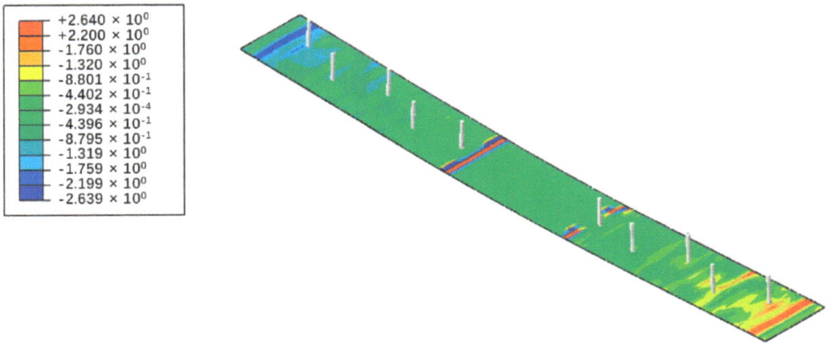

Figure 23. Interfacial shear bond and interfacial stress when rebar yields.

The UHS-4 specimen exhibited the highest ultimate bearing capacity, while the UHS-10 specimen displayed the lowest. At the ultimate state, the UHPC–NC interface stress condition of the UHS-4 specimen should be the most complex, while the opposite was observed for the UHS-10 specimen. The interface stress and stiffness degradation rate of UHS-1 to UHS-10 were extracted, as shown in Figure 24.

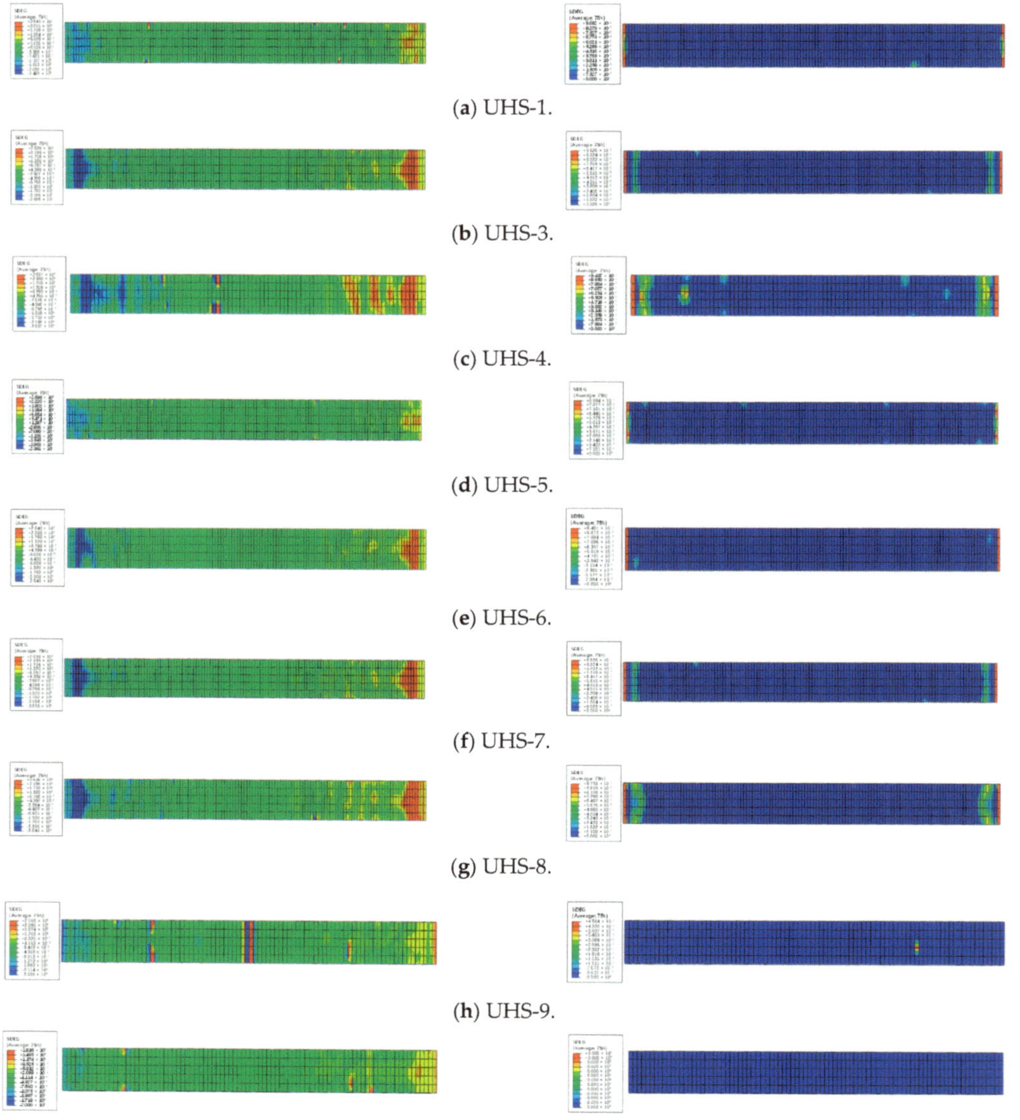

(a) UHS-1.

(b) UHS-3.

(c) UHS-4.

(d) UHS-5.

(e) UHS-6.

(f) UHS-7.

(g) UHS-8.

(h) UHS-9.

(i) UHS-10.

Figure 24. Interfacial stress and stiffness degradation rate of UHS-1~10.

By comparing Figure 22, Figure 23, and Figure 24a to Figure 24c, it can be seen that with the increased v_f in the strengthening layer, the interface peeling on both sides of each specimen became more pronounced. When v_f ranged from 0% to 1.4%, the UHPC–NC interface peeled only at the weak edges on both sides. However, when the v_f reached 2.09%, significant peeling occurred not only at the edges of the interface under the limit state but also in the oblique shear section of the specimen, with considerable peeling, concentrated at the weak points between the shear keys. Overall, the existence of the shear key effectively restricted the further interface peeling.

Compared with the direct increase in the strengthening ratio of high-strength steel wire mesh, increasing the steel fiber volume in UHPC had a relatively minor impact on the interfacial bonding performance under ultimate conditions. However, as the steel fiber volume increased, the debonding phenomenon on both sides of the UHPC–NC interface gradually became more pronounced. Nevertheless, the interface in the oblique shear section still maintained a relatively strong bonding effect.

When the UHS-10 specimen reached the load limit, the UHPC layers on both sides of the specimen were nearly completely peeled off. The stress concentration phenomenon also occurred at the shear keys of the oblique shear section.

In summary, the interface shear bond effectively inhibited the interface debonding phenomenon between the strengthening layer and the original structure. In the absence of interface shear reinforcement, the high-strength steel wire mesh UHPC layer gradually debonded from the strengthened beam as the load increased. However, at the location with the interface shear bond, the interface debonding phenomenon was suppressed. Although the interface between the shear bonds still tended to debond, the presence of the shear bond successfully prevented the debonding from spreading to other locations.

6. Conclusions

(1) The initial cracking strength and ultimate strength of RC beams strengthened with a thin layer of UHPC and high-strength steel wire mesh were significantly improved. The ultimate load capacity of the UHS-4 specimen strengthened with UHPC with three layers of high tensile steel wire mesh increased by 99.4% compared to that of the unstrengthened beam. The significant increase in bearing capacity also reduces the risk of damage to the structure under long-term loads.

(2) The thickness of the UHPC layer and the volume of steel fibers in the strengthening layer had a significant impact on the cracking load of the specimens. Compared to the results for the UHS-10 specimen, an increase in the thickness of the UHPC layer by 20 mm (i.e., UHS-7) resulted in a 41.6% increase in the cracking load. Compared to results for the UHS-5 specimen, an increase in the volume of steel fibers by 3% (i.e., UHS-8) led to a 34.5% increase in the cracking load of the strengthened beam. The increase in cracking load can reduce the occurrence of cracks or control the crack width, effectively preventing the ingress of corrosive media.

(3) The axial stiffness of the high-strength steel wire mesh UHPC layer was identified as the primary factor influencing the crack propagation speed and depth in the strengthened beam. Among various methods, increasing the strengthening ratio of high-strength steel wire mesh within the UHPC layer was the most effective way to enhance its stiffness. By raising the strengthening ratio of high-strength steel wire mesh, the crack propagation speed during loading could be effectively delayed. The crack depth under ultimate load could also be reduced.

(4) An UHPC–RC interface bond–slip model considering cohesion was constructed. Following calculations and analyses, it was determined that the effective bonding of the interfaces was not guaranteed by relying only on the bonding between the UHPC–NC interfaces. The interface debonding could be effectively suppressed and the mechanical performance of the strengthened beams can be ensured by using shear keys to constrain the UHPC–NC interface. The optimal choice is recommended to be a reinforced beam with a UHPC strengthening layer thickness of 30 mm, a steel fiber volume of 2%, and a high-strength steel wire mesh enhancement ratio of 2.09%, as this configuration achieves the best overall performance. Given the differences in steel fiber volume and specimen thickness used in practical reinforcement projects, future research should focus on conducting controlled variable studies on these key parameters.

Author Contributions: Conceptualization, C.Z.; methodology, Z.J.; software, Z.J.; validation, Y.Q. and J.C.; formal analysis, J.Y.; investigation, C.D. and Y.Q.; Resources, Y.L.; data curation, Y.L.; writing—original draft, C.D.; writing—review & editing, C.Z.; supervision, J.Y.; project administration, Z.Z.; funding acquisition, Z.Z. and J.C. All authors have read and agreed to the published version of the manuscript.

Funding: The authors greatly appreciate the financial support from the Science and Technology Research Program of Chongqing Municipal Education Commission (Grant Nos. KJQN202300723 and KJQN202400701) and the Technology Project of the Transportation Department of Guizhou Province (Grant No. 2024-122-001).

Data Availability Statement: The raw data supporting the conclusions of this article will be made available by the authors on request.

Conflicts of Interest: Author Chao Zhu and Chunlin Du are employed by the Chongqing Design Group Co., Ltd. Author Yinbin Li is employed by the Guizhou Provincial Transportation Planning Survey and Design Institute Co., Ltd. The remaining authors declare that the research was conducted in the absence of any commercial or financial relationships that could be construed as a potential conflict of interest.

References

1. Zhu, Y.; Zhang, Y.; Hussein, H.H.; Chen, G. Flexural strengthening of reinforced concrete beams or slabs using ultra-high performance concrete (UHPC): A state of the art review. *Eng. Struct.* **2020**, *205*, 110035. [CrossRef]
2. Zhang, Z.; Pang, K.; Xu, L.; Zou, Y.; Yang, J.; Wang, C. The bond properties between UHPC and stone under different interface treatment methods. *Constr. Build. Mater.* **2023**, *365*, 130092. [CrossRef]
3. Graybeal, B.; Tanesi, J. Durability of an ultrahigh-performance concrete. *J. Mater. Civ. Eng.* **2007**, *19*, 848–854. [CrossRef]
4. Li, Y.; Zhou, H.; Zhang, Z.; Yang, J.; Wang, X.; Wang, X.; Zou, Y. Macro-micro investigation on the coefficient of friction on the interface between steel and cast-in-place UHPC. *Eng. Struct.* **2024**, *318*, 118769. [CrossRef]
5. Wille, K.; El-Tawil, S.; Naaman, A.E. Properties of strain hardening ultra high performance fiber reinforced concrete (UHP-FRC) under direct tensile loading. *Cem. Concr. Compos.* **2014**, *48*, 53–66. [CrossRef]
6. Zhou, J.; Leng, J.; Yang, J.; Zhang, Z.; Du, J.; Zou, Y. Experimental investigation on shear behavior of damaged and acid rain-corroded RC T-beams strengthened with ultra-high-performance concrete. *Eng. Struct.* **2025**, *327*, 119618. [CrossRef]
7. Sun, G.; Shi, J.; Qu, Y. Cracking and yield behavior of reinforced UHPC beams containing steel slag under flexural test. *Eng. Struct.* **2023**, *280*, 115693. [CrossRef]
8. Huang, L.; Yuan, M.; Wei, B.; Yan, D.; Liu, Y. Experimental investigation on sing fiber pullout behaviour on steel fiber-matrix of reactive powder concrete (RPC). *Constr. Build. Mater.* **2022**, *318*, 125899. [CrossRef]
9. Yang, J.; Xia, J.; Zhang, Z.; Zhou, J.; Zou, Y.; Wang, Y.; Shen, X. Mesoscopic shear behavior and strength characteristic of UHPC-NC interface considering the combined effect of mechanical interlocking and dowel action. *Eng. Fract. Mech.* **2024**, *307*, 110306. [CrossRef]
10. Zhang, Y.; Zhang, C.; Zhu, Y.; Cao, J.; Shao, X. An experimental study: Various influence factors affecting interfacial shear performance of UHPC-NSC. *Constr. Build. Mater.* **2020**, *236*, 117480. [CrossRef]
11. Wei, J.; Li, J.; Wu, C.; Hao, H.; Liu, J. Experimental and numerical study on the impact resistance of ultra-high performance concrete strengthened RC beams. *Eng. Struct.* **2023**, *277*, 115474. [CrossRef]
12. Zhang, Y.; Li, X.; Zhu, Y.; Shao, X. Experimental study on flexural behavior of damaged reinforced concrete (RC) beam strengthened by toughness-improved ultra-high performance concrete (UHPC) layer. *Compos. Part B Eng.* **2020**, *186*, 107834. [CrossRef]
13. Ke, L.; Guo, J.; Yan, B.; Cheng, H.; Feng, Z.; Zhou, J.; Yoo, D.-Y. Shear performance evaluation of damaged RC beams strengthened with cast-in-place U-shaped UHPFRC shell. *Structures* **2023**, *58*, 105530. [CrossRef]
14. Nadir, W.; Kadhim, M.M.A.; Jawdhari, A.; Fam, A.; Majdi, A. RC beams strengthened in shear with FRP-Reinforced UHPC overlay: An experimental and numerical study. *Structures* **2023**, *53*, 693–715. [CrossRef]
15. Mirdan, D.; Saleh, A.R. Flexural performance of reinforced concrete (RC) beam strengthened by UHPC layer. *Case Stud. Constr. Mater.* **2022**, *17*, e01655. [CrossRef]
16. Yoo, D.Y.; Oh, T.; Choi, J.; Min, K.-H.; Shin, H.-O. Structural strengthening of reinforced concrete beams using ultra-high-performance concrete with various fiber volume fractions and layer thicknesses. *Structures* **2024**, *70*, 107735. [CrossRef]
17. Zhang, W.; Zheng, D.; Huang, Y.; Kang, S. Experimental and simulative analysis of flexural performance in UHPC-RC hybrid beams. *Constr. Build. Mater.* **2024**, *436*, 136889. [CrossRef]

18. Yu, J.; Zhang, B.; Chen, W.; Liu, H.; Li, H. Multi-scale study on interfacial bond failure between normal concrete (NC) and ultra-high performance concrete (UHPC). *J. Build. Eng.* **2022**, *57*, 104808. [CrossRef]
19. Krishna, A.; Sreekumaran, S.; Kaliyaperumal, S.R.M.; Shreemathi. Performance evaluation of axially loaded high strength ferrocement confined fibre reinforced concrete columns. *Innov. Infrastruct. Solut.* **2024**, *9*, 68. [CrossRef]
20. Paschalis, S.A.; Lampropoulos, A.P.; Tsioulou, O. Experimental and numerical study of the performance of ultra high performance fiber reinforced concrete for the flexural strengthening of full scale reinforced concrete members. *Constr. Build. Mater.* **2018**, *186*, 351–366. [CrossRef]
21. Shafieifar, M.; Farzad, M.; Azizinamini, A. A comparison of existing analytical methods to predict the flexural capacity of Ultra High Performance Concrete (UHPC) beams. *Constr. Build. Mater.* **2018**, *172*, 10–18. [CrossRef]
22. Zhao, H. Axial compressive behaviour of concrete strengthened with steel rings, wire mesh and modified high strength mortar (MHSM). *Constr. Build. Mater.* **2020**, *250*, 118938. [CrossRef]
23. Ma, S.; Xu, M.; Wu, C.; Huang, Z.; Zhang, F. Experimental study on bending fatigue performance of ambient-cured ultra-high-performance concrete thin plate with embedded steel wire mesh. *Constr. Build. Mater.* **2024**, *449*, 138276. [CrossRef]
24. *GB/T 31387-2015*; Reactive Powder Concrete. Ministry of Housing and Urban-Rural Development, PRC: Beijing, China, 2015.
25. *GB/T 1499.2-2018*; Steel for Reinforced Concrete. Ministry of Housing and Urban-Rural Development, PRC: Beijing, China, 2018.
26. Wille, K.; Kim, D.J.; Naaman, A.E. Strain-hardening UHP-FRC with low fiber contents. *Mater. Struct.* **2011**, *44*, 583–598. [CrossRef]
27. Yang, J.; Yu, J.; Zhang, Z.; Zou, Y.; Chen, R.; Zhou, J.; Li, B. Flexural behavior of 15-year-old full-scale hollow slab beams strengthened with fiber-reinforced composites. *Case Stud. Constr. Mater.* **2023**, *19*, e02545. [CrossRef]
28. *JGJ 55-2011*; Specification for Mix Proportion Design of Ordinary Concrete. Ministry of Housing and Urban-Rural Development, PRC: Beijing, China, 2011.
29. *GB/T 50081-2002*; Standard for Test Method of Mechanical Properties on Ordinary Concrete. Ministry of Construction, PRC: Beijing, China, 2002.
30. *GB/T 50367-2013*; Code for Design of Strengthening Concrete Structure. Ministry of Housing and Urban-Rural Development, PRC: Beijing, China, 2013.
31. *GB/T 50010-2010*; Code for Design of Concrete Structures. Ministry of Housing and Urban-Rural Development, PRC: Beijing, China, 2010.
32. Chen, R.; Zhang, Z.; Zou, Y.; Yang, J.; Zhou, J.; Kuang, Y.; Wang, Y. In-situ evaluation on existing RC beam strengthened with GFRP-reinforced UHPC overlay. *Constr. Build. Mater.* **2024**, *429*, 136363. [CrossRef]

Disclaimer/Publisher's Note: The statements, opinions and data contained in all publications are solely those of the individual author(s) and contributor(s) and not of MDPI and/or the editor(s). MDPI and/or the editor(s) disclaim responsibility for any injury to people or property resulting from any ideas, methods, instructions or products referred to in the content.

buildings

Article

Experimental Study on Flexural Behavior of Retard-Bonded Prestressed UHPC Beams with Different Reinforcement Ratios

Lingfeng Du [1], Dongchang Wu [2], Jun Wang [1,*], Shaowei Wang [2], Boyi Zhao [1] and Xiufeng Tang [2]

1 College of Civil Engineering, Henan University of Technology, Zhengzhou 450001, China; dulingfeng2022@stu.haut.edu.cn (L.D.); iszhaoby@stu.haut.edu.cn (B.Z.)
2 China Construction Seventh Engineering Division Corp. Ltd., Zhengzhou 450004, China; 20243435@cscec.com (D.W.); wangshaowei@cscec.com (S.W.); tangxiufeng368@163.com (X.T.)
* Correspondence: wangj@haut.edu.cn

Abstract: Ultra-high performance concrete (UHPC), functioning as a next-generation cementitious engineering material, demonstrates marked superiority over conventional concrete in critical performance metrics, with its groundbreaking characteristics primarily manifested through exceptional strength and enhanced durability parameters. To address structural demands for reduced self-weight, material efficiency, and simplified construction processes in large-span durable structures, this study proposes a retard-bonded prestressed UHPC (RBPU) beam combining UHPC with retard-bonded prestressing tendons (RBPTs). Three RBPU beam specimens, with varying reinforcement ratios, underwent flexural testing to quantitatively assess their bending performance characteristics, providing foundational references for design optimization. The test results show that the failure mode of RBPU beams is typical flexural failure, demonstrating good load-bearing capacity and excellent ductility. As the reinforcement ratio increases, the cracking moment (M_{cr}) is improved to some extent, while the ultimate moment (M_{u}) and ductility are significantly enhanced. The plastic influence coefficient of the section modulus (γ) in the calculation of the M_{cr} was revised, and the ultimate moment M_u was subsequently calculated. The comparison demonstrates good agreement between the experimental values and computational results. This study provides both experimental and theoretical references for further in-depth research and practical applications of RBPU beams.

Keywords: retard-bonded prestress; UHPC; flexural behavior; reinforcement ratio; load-bearing capacity calculation method

1. Introduction

Ultra-high performance concrete (UHPC), an advanced cementitious composite material, is characterized by a dense cementitious matrix uniformly reinforced with steel fibers [1]. It exhibits exceptional mechanical properties, including a superior compressive strength, elevated elastic modulus, high tensile performance, and enhanced durability [2–4]. The implementation of UHPC enables a significant reduction in structural member cross-sections while simultaneously decreasing the overall structural self-weight. These superior material properties have garnered substantial attention within the architectural and structural engineering communities [5–7].

The retard-bonded prestressing system is a specialized structural technique characterized by a sheathing with transverse ribs analogous to corrugated ducts, wherein the annular space between the prestressing strand and sheathing is densely filled with a newly developed retarded adhesive material [8]. The retarder-bond adhesive cures at ambient

https://doi.org/10.3390/buildings15060887

temperature, eliminating the need for on-site heating equipment and preventing thermal impacts on formwork and concrete, while significantly reducing energy consumption and associated costs. The structural configuration of an RBPT is delineated in the schematic diagram provided in Figure 1. During the construction phase, the retarded adhesive remains in an uncured state, resulting in an unbonded interaction between the strand and retarded-bond material within retard-bonded prestressing tendons (RBPTs). This configuration permits unconstrained strand movement within the sheathing. Distinct from unbonded prestressing systems that require dedicated duct reservation, this technology minimizes cross-sectional weakening, thereby enhancing the structural load-bearing capacity to a certain extent. Furthermore, it eliminates the grouting process essential in bonded prestressing systems, effectively mitigating potential quality defects associated with grouting operations. Upon the commencement of service post-construction, the complete polymerization of the retarding adhesive enables RBPTs to achieve mechanical performance parameters equivalent to those of conventional bonded prestressing tendons [9]. This innovative technology amalgamates the inherent construction expediency of unbonded prestressed systems with the superior structural characteristics of bonded prestressed concrete. It optimizes the construction process when employing prestressing techniques and is particularly suitable for high-performance buildings and structures that demand enhanced performance specifications [10,11].

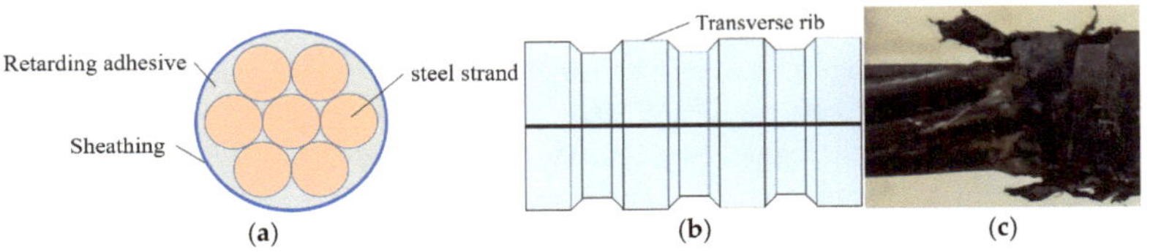

Figure 1. Details of retard-bonded prestressing steel strand: (**a**) cross-section; (**b**) external appearance; (**c**) photo.

The integration of ultra-high performance concrete (UHPC) with prestressing technology has been a significant advancement in structural engineering, driven by UHPC's exceptional compressive strength and superior crack control capabilities [12–14]. Bahmani et al. [15] demonstrated that UHPC with 50% steel slag replacement of silica sand achieves significant improvements in compressive, flexural, and tensile strengths (by 13.5%, 20.8%, and 6.5%, respectively), while achieving a 5.1% reduction in carbon emissions, thereby validating steel slag's potential as a sustainable aggregate in engineering applications. Tian et al. [16] established a computational framework employing 3D nonlinear FE modeling to replicate the bending responses of post-tensioned UHPC members, deriving an analytical protocol for failure mode identification. El-Helou et al. [17] conducted displacement-controlled destructive tests on prestressed UHPC bridge beams to assess their flexural behavior, subsequently developing a design methodology rooted in equilibrium and strain compatibility principles for such structural components. Through the full-scale failure testing of web-reinforcement-free prestressed UHPC-RC composite beams, Sun et al. [18] systematically outlined the critical reinforcement ratio for hybrid-reinforced members, developing a computational model for determining the minimum reinforcement ratio that incorporates UHPC's tensile contribution. Sim et al. [19], based on experimental validation and numerical examples, recommended prioritizing UHPC's tensile performance over its ultra-high compressive strength in flexural design, achieving optimal performance and economic efficiency while reducing the need for shear reinforcement. Li et al. [20] conducted experimental investigations on large-scale prestressed UHPC beams through bending tests, proposing a modified UHPC tensile constitutive model that incorporates UHPC's tensile strength contribution for the accurate prediction of flexural performance. Leutbecher et al. and Yang et al. [21,22]

considered fiber-bearing effects in the flexural cracking of prestressed UHPC beams, developing design methods that effectively reflect experimental bending capacities. Deng et al. [23] experimentally investigated the flexural behavior of prestressed UHPC beams reinforced with hybrid fibers, calculating cracking and ultimate moments based on the plane-section assumption while considering short fiber bridging effects, with good agreement between the experimental and calculated values. Through the computational modeling of unbonded prestressed UHPC beams, Dogu et al. [24] developed a theoretical framework capable of simulating full-range moment–curvature relationships. Their sensitivity analysis revealed that structural ductility exhibited strong dependency on two key variables: the maximum attainable tensile strain in the UHPC and specific load application patterns. Zhang et al. [25,26] developed a UHPC reinforcement system using pre-tensioned layers, with four-point bending tests investigating flexural performance. Their numerical framework accurately predicts cracking and the ultimate load in strengthened beams. Li et al. [27] investigated the performance of UHPC-NC composite beams through experimental analysis, focusing on three critical parameters: UHPC thickness, prestressing level, and normal concrete compressive strength. Their finite strip model demonstrated reliable accuracy in predicting flexural responses. Mohebbi et al. [28] developed data-driven models for predicting creep and shrinkage behavior in UHPC-like materials, enabling the accurate estimation of prestress losses in UHPC beams. Joh et al. [29] measured the prestress transfer length, development length, and prestress losses in UHPC-cast beams, demonstrating that creep and shrinkage contributed minimally to prestress losses. These studies collectively advance the understanding and application of prestressed UHPC in structural engineering, providing robust methodologies for design and analysis. Zheng et al. [30] demonstrated that the combination of external prestressing and UHPC materials can significantly enhance crack resistance in negative moment regions (with cracking load increased by 118.8% to 148%), and established an ultimate bearing capacity calculation formula that provides a theoretical basis for the engineered design. Feng et al. [31] systematically evaluated the reliability of predictive models as regards flexural capacity in both ordinary reinforced UHPC beams and prestressed UHPC beams across various design codes, while calibrating material partial safety factors and structural safety factors through comprehensive analyses.

The novel structural system combining retard-bonded prestressing tendons (RBPTs) and UHPC can fully exploit the properties of these two high-strength materials; significantly reduce structural self-weight; save construction materials; and greatly enhance the performance of structures with long spans, heavy loads, or high importance. Current research on retard-bonded prestressed structures primarily focuses on their combination with normal concrete. Cao et al. [32,33] compared the mechanical properties of retard-bonded prestressed concrete (RBPC) beams with those of traditional bonded prestressed concrete (BPC) beams. They found that the mechanical properties of the two are similar, with RBPC beams even slightly outperforming BPC beams. Wang et al. [34] selected a retard-bonded prestressed concrete beam at a construction site as a test subject for monitoring, revealing the stress transfer mechanism in the concrete of the test beam. They analyzed the critical sections of the test beam using an ANSYS model to provide early warnings. However, research on the combination of RBPTs and UHPC remains limited. Xiong et al. [35] conducted experimental research on a large-scale variable-section π-shaped beam using RBPTs and UHPC. The results showed that the beam exhibited excellent load-bearing capacity and ductility.

Currently, research on the flexural behavior of retard-bonded prestressed UHPC (RBPU) beams remains scarce, which hinders the accurate elucidation of the bending mechanism of this novel system. To investigate the flexural performance of RBPU beams and study the influence of different reinforcement ratios on their flexural behavior, thereby revealing the flexural mechanism of this novel system, this research designed three RBPU beams with varying reinforcement ratios and conducted four-point bending tests. During the loading process, parameters such as load, strain, and crack development were monitored. Based on the material properties of both UHPC and RBPTs, we calculated the cracking moment and ultimate moment of RBPU beams. The results of this study validate the flexural performance of beams combining RBPTs and UHPC, providing valuable insights and theoretical support for the design and analysis of RBPU beams.

2. Experimental Program

2.1. Test Specimens

Three RBPU beam specimens were fabricated for flexural tests, with their structural configurations and parameters visually presented in Figure 2 and tabulated in Table 1, respectively. The beam span was 2000 mm, the effective length was 1800 mm, and the cross-sectional dimensions were 180 mm × 250 mm. All the beam specimens were made with retard-bonded prestressing steel strands of a 15.2 mm diameter and featuring a transverse rib thickness of 1.2 mm, transverse rib width of 9 mm, transverse rib height ≥ 1.2 mm, and transverse rib spacing of 13 mm, incorporating two 8 mm diameter erection reinforcements on the compression side and 10 mm diameter stirrups arranged at 100 mm intervals along the beam length. At the positions where tension force was applied at both ends of the beam, steel plates with dimensions of 180 mm × 140 mm and a thickness of 20 mm were embedded. After tensioning, low-shrinkage anchorage was used for anchoring.

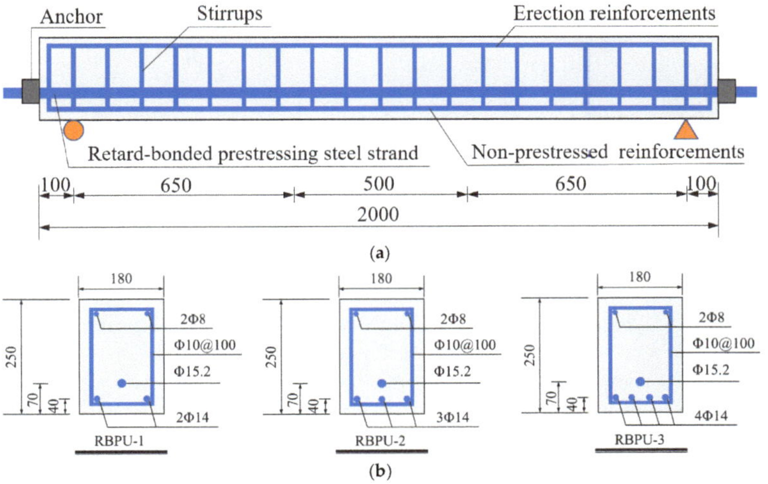

Figure 2. Structural configuration and dimensional parameters of RBPU beam specimens (unit: mm): (**a**) arrangement of reinforcements and prestressing strands; (**b**) cross-section.

Table 1. The design parameters of the RBPU beams.

Specimen	Dimensions (mm)	Non-Prestressed Reinforcements	Stirrups	Number of Steel Strands	Prestressing Level (%)	σ_{con} (MPa)
RBPU-1 RBPU-2 RBPU-3	2000 × 180 × 250	2Φ14 3Φ14 4Φ14	Φ10@100	1	55	1023

Note: The prestressing level is the ratio of the controlled tensioning stress to the standard value of the tensile strength of the prestressing steel strand; σ_{con} denotes the controlled tensioning stress.

Figure 3 illustrates the fabrication process of the RBPU beam. In the first step, the reinforcement cage was tied. Upon the completion of the rebar tying process, the assembled reinforcement cage and end-plate system were positioned within the timber formwork, and the retard-bonded prestressing steel strands were fixed at the designated positions. Finally, the different components of UHPC were poured into the concrete mixer according to the mix design, and after mixing for the appropriate duration, the UHPC was poured into the mold for casting. After casting was finished, a plastic film was used to cover the surface of the beam specimen. The specimens then underwent curing under laboratory conditions (the ambient temperature was maintained at 25–30 °C) throughout the 28-day hardening period. Subsequently, within the tensioning application period, an electric oil pump was used to tension and anchor the prestressed steel strands. After the retarding adhesive had fully cured, the bending test was conducted on the specimen.

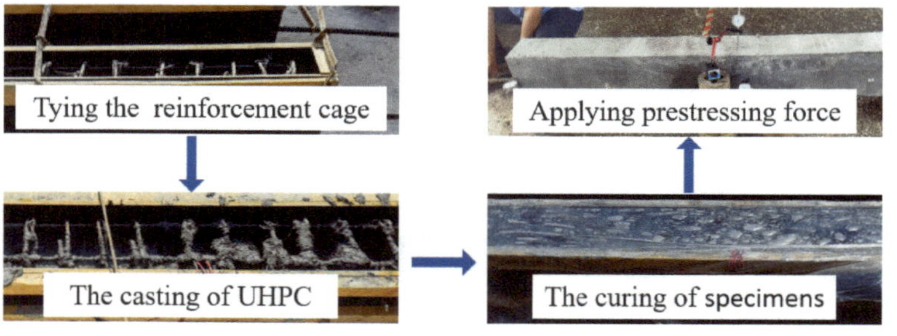

Figure 3. The fabrication process of the RBPU beam.

2.2. Material Properties

This investigation employed brass-coated steel fibers at a 2% volumetric integration within the UHPC matrix, comprising equal proportions (50% each) of straight and hooked-end fibers measuring 13 mm and 14 mm in length, respectively, both maintaining a 0.2 mm diameter. The constituent materials, including specialized pre-blended UHPC powder and steel fibers, were sourced from Zhejiang Hongri Tenacal New Material Technology Co., Ltd. (Zhejiang, China), with the manufacturer's proprietary formulation details tabulated in Table 2 [1]. The steel fibers demonstrated a characteristic tensile strength of 2000 MPa, with a corresponding elastic modulus reaching 200 GPa. Consistent with RBPU beam curing protocols, triplicate sets of 100 mm^3 cubic molds and standardized dog-bone specimens were cast from identical UHPC batches. Mechanical evaluation was conducted in compliance with T/CECS864-2021 [36], with the dog-bone specimen geometry specifications illustrated in Figure 4. Post 28-day curing, the material demonstrated a characteristic compressive strength of 121.6 MPa and tensile capacity of 6.9 MPa. The mean values from the mechanical characterization tests are systematically compiled in Table 3.

Table 2. Mix proportions of UHPC matrix (kg/m^3).

Cement	Silica Fume	Grinding Quartz Powder	Silica Sand	Water	Superplasticizer
745	223.5	223.5	998.3	179.0	13.1

The grades of the reinforcement and steel strand are HRB400 and 1860, respectively. Uniaxial tensile testing protocols were implemented in strict accordance with GB/T 228.1-2021 [37] (reinforcement) and GB/T 21839-2019 [38] (steel strand) to characterize their mechanical performance. Table 4 presents the experimentally determined mechanical characteristics of reinforcement and prestressing strands of various diameters.

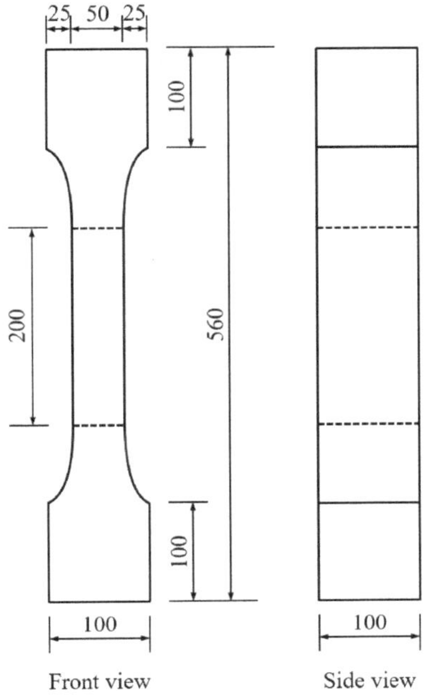

Figure 4. Detailed dimensions of dog-bone specimen (unit: mm).

Table 3. Mechanical properties of UHPC.

f_{cu} (MPa)	f_c (MPa)	f_t (MPa)	E_c (GPa)
121.6	94.2	6.9	43.0

Note: f_{cu} denotes the cubic compressive strength of UHPC; f_c denotes the axial compressive strength of UHPC; f_t denotes the axial tensile strength of UHPC; E_c denotes the elastic modulus of UHPC.

Table 4. Mechanical properties of steel.

Type	Grade	Nominal Diameter	f_y (MPa)	f_u (MPa)	E_c (GPa)
Erection reinforcements	HRB400	8	448.6	635.2	204.5
Stirrups	HRB400	10	445.2	625.4	205.4
Non-prestressed reinforcements	HRB400	14	490.4	668.8	205.3
Steel strands	1860	15.2	1771.3	1970.1	196.7

Note: f_y denotes the yield strength, and the yield strength of a steel strand is defined as the stress corresponding to a residual strain of 0.2%; f_u denotes the ultimate strength; E_c denotes the elastic modulus.

2.3. Test Setup and Instrumentation

The experimental configuration and measurement devices are illustrated in Figure 5. The test beams rested on simple supports and were subjected to four-point bending evaluation using a 12,000 kN electro-hydraulic servo system. As demonstrated in Figure 5b, the specimens featured a 500 mm pure bending segment and 650 mm shear zone, achieving a shear span–depth ratio of 3.1. Initial calculations were performed to determine the theoretical cracking and failure loads before formal testing commenced. Following the code of GB/T50152-2012 [39], preliminary loading at 5% of the predicted failure load preceded formal testing to verify the equipment's functionality and eliminate system clearance. The bending protocol implemented phased loading strategies: a pre-yield phase that utilized force-controlled increments (20 kN/step) at 6 kN/min, with stabilization periods for crack documentation using the HC-CK103 crack-monitoring device (Beijing Haichuang High Technology Co., Ltd. (Beijing, China), ±0.01 mm precision), and a post-yield transition to displacement-controlled loading at 0.6 mm/min with 2 mm displacement increments per phase.

The instrumentation layout is detailed in Figure 5b. Ten linear variable differential transformers (LVDTs) were positioned at the beam's mid-span, loading points, and support regions to record the vertical displacement responses under load. Strain measurements were obtained through gauges mounted on the UHPC surface (top, bottom, and lateral faces) and targeted longitudinal reinforcement. Figure 5c shows the strain gauge distribution: labels U1–U19 correspond to UHPC measurement points, while R1–R4 identify reinforcement strain-monitoring locations.

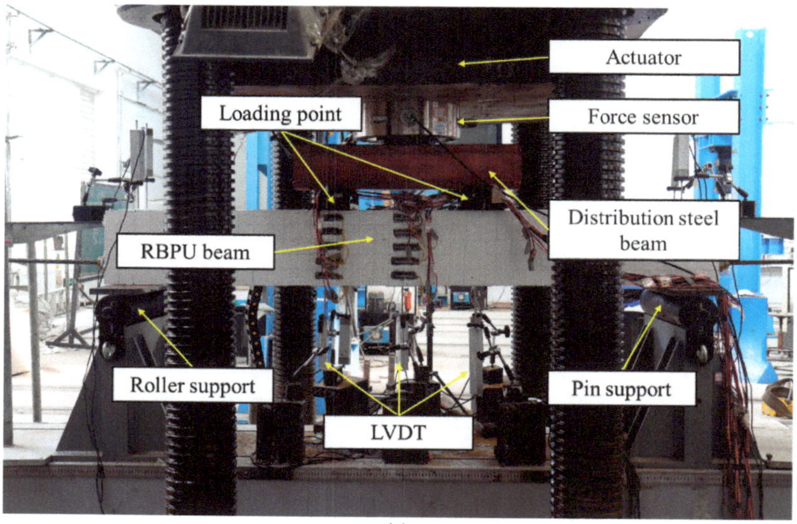

(**a**)

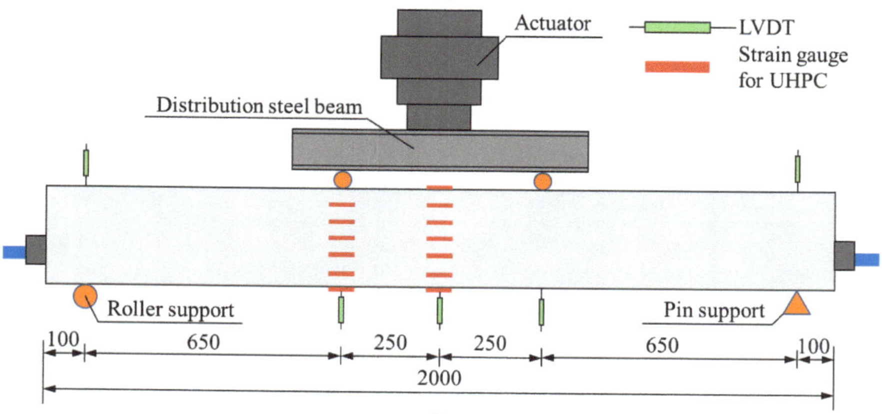

(**b**)

Figure 5. *Cont.*

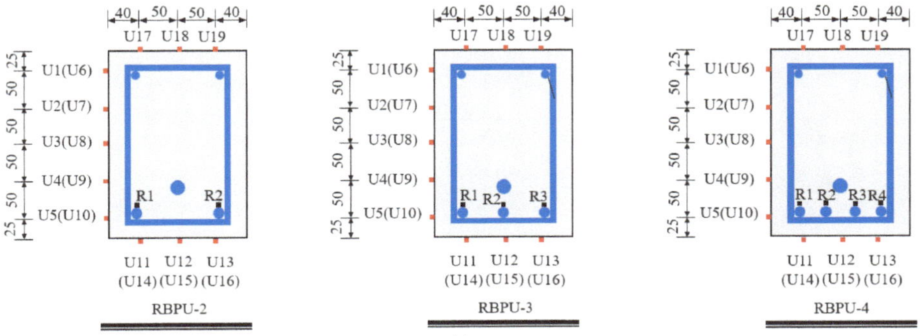

(c)

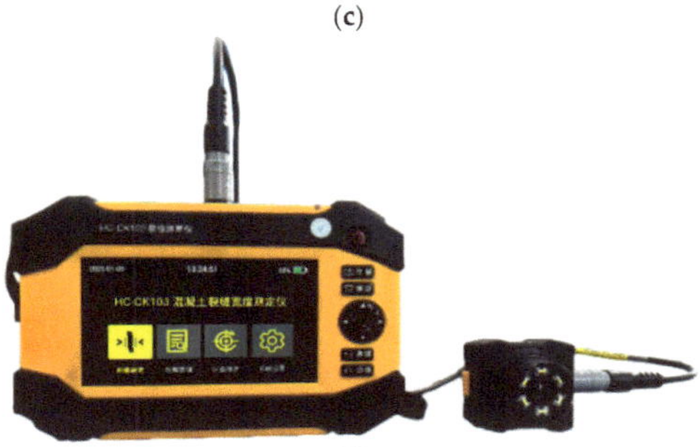

(d)

Figure 5. Test setup and measuring point arrangement (unit: mm): (**a**) photo of test. (**b**) Arrangement of LVDTs. (**c**) Arrangement of strain gauges. (**d**) Photo of crack-measuring instrument, HC-CK103.

3. Experimental Results and Discussion

3.1. Failure Modes and Crack Patterns

The failure characteristics and fracture distributions are presented in Figure 6. All three specimens exhibited flexural-dominated failure patterns consistent with conventional bending failure mechanisms, featuring compressive failure in the upper UHPC zone coupled with the simultaneous yielding of non-prestressed reinforcements and prestressed reinforcements.

During the initial loading phase, the specimens remained uncracked under small bending moments, demonstrating linear elastic behavior with a stable neutral axis position. During the post-cracking to pre-yielding phase, progressive loading triggered crack initiation across the pure bending segment, accompanied by upward neutral axis migration. Three key crack parameters—number, width, and propagation length—progressively increased with the applied load. The transition to the yielding phase manifested through dominant macrocracks (typically 1–2 primary fractures) within the bending region, exhibiting accelerated width expansion and vertical penetration toward the compressive failure zone. The synergistic interaction between steel fiber bridging mechanisms and prestressing technology in RBPU beams significantly retards macrocrack propagation while enhancing crack distribution homogeneity compared to conventional reinforced concrete systems [23]. Following peak load attainment, crack stabilization occurred as fracture propagation ceased. The post-peak degradation phase manifested through fiber extraction in tensile regions concurrent with compressive zone deterioration involving UHPC fragmentation, ultimately culminating in structural failure. At this point, the compression zone of the RBPU beam only experienced slight UHPC spalling and did not exhibit significant spalling or bulging as seen in ordinary reinforced concrete. The ultra-high strength of

UHPC, combined with the bridging effect of steel fibers, enables it to withstand higher loads and reduces its susceptibility to cracking, resulting in a unique behavioral pattern that differs from ordinary reinforced concrete systems.

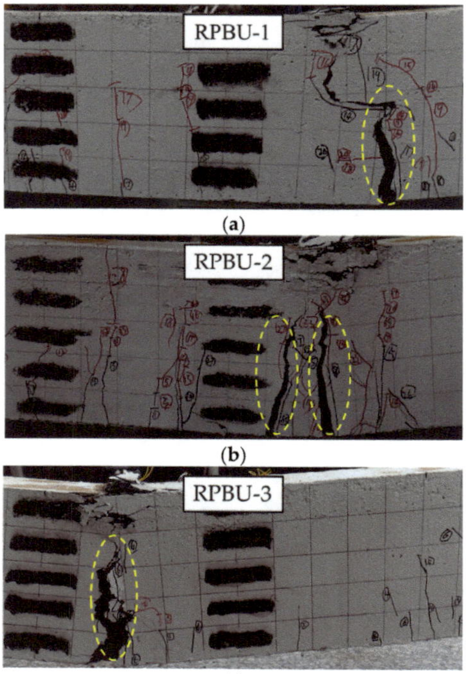

(a)

(b)

(c)

Figure 6. Failure modes and crack patterns: (**a**) RPBU-1; (**b**) RPBU-2; (**c**) RPBU-3.

3.2. Load–Deflection Relationship

The mid-span load–deflection response, as presented in Figure 7, reveals three distinct phases: (1) a linear-elastic regime preceding the tensile cracking of UHPC; (2) a nonlinear hardening phase post-cracking with nonlinear stiffness evolution, and (3) a post-yielding degradation phase marked until specimen beam failure.

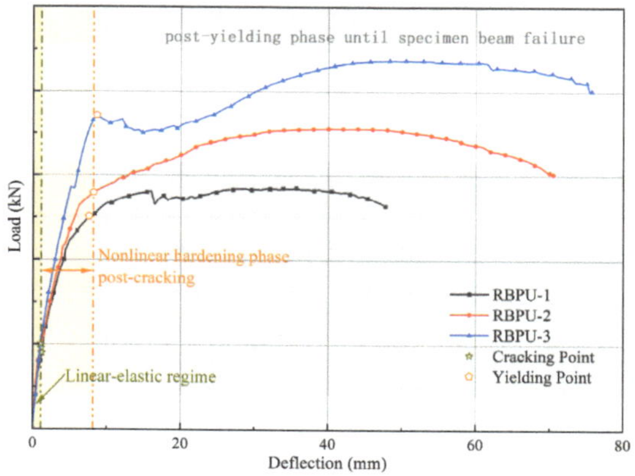

Figure 7. Load–deflection curve.

In the linear-elastic regime preceding tensile UHPC cracking, the bending moments remain below critical thresholds, with the stress levels in both the tensile reinforcement and UHPC maintained within elastic limits. The beam's stress condition resembles that of a homogeneous elastic beam, with the load–displacement curve ascending almost linearly. Following the initiation of UHPC cracking, the load–displacement response transitions to a nonlinear hardening phase. The slope of the curve does not decrease significantly, and the inflection point is not pronounced, with the load–displacement curve of the specimen still approximating a linear progression. This behavior likely stems from the crack-inhibiting mechanism of the retard-bonded prestressing steel strands, resulting in an insignificant reduction in the beam's stiffness, thereby minimizing the impact of UHPC cracking on the overall stiffness of the test beam. As the curve ascends, a gradual divergence in the curves can be observed, with specimens exhibiting higher reinforcement ratios demonstrating greater stiffness during this phase. Furthermore, there is a strong positive correlation between the reinforcement ratio and flexural capacity enhancement in RBPU beams, demonstrating parametric sensitivity of the reinforcement ratio to the bending resistance characteristics.

When the load–deflection curve reaches a certain level, the curve noticeably decelerates, and the specimen's stiffness decreases significantly, signifying the specimen's entry into the yielding phase. This stage exhibits rapid strain escalation in non-prestressed reinforcement, coordinated with progressive growth in mid-span deflection and neutral axis elevation. A slight decline in the curve of specimen RBPU-3 is observed before it continues to rise slowly, a phenomenon attributed to the internal force redistribution within the specimen.

With continued loading, the load increases gradually. The load attains its maximum prior to UHPC spalling initiation within the beam's pure bending segment, followed by a progressive decline in load-bearing capacity. At a mid-span deflection of 47.73 mm, the non-prestressed reinforcement in specimen RBPU-1 fractures. At this juncture, the mid-span displacement has exceeded the $1/50L_0$ stipulated by the code of GB/T 50152-2012 [39], and while the load decrease is not pronounced in any of the test beams, the other two specimens can still sustain further loading. This indicates that RBPU beams configured in this manner possess commendable deformation capacity and ductility.

3.3. Flexural Capacity and Ductility

The experimental testing results are detailed in Table 5, including the characteristic load and corresponding mid-span deflection. Structural ductility in concrete systems refers to a sustained deformation capability post-yielding while load resistance is maintained, indicative of energy absorption potential. This property is quantified through the displacement ductility coefficient (μ), expressed as follows:

$$\mu = \frac{\delta_u}{\delta_y} \quad (1)$$

where δ_u represents the failure load-induced deflection. Most of the RBPU beams did not exhibit significant load reduction. For the analysis of the RBPU beams, the failure load was defined at 95% peak load magnitude [40], enabling parametric influence quantification. δ_y denotes the deflection corresponding to the yield load, which was calculated using the farthest point method as described in ref. [41]. The calculated displacement ductility factors for all the tested beams are presented in Table 5.

Table 5. The characteristic loads and deflections and displacement ductility factors.

Specimen	P_{cr} (kN)	δ_{cr} (mm)	P_y (kN)	δ_y (mm)	P_p (kN)	δ_p (mm)	P_u (kN)	δ_u (mm)	μ
RBPU-1	90.0	1.22	250.6	7.6	285.6	35.68	272.1	46.45	6.1
RBPU-2	95.4	1.26	279.6	8.21	355.8	40.51	336.7	59.01	7.2
RBPU-3	100.1	1.42	371.2	8.65	437.1	50.49	411.8	73.56	8.5

Note: P_{cr}, P_y, P_p, and P_u denote the cracking load, yield load, peak load, and ultimate load, respectively; δ_{cr}, δ_y, δ_p, and δ_u denote the deflections corresponding to the cracking load, yield load, peak load, and ultimate load, respectively; μ denotes the displacement ductility factor.

Figure 8 demonstrates how varying the reinforcement ratios affected bending performance. According to the experimental results recorded in Table 5 and visualized in Figure 8, progressive increases in the total reinforcement ratio from 1.28% to 2.10% elevated the cracking load by 11.22%, yield load by 48.11%, and maximum load capacity by 53.04%. The limited impact of reinforcement ratio elevation on cracking load originates from its inability to augment UHPC's inherent tensile properties. Elevated reinforcement ratios substantially strengthen the total tensile resistance within the beam's tension regions, which drives significant improvements in both yield load thresholds and maximum load capacities. This mechanical enhancement directly contributes to superior flexural performance in RBPU beam systems.

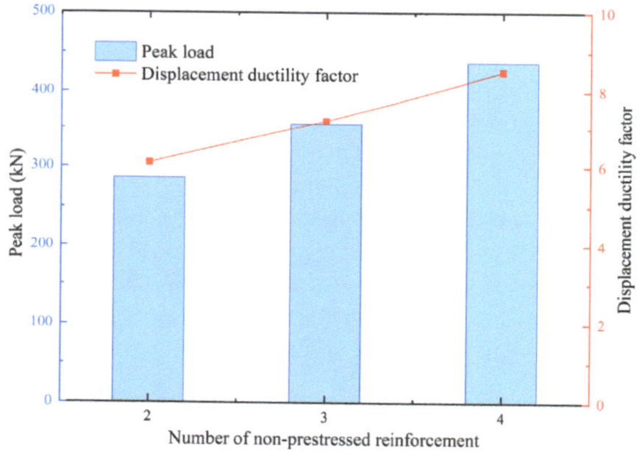

Figure 8. Comparison of flexural capacity and displacement ductility factors of different specimens.

Furthermore, the experimental data in Table 5 and Figure 8 reveal displacement ductility factors exceeding 6.0 across all the specimens, demonstrating superior deformation performance. When the total ratio of reinforcement increases from 1.28% to 2.10%, the displacement ductility factors of the specimens increase by 39.34%. This suggests that, within this configuration, a higher ratio of reinforcement results in greater deformation capacity of the specimens. This conclusion is similar to that of Shao et al. [42], indicating that increasing the reinforcement ratio within the range of 0.96% to 2.1% can significantly enhance ductility.

3.4. UHPC Strain of Mid-Span Section

Strain gauge placement (Figure 5) enabled the measurement of the UHPC's strain distribution under loading (Figure 9). The linear strain profile across the mid-span sections validates the plane-section hypothesis, demonstrating consistent compliance with fundamental beam theory. The analysis indicates that the neutral axis remained stable until the tensile-zone concrete cracked. Once the load surpassed the cracking point, the neutral axis began to move upward, accelerating significantly after the specimen yielded.

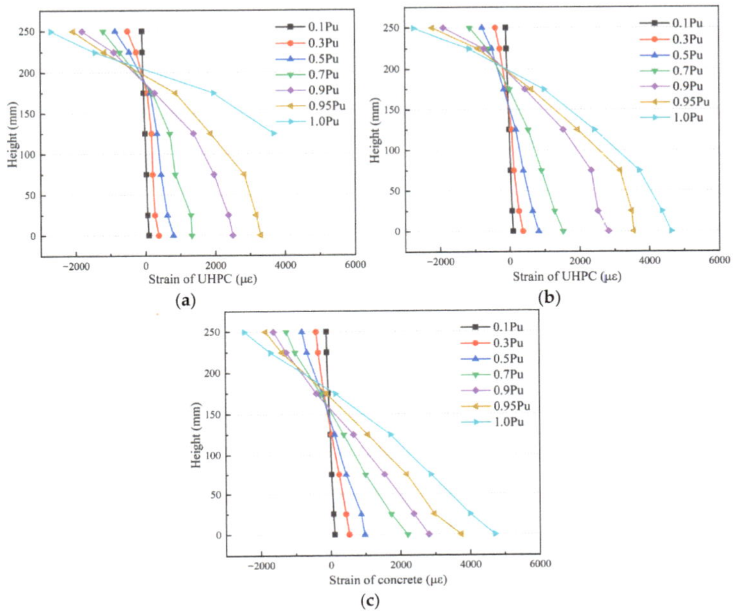

Figure 9. The average strain of UHPC along the cross-sectional height of each test beam: (**a**) RBPU-1; (**b**) RBPU-2; (**c**) RBPU-3.

4. Theoretical Analysis of Flexural Capacity

4.1. Basic Assumptions

Firstly, it is postulated that the RBPU beam adheres to the plane-section hypothesis. The cross-sections perpendicular to the beam axis (i.e., the cross-sections of the member) subjected to pure bending remain planar after deformation and orthogonal to the deformed member axis. Each cross-section undergoes rotation, and the rotation of each cross-section can be determined by two rotational angles. Secondly, the slip between the tensile reinforcement, the slowly bonded prestressing steel strands, and the UHPC is neglected.

4.2. Calculation of Prestressing Losses

The retard-bonded prestressed UHPC beam specimen can be regarded as a post-tensioned construction element, and its prestress loss can be calculated in two stages. In the first stage, the prestress loss includes the following: the loss due to the deformation of the anchorage at the tensioning end and retraction of the retard-bonded prestressing steel strand (σ_{l1}), as well as the loss caused by friction between the prestressing steel strand and the sheathing wall (σ_{l2}). In the second stage, the prestress loss consists of the stress relaxation of the prestressing steel strand (σ_{l4}) and the shrinkage and creep of UHPC (σ_{l5}). The calculation methods for each type of prestress loss are based on the definitions provided in the code of JGJ387-2017 [43], as shown in Equations (2)–(5), and the calculation results are presented in Table 6.

$$\sigma_{l1} = \frac{a}{l} E_{\mathrm{p}} \tag{2}$$

$$\sigma_{l2} = \sigma_{\mathrm{con}}(\kappa x + \mu \theta) \tag{3}$$

$$\sigma_{l4} = 0.125\left(\frac{\sigma_{con}}{f_{ptk}} - 0.5\right)\sigma_{con} \quad (4)$$

$$\sigma_{l5} = \frac{55 + 300\frac{\sigma_{pc}}{f_{cu}}}{1 + 15\rho} \quad (5)$$

where a represents the deformation of the anchorage and the retraction value of the prestressing tendon, which can be taken as 5 mm; l denotes the distance between the tensioning end and the anchoring end; E_p is the elastic modulus of the prestressing tendon; κ is the friction coefficient accounting for the local deviation per meter length of the sheath wall of the retard-bonded prestressing steel strand, which can be taken as 0.006; x is the length from the tensioning end to the calculated section; θ is the radian value corresponding to the angle between the tangent of the curve and the line from the tensioning end to the calculated section; σ_{pc} represents the normal compressive stress of the UHPC at the resultant force point of the prestressing tendon in the tensile zone; and ρ is the total reinforcement ratio of the retard-bonded prestressing steel strand and non-prestressed reinforcement in the tensile zone.

Table 6. The calculation results of the prestress loss for the test beam (MPa).

Specimens	σ_{l1}	σ_{l2}	σ_{lI}	σ_{l4}	σ_{l5}	σ_{lII}	σ_{l}	σ_{pe}	σ_{pc}
RBPU-1	491.75	6.14	497.89	6.39	53.10	59.49	557.37	465.63	2.01
RBPU-2	491.75	6.14	497.89	6.39	50.72	57.11	555.00	468.00	1.87
RBPU-3	491.75	6.14	497.89	6.39	48.53	54.92	552.80	470.20	1.71

4.3. Calculation of Cracking Moment

According to the code of JGJ 369-2016 [44], the cracking moment of prestressed concrete beams can be calculated using Equations (6) and (7).

$$M_{cr} = (\sigma_{pc} + \gamma f_{tk})W_0 \quad (6)$$

$$\gamma = \left(0.7 + \frac{120}{h}\right)\gamma_m \quad (7)$$

where σ_{pc} represents the compressive stress in the concrete at the edge of the crack-checking section after deducting all the prestress losses; γ denotes the plastic influence coefficient of the section modulus of the concrete member; W_0 is the elastic section modulus at the edge of the transformed section for crack checking; h is the section height, and when h < 400 mm, it is taken to be 400 mm; and γ_m is the basic value of the plastic influence coefficient of the section modulus for concrete members, which is taken to be 1.55 for a rectangular section.

The crack-inhibiting mechanism of steel fibers, through their bridging effect, necessitates the introduction of the correction factor α_{cr} in Equation (8) to differentiate RBPU beam behavior from conventional prestressed concrete systems.

$$\gamma = \alpha_{cr}\left(0.7 + \frac{120}{h}\right)\gamma_m \quad (8)$$

where α_{cr} is the correction factor.

By substituting the experimental data and [22] into Equations (6) and (8), the average value of α_{cr} was found to be 1.131, with a standard deviation of 0.149 and a coefficient of variation of 0.131, as shown in Table 7.

Table 7. Calculation of correction factor.

Specimen	$M_{cr,t}$ (kN · m)	σ_{pc} (MPa)	f_t (MPa)	α_{cr}
SB80-2 [22]	57.72	3.69	4.3	1.067
SB90-2 [22]	58.15	4.28	4.3	0.999
SB95-2 [22]	63.68	4.58	4.3	0.890
SB90-3 [22]	81.03	6.4	4.3	1.062
RBPU-1	40.51	2.01	6.9	1.198
RBPU-2	42.95	1.87	6.9	1.258
RBPU-3	45.04	1.71	6.9	1.306
Average value	-	-	-	1.131
Coefficient of variation	-	-	-	0.131

Note: $M_{cr,t}$ represents the experimental value of the cracking moment.

Taking $\alpha_c = 1.131$, the calculated cracking moments and the measured cracking moments are presented in Table 8. As shown in Table 8, the ratio of the experimental value to the calculated value of the cracking moment is 1.111, with a standard deviation of 0.044 and a coefficient of variation of 0.039. The results indicate that the introduction of the correction factor α_{cr} significantly improves the accuracy of the calculated cracking moments for the RBPU beams tested in this study.

Table 8. Calculated and experimental values of the cracking moment and ultimate bearing moment.

Specimen	$M_{cr,c}$ (kN · m)	$M_{cr,t}$ (kN · m)	$M_{cr,t}/M_{cr,c}$	$M_{u,c}$ (kN · m)	$M_{u,t}$ (kN · m)	$M_{u,t}/M_{u,c}$
RBPU-1	38.56	40.51	1.066	99.16	92.82	0.936
RBPU-2	39.13	42.95	1.114	112.20	115.64	1.031
RBPU-3	39.65	45.04	1.153	125.02	142.06	1.136
Average value	-	-	1.111	-	-	1.034
Coefficient of variation	-	-	0.039	-	-	0.097

Note: $M_{cr,c}$ and $M_{cr,t}$ represent the calculated value and the experimental value of the cracking moment, respectively; $M_{u,c}$ and $M_{u,t}$ denote the calculated value and the experimental value of the ultimate moment, respectively.

4.4. Calculation of Ultimate Moment

RBPU beams exhibit a distinct behavior compared to traditional prestressed concrete beams. Specifically, under ultimate moment conditions, the tensile-zone concrete in RBPU beams can effectively utilize the tensile strength of UHPC. The concrete in the compressive zone partially enters plasticity, and both non-prestressed and prestressed steel strands reach their yield states, aligning with the ultimate limit state behavior of traditional prestressed concrete beams. To simplify the calculations, the stress distributions in the concrete's compressive and tensile zones were idealized as rectangular stress blocks. Figure 10 illustrates the simplified diagram for determining the ultimate flexural state of RBPU beams.

Using the horizontal force equilibrium equation and cross-section geometry, the following expressions are obtained:

$$\alpha_1 f_c b x = f_s A_s + f_p A_p + \sigma_{te} b x_t \tag{9}$$

$$x_t = h - x/\beta \tag{10}$$

$$\sigma_{te} = k f_t \tag{11}$$

where α_1 and β are coefficients for the equivalent rectangular stress block in the compressive zone, taken as 1.0 and 0.8 [23], respectively; k is the ratio of the stress value in the rectangular stress block of the UHPC tensile zone to the axial tensile strength of the UHPC, taken as 0.9 [45]; b and h are the width and height of the calculated cross-section, respectively; x_t and x are the equivalent heights of the tensile and compressive zones, respectively; σ_{te} is the equivalent tensile stress in the UHPC tensile zone; f_y and f_{py} are the yield strength of the non-prestressed reinforcement and prestressed reinforcement, respectively; and A_s and A_p are the cross-sectional areas of the non-prestressed reinforcement and prestressed reinforcement, respectively.

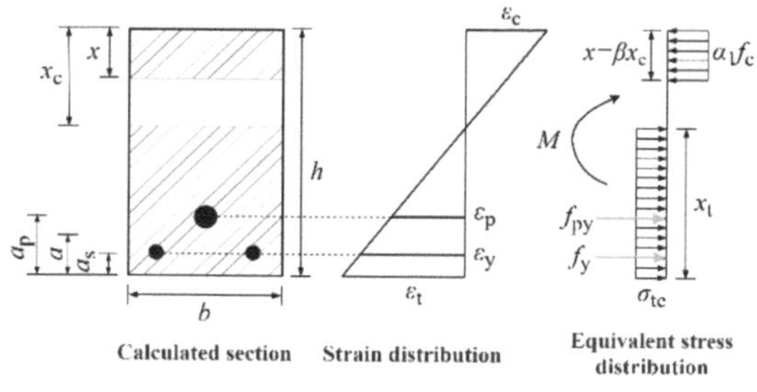

Figure 10. Calculation of flexural bearing capacity in ultimate state.

The calculated equivalent compressive zone height x for UHPC must satisfy the following condition:

$$2a' \leq x \leq \xi_b h_0 \tag{12}$$

where a' is the distance from the compressive reinforcement to the upper edge; ξ_b is the limiting relative compressive zone height.

According to the code of T/CCPA35-2022 [46], the value of ξ_b was taken as 0.23. Verification showed that all the RBPU beam specimens in this experiment met the required criteria.

Based on the principle of moment equilibrium, by taking moments about the resultant force point of the tensile reinforcement, the ultimate bearing capacity can be calculated using Equation (13).

$$M_u = \alpha_1 f_c b x (h_0 - x/2) - \sigma_{te} b x_t (x_t/2 - a) \tag{13}$$

where f_c represents the axial compressive strength of UHPC; a denotes the distance from the resultant force point of the tensile reinforcement to the bottom of the beam, which can be calculated by $a = (f_{py}A_p a_p + f_y A_s a_s)/(f_{py}A_p + f_y A_s a_s)$; and a_p and a_s are the distances from the resultant force points of the prestressed reinforcement and non-prestressed reinforcement, respectively, to the bottom of the tensile zone, measured to the bottom of the beam.

Table 8 summarizes the ultimate moment calculations, revealing a mean $M_{u,t}/M_{u,c}$ ratio of 1.034 with a 0.097 coefficient of variation, demonstrating strong correlation between the theoretical predictions and experimental results. Additionally, it can be observed that, for lower reinforcement ratios, the calculated M_u values are smaller than the measured values. As the reinforcement ratio increases, the ratio of the calculated M_u to the measured value gradually rises and exceeds 1. This suggests that, for RBPU beams with this configuration, the use of Equation (13) for calculating M_u becomes increasingly conservative as the reinforcement ratio increases, thereby providing a greater margin of safety.

5. Conclusions

This study introduces a new retard-bonded prestressed UHPC (RBPU) beam made by combining retard-bonded prestressed technology with UHPC. The failure modes, crack distribution and propagation, characteristic loads, and ductility of the beams were analyzed under varying reinforcement ratios (2, 3, and 4 steel strands). The main conclusions are as follows:

(1) The RBPU beams in this experiment exhibited typical flexural failure modes. Compared to that of conventional prestressed concrete beams, the crack distribution was more uniform, which can be attributed to the retard-bonded prestressed technology and the bridging effect of steel fibers in UHPC. Additionally, compared to conventional concrete beams, the phenomena of concrete spalling and bulging at the late loading stage were relatively mild, and there was no significant drop in load.
(2) As the reinforcement ratio was increased from 1.28% to 2.10%, the cracking load saw an 11.22% increase, suggesting that higher reinforcement ratios can improve crack resistance. Additionally, the ultimate load capacity and ductility coefficient grew by 53.04% and 39.34%, respectively, under the same conditions. This indicates that increasing the reinforcement ratio enhances both the load-bearing capacity and ductility of the component.
(3) Taking into account the bridging effect of steel fibers in UHPC, the plastic influence coefficient γ for the section's moment resistance was adjusted. Post-modification, the computed cracking moment values aligned closely with the experimental data. Likewise, the calculated ultimate moment values demonstrated strong agreement with the experimental findings.
(4) While this study investigates the flexural behavior of retard-bonded prestressed UHPC (RBPU) beams with varying reinforcement ratios, there are certain limitations that warrant further exploration in the following aspects:
    (i) A parametric influence analysis of RBPU beams: experimental studies should be conducted to examine the effects of critical parameters such as prestressing levels and steel fiber volume content on the flexural performance of RBPU beams;
    (ii) High-precision numerical modeling: refined finite-element models should be developed to elucidate the underlying mechanisms of flexural behavior in RBPU beams.

Author Contributions: Conceptualization, J.W. and L.D.; methodology, L.D.; software, L.D.; validation, D.W. and S.W.; formal analysis, B.Z.; investigation, X.T.; resources, J.W.; data curation, L.D.; writing—original draft preparation, L.D.; writing—review and editing, L.D. and D.W.; visualization, D.W.; supervision, S.W.; project administration, J.W.; funding acquisition, D.W. All authors have read and agreed to the published version of the manuscript.

Funding: This work was supported by the National Natural Science Foundation of China [Grant No. 52208242], Key R&D projects in Henan Province [Grant No. 241111322600], the China Postdoctoral Science Foundation [Grant No. 2023M733356], and the Scientific and Technological Research and Development Projects of China Construction Seventh Engineering Division Corp. Ltd. (047H2023TM047).

Data Availability Statement: All the data generated and analyzed during this study are included within this article.

Conflicts of Interest: Authors Dongchang Wu, Shaowei Wang and Xiufeng Tang were employed by China Construction Seventh Engineering Division Corp. Ltd. The remaining authors declare that the research was conducted in the absence of any commercial or financial relationships that could be construed as a potential conflict of interest.

References

1. Gao, X.L.; Shen, S.Y.; Wan, Y.; Qin, S.W. Experimental Study on Bond Behavior of Steel Bar Embedded in Thin UHPC. *J. Build. Eng.* **2024**, *86*, 108865. [CrossRef]
2. Shao, Y.; Hung, C.C.; Billington, S.L. Gradual Crushing of Steel Reinforced HPFRCC Beams: Experiments and Simulations. *J. Struct. Eng.* **2021**, *147*, 04021114. [CrossRef]

3. Shafieifar, M.; Farzad, M.; Azizinamini, A. A Comparison of Existing Analytical Methods to Predict the Flexural Capacity of Ultra High Performance Concrete (UHPC) Beams. *Constr. Build. Mater.* **2018**, *172*, 10–18. [CrossRef]
4. Krahl, P.A.; Carrazedo, R.; El Debs, M.K. Mechanical Damage Evolution in UHPFRC: Experimental and Numerical Investigation. *Eng. Struct.* **2018**, *170*, 63–77. [CrossRef]
5. Yan, W.; Zhang, R.; Sushant, S.; Ashour, A.; Fu, S.; Qiu, L.; Zhang, Z.; Ge, W. Experimental Investigation on Flexural Performance of UHPC Beams Reinforced with Steel-FRP Bars. *Arch. Civ. Mech. Eng.* **2024**, *24*, 132. [CrossRef]
6. Hung, C.; El-Tawil, S.; Chao, S. A Review of Developments and Challenges for UHPC in Structural Engineering: Behavior, Analysis, and Design. *J. Struct. Eng.* **2021**, *147*, 03121001. [CrossRef]
7. Zhou, M.; Lu, W.; Song, J.; Lee, G. Application of Ultra-High Performance Concrete in Bridge Engineering. *Constr. Build. Mater.* **2018**, *186*, 1256–1267. [CrossRef]
8. Li, D.B.; Li, W.L. Development of retard-bonded prestressed technique in China. *Build. Sci.* **2022**, *38*, 1–7. [CrossRef]
9. Zhang, X.; Fan, X.; Jiang, F.; Liu, F.; Yin, Y.; Jin, H. Effect of Aging and Temperature on the Viscosity of the Adhesive Used for Retard-Bonded Prestressed Systems. *J. Test. Eval.* **2019**, *47*, 1848–1863. [CrossRef]
10. Xu, J.Q.; Zhang, W.; Chen, S.Z. Design and analysis of slowly-bonded prestressed structure of Ganzhou West Railway Station. *Build. Sci.* **2020**, *50*, 132–137. [CrossRef]
11. Song, X.F.; Li, J.A.; Zhu, S.W.; Zhu, S. Construction of Retard-bonded Prestressed in the Shuangliao Station. *Build. Sci.* **2011**, *27*, 167–168+171. [CrossRef]
12. Pham, H.D.; Khuc, T.; Nguyen, T.; Cu, H.; Le, D.B.; Trinh, T.P. Investigation of Flexural Behavior of a Prestressed Girder for Bridges Using Nonproprietary UHPC. *Adv. Concr. Constr.* **2020**, *10*, 71–79. [CrossRef]
13. Chen, L.; Graybeal, B.A. Modeling Structural Performance of Ultrahigh Performance Concrete I-Girders. *J. Bridge Eng.* **2012**, *17*, 754–764. [CrossRef]
14. Steinberg, E. Structural Reliability of Prestressed UHPC Flexure Models for Bridge Girders. *J. Bridge Eng.* **2010**, *15*, 65–72. [CrossRef]
15. Bahmani, H.; Mostafaei, H.; Santos, P.; Fallah Chamasemani, N. Enhancing the Mechanical Properties of Ultra-High-Performance Concrete (UHPC) Through Silica Sand Replacement with Steel Slag. *Buildings* **2024**, *14*, 3520. [CrossRef]
16. Tian, X.; Fang, Z.; Shao, Y. Flexural Behavior and Design of Prestressed Ultra-High Performance Concrete (UHPC) Beams: Failure Mode and Ductility 2024. *arXiv* **2024**, arXiv:2411.01113.
17. El-Helou, R.G.; Graybeal, B.A. Flexural Behavior and Design of Ultrahigh-Performance Concrete Beams. *J. Struct. Eng.* **2022**, *148*, 04022013. [CrossRef]
18. Sun, X.; Ma, Y.; Jiang, F.; Fan, X.; Wu, H. Bending Resistance Mechanism of Prestressed Ultra-High Performance Concrete—Reinforced Concrete Beam Based on a Full-Scale Experiment. *Adv. Struct. Eng.* **2024**, *27*, 1746–1761. [CrossRef]
19. Sim, C.; Tadros, M.; Gee, D.; Asaad, M. Flexural Design of Precast, Prestressed Ultra-High-Performance Concrete Members. *PCI J.* **2020**, *65*, 35–61. [CrossRef]
20. Li, H.; Li, L.; Fan, X.; Ye, M.; Shao, X.; Yi, L. Experimental and Numerical Investigation on the Flexural Behavior of a Large-Scale Prestressed UHPC T-Shaped Girder. *Eng. Struct.* **2022**, *272*, 115027. [CrossRef]
21. Leutbecher, T.; Heck, L.; Metje, K.; Riedel, P. Flexural design of UHPC girders with mixed reinforcement. *Beton-Stahlbetonbau* **2022**, *117*, 863–877. [CrossRef]
22. Yang, I.H.; Joh, C.; Kim, B.S. Flexural Strength of Large-Scale Ultra High Performance Concrete Prestressed T-Beams. *Can. J. Civ. Eng.* **2011**, *38*, 1185–1195. [CrossRef]
23. Deng, Z.; Li, Q.; Tuladhar, R.; Shi, F. Flexural Performance of Prestressed UHPC Beams with Different Prestressing Degrees and Levels. *Comput. Concr.* **2024**, *34*, 379–391. [CrossRef]
24. Dogu, M.; Menkulasi, F. Moment-Curvature-Deformation Response of Post-Tensioned Ultra-High-Performance Concrete Beams. *ACI Struct. J.* **2022**, *119*, 301–315. [CrossRef]
25. Zhang, Y.; Huang, S.; Zhu, Y.; Hussein, H.H.; Shao, X. Experimental Validation of Damaged Reinforced Concrete Beam Strengthened by Pretensioned Prestressed Ultra-High-Performance Concrete Layer. *Eng. Struct.* **2022**, *260*, 114251. [CrossRef]
26. Zhang, Y.; Huang, S.; Liu, Y.; Fan, W.; Shao, X. Flexural Behavior of Damaged RC Beams Strengthened with Prestressed UHPC Layer. *Eng. Struct.* **2023**, *283*, 115806. [CrossRef]
27. Li, W.; Ji, W.; An, M.; Zhu, L.; Wang, J. Flexural Performance of Composite Prestressed UHPC-NC T-Girders. *J. Bridge Eng.* **2020**, *25*, 04020064. [CrossRef]
28. Mohebbi, A.; Graybeal, B. Prestress Loss Model for Ultra-High Performance Concrete. *Eng. Struct.* **2022**, *252*, 113645. [CrossRef]
29. John, E.E.; Ruiz, E.D.; Floyd, R.W.; Hale, W.M. Transfer and Development Lengths and Prestress Losses in Ultra-High-Performance Concrete Beams. *Transp. Res. Rec.* **2011**, *2251*, 76–81. [CrossRef]
30. Zheng, Y.; Zhu, J.; Yang, H.; Liang, Z.; Li, W. Flexural Behavior of Externally Prestressed Steel-UHPC-NC Composite Beams in Negative Moment Regions. *SSRN* 2025, *preprint*. [CrossRef]

31. Feng, J.; Shao, X.; Qiu, M.; Li, H.; Gao, X.; Huang, Z. Reliability Evaluation of Flexural Capacity Design Provision for UHPC Beams Reinforced with Steel Rebars/Prestressing Tendons. *Eng. Struct.* **2024**, *300*, 117160. [CrossRef]
32. Cao, G.H.; Hu, J.X.; Xiao, M.; Zhang, K. Ultimate Bearing Capacity Test of Retard-Bonded Prestressed Concrete T-Beams. *Appl. Mech. Mater.* **2013**, *256*, 769–774. [CrossRef]
33. Cao, G.; Zhang, W.; Hu, J.; Zhang, K. Experimental Study on the Long-Term Behaviour of RBPC T-Beams. *Int. J. Civ. Eng.* **2018**, *16*, 887–895. [CrossRef]
34. Wang, Y.; Feng, X.; He, P.; Wang, L.; Chen, Z. Experimental Study and Finite Element Analysis on Stress Transfer Mechanism of Retard-Bonded Prestressed Concrete Beams. In Proceedings of the 2020 International Conference on Intelligent Transportation, Big Data & Smart City (ICITBS), Vientiane, Laos, 11–12 January 2020; pp. 288–293.
35. Xiong, X.; He, L.; Bao, L. Experimental Study on Flexural Behavior of Full-Scale Retard-Bonded Prestressed UHPC Variable Section Cantilever Beams. *Eng. Struct.* **2023**, *284*, 115943. [CrossRef]
36. *T/CECS 864-2021*; Standard for Test Method of Ultra-High Performance Concrete. China Architecture & Building Press: Beijing, China, 2021.
37. *GB/T 228.1-2021*; Metallic Materials: Tensile Testing. Standards Press of China: Beijing, China, 2021.
38. *GB/T 21839-2019*; Test Methods of Steel for Prestressing Concrete. Standards Press of China: Beijing, China, 2019.
39. *GB/T 50152-2012*; Standard for Test Method of Concrete Structures. China Architecture & Building Press: Beijing, China, 2012.
40. Gao, X.; Wang, M.; Guo, J.; Li, H. Flexural Behaviors of a Novel Precast Hollow UHPC Composite Beam Reinforced with Inverted T-Shaped Steel: Experimental Investigation and Theoretical Analysis. *J. Build. Eng.* **2024**, *86*, 108893. [CrossRef]
41. Feng, P.; Cheng, S.; Bai, Y.; Ye, L. Mechanical Behavior of Concrete-Filled Square Steel Tube with FRP-Confined Concrete Core Subjected to Axial Compression. *Compos. Struct.* **2015**, *123*, 312–324. [CrossRef]
42. Shao, Y.; Billington, S.L. Impact of UHPC Tensile Behavior on Steel Reinforced UHPC Flexural Behavior. *J. Struct. Eng.* **2022**, *148*, 04021244. [CrossRef]
43. *JGJ 387-2017*; Technical Specification for Concrete Structures Prestressed with Retard-Bonded Tendons. China Architecture & Building Press: Beijing, China, 2017.
44. *JGJ 369-2016*; Code for Design of Prestressed Concrete Structures. China Architecture & Building Press: Beijing, China, 2016.
45. *SIA 2052*; Recommendation: Ultra-high Performance Fiber Reinforced Cement-based Composites (UHPFRC) Construction Material, Dimensioning and Application. MCS-EPFL: Lausanne, Switzerland, 2017.
46. *T/CCPA 35-2022*; Specification for Design of Ultra-High Performance Concrete Structures. China Building Material Council: Beijing, China, 2022.

Disclaimer/Publisher's Note: The statements, opinions and data contained in all publications are solely those of the individual author(s) and contributor(s) and not of MDPI and/or the editor(s). MDPI and/or the editor(s) disclaim responsibility for any injury to people or property resulting from any ideas, methods, instructions or products referred to in the content.

buildings

MDPI

Article

A Study on the Calculations of the Bottom Void Range of an Underground Pipe Gallery Structure Under the Action of Ground Fissure Dislocations

Xintao Yu [1], Baoan Han [1], Yubo Zhao [2], Botuan Deng [3,*], Kang Du [3] and Haosheng Liu [3]

1 China Construction Silk Road Construction Investment Co., Ltd., Xi'an 710000, China; yxt@stu.xust.edu.cn (X.Y.); hba@stu.xust.edu.cn (B.H.)
2 China Construction No. 6 Bureau Construction Engineering Co., Ltd., Tianjin 300000, China; zyb@stu.xust.edu.cn
3 School of Architecture and Civil Engineering, Xi'an University of Science and Technology, Xi'an 710000, China; kangdu@stu.xust.edu.cn (K.D.); haoshengliu@stu.xust.edu.cn (H.L.)
* Correspondence: botuandeng@xust.edu.cn or deng13720412121@139.com

Abstract: Ground fissures are extraordinary urban geological disasters, and their harmful effects on underground structures have been highlighted in many cities. Differential settlements between strata can cause a void phenomenon at the bottom of a pipe gallery structure, significantly threatening the project's construction and operation. This study analyzes the void phenomenon at the bottom of a pipe gallery structure, and a calculation method for the bottom void range is proposed. Through a model test, the stress and deformation laws of the pipe gallery structure under the conditions of orthogonal (90°) and oblique (45°) ground fissure displacements are analyzed. The results show that, owing to the dislocation of the ground fissure, the bottom void range of the pipe gallery is 2.87–3 times the length of the bottom edge of the pipe gallery section under the orthogonal condition and 3.125–3.5 times the length under the oblique condition. Under the dislocation of the ground fissure, the top plate of the structure is under tension; the bottom plate is under compression, and the strains on the side plates are significantly less than those on the top and bottom plates. The maximum contact pressure between the structure and the surrounding soil is distributed on the top plate of the hanging wall and the bottom plate of the footwall near the ground fissure. This study provides a theoretical basis for the optimal design of pipe gallery structures crossing ground fissures and has theoretical significance and application value.

Keywords: ground fissures; underground pipe gallery; bottom void area; elastic foundation beam; model test

1. Introduction

An underground pipe gallery is a modern, scientific, and intensive urban infrastructure formed by centrally setting up two or more pipelines, such as electricity, communication, and water supply, in the same underground space. It has been built on a large scale in many large cities [1,2]. As the core city in Northwest China, under the background of rapid economic and population growth, underground space development has become the best choice for the urban expansion of Xi'an [3]. In recent years, the planning and construction of urban underground pipe galleries in Xi'an has been in full swing, although the existence of ground fissures has become an essential obstacle in their construction [4–6]. As a type of geological disaster on a slow variation time scale, ground fissures have appeared in many countries and regions [7–10]. The stratum dislocation caused by ground fissure activity is similar to fault activity, which leads to damage to underground structures and threatens the operational safety of underground tunnels and pipelines. An urban underground pipe gallery is a typical linear structure that inevitably crosses ground fissures when it is built. A pipe gallery structure crossing ground fissures is affected by the dislocation of ground fissures for a long time during its operation [6,8].

At present, significant progress has been made in the research on the mechanical behavior of underground structures under fault dislocation at home and abroad, which can provide a reference for the study of the bottom void phenomenon of underground pipe gallery structures under the action of ground fissures [5,11–19]. Huang et al. [5] and Liu et al. [20] studied the denaturing characteristics, structural stress, earth-pressure variation law, and failure mode of water supply pipelines under the action of ground fissures through full-scale tests. Many studies have been conducted on the mechanism of the interaction between subway tunnels and ground fissures at different angles and the failure mode of tunnel structures [21–31]. Yan et al. [30,31] and Deng et al. [21,22] used finite element software packages, model tests, and theoretical analyses to analyze the mechanical properties, influence range of ground fissures, and failure mode of an underground pipe gallery in Xi'an, China. The study factors are different dislocations and angles between the comprehensive pipe gallery and the ground fissure. According to the characteristics of ground fissures in Xi'an, China, some prevention measures for underground structures have been proposed [32–34]. Maqsood, Z. [35], by loading GMS materials using different strain rates, found that the strength as well as the deformation of the material is affected by the loading rate. Li [36,37], Wang [34], and others took the research and development of equipment and materials for underground engineering model tests as a breakthrough and carried out research work on the loading and unloading devices for model tests and new similar materials based on geomechanical models and the coupling of flow–solid in underground engineering. The application of these new means and new materials has led to a large number of important results in reducing boundary effects and improving test accuracy in underground engineering model tests.

In our previous study on the stress deformation of a pipe gallery under the action of ground fissure dislocations [22], it was found that near the ground fissure, a stress reduction area appears at the bottom of the hanging wall of the pipe gallery structure along with the bottom void phenomenon. When the bottom of the pipe gallery is void, the foundation of the voiding part loses its support to the structure, changing the stress state of the structure. Several studies have been conducted on the phenomenon of the bottom void of the underground structure when it passes through ground fissures. Li et al. [38] conducted a model test of the Xi'an metro tunnel crossing a ground fissure zone and considered that with an increase in dislocation, the expansion process of the void area at the bottom of the pipe gallery can be divided into three stages: the simultaneous deformation stage, the critical void stage, and the void development stage. Pang et al. [39] studied the voiding distance and influence range of a subway tunnel under the action of ground crack dislocations. Hu et al. [40] and Zhang et al. [41] studied the deformation failure mode of a pipe gallery structure under oblique conditions through model tests. They concluded that the stress on the pipe gallery structure was primarily due to the constrained torsion of the thin-walled members. The reason for the torsion of the pipe gallery structure was that the resultant force of the shear flow in the void area did not pass through the shear center of the structural section. Mei et al. [42] studied the stress deformation and bottom voiding phenomenon under the intersection of pipe gallery structures and ground fissures at different angles through a numerical simulation and proposed corresponding early warning indicators and control measures. Yan et al. [30] studied the structure at an angle of 30° with the ground fissure and concluded that the angle between the ground fissure and the pipe gallery structure has a significant influence on the void area formation at the bottom of the pipe gallery structure.

Several studies have been conducted on the bottom void phenomenon of structures under the action of ground fissure dislocations; however, very few researchers have performed relevant research on the prediction of the formation range of the void area at the bottom of the pipe gallery structure. The determination of the void area directly affects the determination of the structural stress state and damage range. Therefore, this study considers an underground pipe gallery crossing a ground fissure in Xi'an, China, as the research background, using a method combining theoretical calculations and model tests, and attempts to develop a calculation method for the void area range. Thus, the void area range is obtained, which can provide a theoretical basis for the design optimization of pipe gallery structures crossing ground fissures.

2. Bottom Void Mechanism of Pipe Gallery

2.1. Mechanism Analysis of Structural De-Voiding in Pipeline Corridors in Region of Ground Cracks

During ground crack action, the upper disk descends and the lower disk is fixed relative to the upper disk. During the descent of the upper disk, there is movement in the soil body of the upper disk in two directions, vertically and horizontally, due to the existence of an 80° inclination angle in the ground crack. In the moving process in the ground crack, the upper disk of the soil body (the top of corridor A and the bottom of corridor B) and the corridor structure in the contact surface between the upper and lower disks of the soil body (the top of corridor C and the bottom of corridor D) produce friction in the overlaying load. Additionally, friction is generated under the joint action of the upper disk of the corridor and soil with the settling of deformation. In the process of settling, the soil body's settlement displacement is greater than that of the corridor. In the settlement process, the soil's settlement displacement is larger than the pipe corridor's bending deformation displacement, and in the upper coil pipe corridor structure near the ground cracks, dehollowing appears. In the whole process, the upper disk corridor structure is subjected to drag, and the lower disk corridor structure is always in contact with the soil and is subjected to the anchoring effect of the soil. Therefore, the upper disk of the corridor can be regarded as a loaded section, and the lower disk of the corridor can be regarded as an anchored section by the soil constraints (see Figure 1) (longitudinal section).

From the ground cracks under the action of the pipe corridor structure and soil interaction mechanism and the existing pipe corridor through the ground cracks, a model test can be seen. In the early stage of upper plate settlement, the pipe corridor's overlying load is uniformly distributed, and with the increase in the amount of the upper plate that is misaligned, the top of the pipe corridor's contact pressure changes, the performance of the upper plate improves, and that of the lower plate decreases. Thus, it is necessary to explain the following: only the earth pressure acting on the pipe corridor is changed, and the contact pressure at the bottom of the pipe corridor is opposite to that at the top. The deformation of the pipe corridor mainly shows that the bottom of the lower disk is compressed and the bottom of the upper disk is strained. The lower coil corridor is deformed greatly near the ground crack, and the lower coil corridor remains unchanged away from the ground crack. The upper coil corridor forms a local hollow at the bottom of the pipe corridor due to the friction caused by the soil settlement. A schematic diagram of the structural deformation force of the tube corridor is shown in Figure 2.

To summarize, the reason for the bottom hollowing of the underground integrated pipe gallery in the ground fracture area is that the pipe gallery structure and soil deformation are not coordinated and inconsistent when the upper wall of the ground fracture falls.

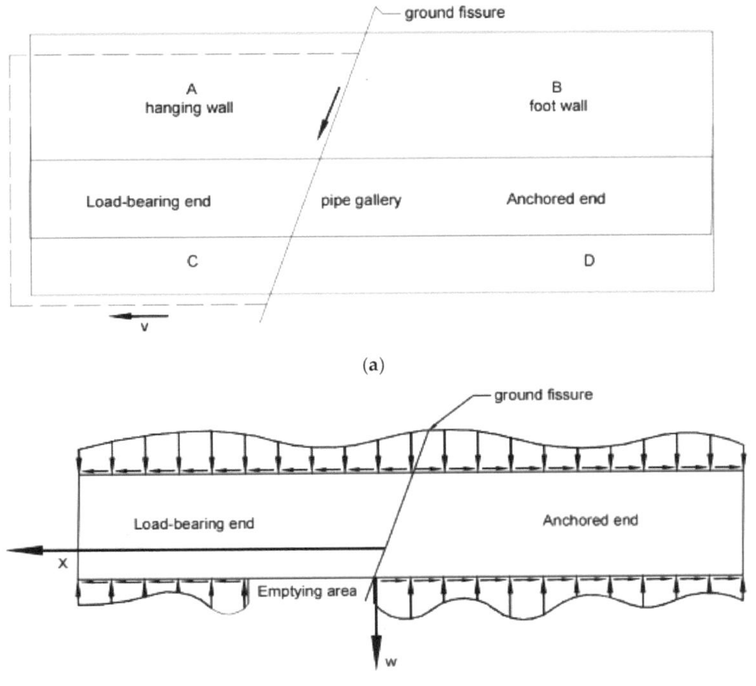

Figure 1. A schematic diagram of the pipe gallery–soil interaction under the action of ground cracks. (**a**) The effect of pipe corridor on soil; (**b**) The effect of soil on the pipe corridor.

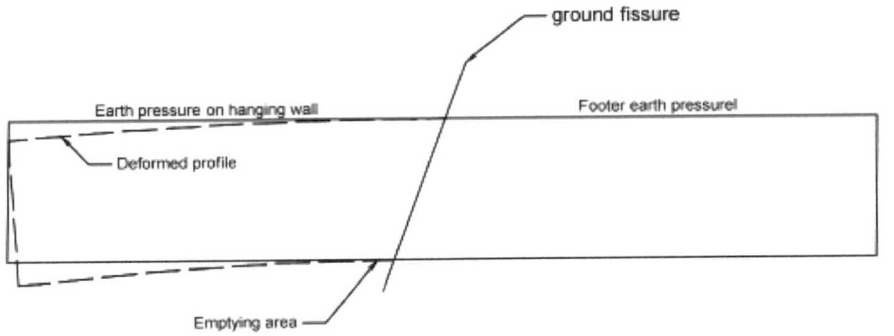

Figure 2. Longitudinal force diagram of pipe corridor structure.

2.2. The Process of Dehollowing at the Bottom of the Pipeline Corridor

Ground crack misalignment is a slow-change geological disaster which occurs when the bottom of the corridor at all stages of the dehollowing phenomenon is not the same. According to the degree of contact between the corridor structure and the soil, dehollowing can occur at the bottom of the structure, and the process is divided into three stages.

The first stage is the common deformation stage. In this stage, the amount of ground crack misalignment is small, the bottom of the pipe corridor structure is in close contact with the foundation, the bottom of the upper and lower coil pipe corridors are subjected to the same base pressure at the ground crack, and the displacement does not change with the dislocation of the ground crack. Dehollowing does not occur during this stage, as shown in Figure 3a.

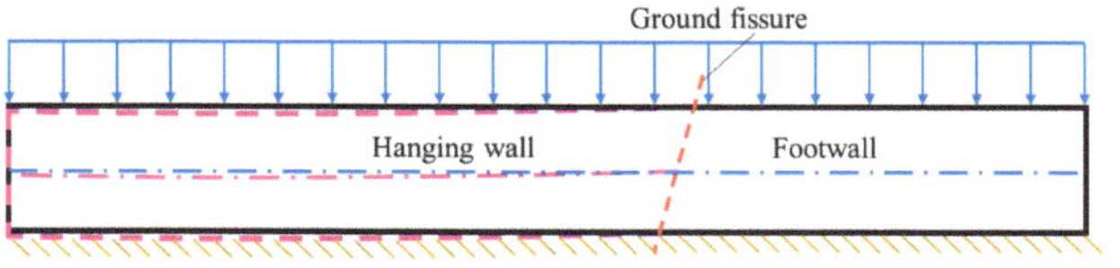

(a) Simultaneous deformation stage

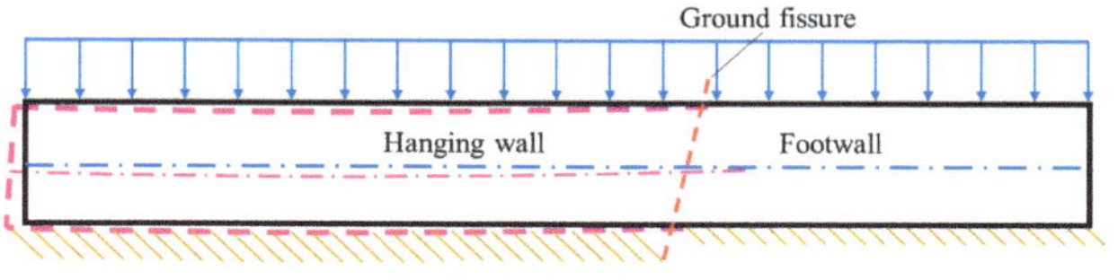

(b) Critical void stage

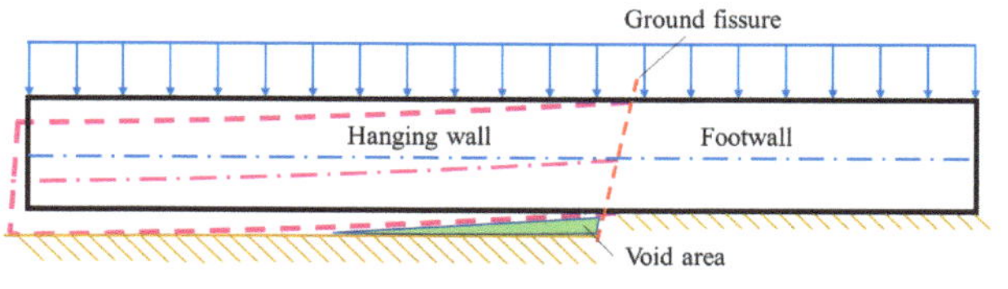

(c) Void development stage

Figure 3. Bottom void stage diagrams of a pipe gallery structure.

The second stage is the critical dehollowing stage. In this stage, the amount of ground crack misalignment increases, and the bottom of the upper and lower disks of the corridor are in contact with the ground cracks at the beginning, causing a difference in the base pressure. The upper disk depressurizes, and the soil under the pipe springs off. The bottom plate is pressurized, but its bottom still touches the ground. This stage lasts for a short time and is a prelude to the emergence of the dehollowing phenomenon, as shown in Figure 3b.

The third stage is the dehollowing development stage in which the amount of misalignment with the ground cracks continues to increase, the base pressure on the bottom of the upper coil corridor at the ground crack decreases to zero, and the detachment of the bottom of the corridor structure from the foundation begins to occur. A dehollowing region appears at this stage, and after this stage, the dehollowing region continues to increase, as shown in Figure 3c.

3. Theoretical Calculation of Void Area Range

Based on the mechanical characteristics and deformation mechanisms of the pipe gallery structure, a mechanical abstract model for calculating the void range at the bottom of the pipe gallery under the action of a ground fissure was established using the rigid bar method in the elastic foundation beam theory and selecting a semi-infinite elastic foundation and Euler–Bernoulli beam.

3.1. Establishing the Calculation Model

By referencing the calculation model of a tunnel crossing a ground fissure [43], the calculation model of the pipe gallery structure was established according to its mechanical characteristics. When the pipe gallery structure crosses the ground fissure, with the subsidence of the hanging wall at the ground fissure, the footwall of the pipe gallery structure is always in contact with the foundation, and the deformation of the pipe gallery near the ground fissure is the largest, whereas that far away from the ground fissure is close to zero. Therefore, the constraint on the footwall soil layer at the end of the pipe gallery far from the ground fissure can be assumed as a fixed end. When the hanging wall dislocation at the ground fissure reaches a certain value, the hanging wall of the pipe gallery is deformed along with the soil. Owing to the disharmony between the pipe gallery and the surrounding soil, a void area is formed between the bottom of the pipe gallery and the hanging wall. Therefore, the constraint on the soil layer at the end of the pipe gallery in contact with the hanging wall can be assumed as a directional support. To simplify the calculations, the variation in the contact pressure at the top of the pipe gallery is not considered, and it is calculated as the overburden load, q_0. The calculation model is shown in Figure 4.

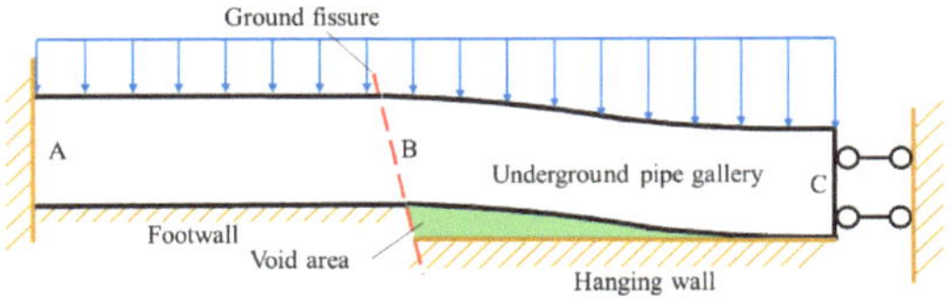

Figure 4. The calculation model of a pipe gallery structure under the action of a ground fissure.

The pipe gallery was simplified as a Euler–Bernoulli beam in the Boussinesq foundation, referred to as the Euler–Bernoulli–Boussinesq model. According to the principle of the rigid bar method, n rigid bars were used to connect parts of the pipe gallery to the foundation. Here, c denotes the distance between adjacent chain rods; a_k denotes the distance between the first bar and the kth bar; p_n denotes the interaction force between the pipe gallery and the foundation; Δ_1 denotes the end displacement of the hanging wall pipe corridor structure; and Δ_2 denotes the dislocation of the ground fissure. The calculation model is shown in Figure 5a. As the footwall of the pipe gallery is always in contact with the foundation, only the hanging wall of the pipe gallery is considered in the calculation of the bottom void range of the pipe corridor. Therefore, the calculation model can be simplified to the basic system of a cantilever beam structure, with point B as the fixed end, as shown in Figure 5b.

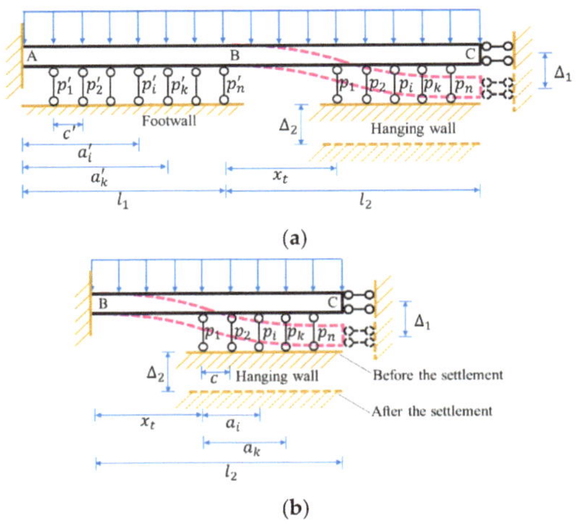

Figure 5. Calculation diagrams of the pipe gallery structure under the action of the ground fissure.

By combining this study's results with the existing literature on the relationship between the relative displacement of the underground tunnel structure and soil under the action of ground fissures [23,38–40,42], we can conclude that the vertical displacement of the hanging wall of the pipe gallery (Δ_1) and dislocation of the ground fissure (Δ_2) have the following relationship: ① With an increase in the dislocation of the ground fissure (Δ_2), the vertical displacement at the end of the hanging wall of the pipe gallery (Δ_1) exhibits an increasing trend. ② The bending stiffness of the pipe gallery structure has a tendency to hinder the increase in the end displacement of the pipe gallery (Δ_1). The hyperbolic tangent function has been used in the study of spatial effects in underground engineering [44]; therefore, the relationship between Δ_1 and Δ_2 can be fitted using a hyperbolic tangent function. A graph depicting the functional relationship between Δ_1 and Δ_2 is shown in Figure 6.

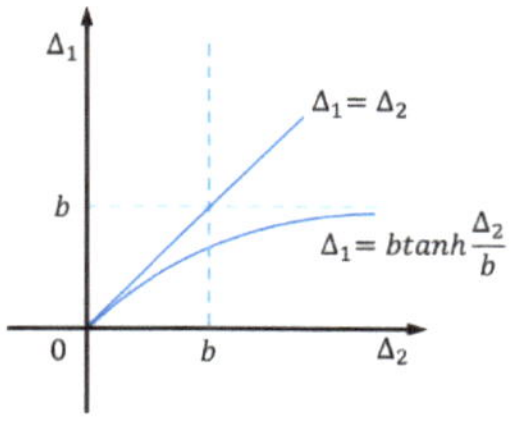

Figure 6. A graph of the functional relationship between Δ_1 and Δ_2.

The functional relationship is expressed as Equation (1).

$$\Delta_1 = b\tanh\left(\frac{\Delta_2}{b}\right), \quad (1)$$

where b denotes the maximum ground fissure dislocation; Δ_1 denotes the vertical displacement at the end of the hanging wall of the pipe gallery; and Δ_2 denotes the ground fissure dislocation.

3.2. Calculation Method for Pipe Gallery's Bottom Void Range

The calculation concept of a general elastic foundation beam can be used for the calculation and derivation of the pipe gallery's bottom void range. First, the finite statically indeterminate structure shown in Figure 5b was transformed into a cantilever beam and used as the basic system to determine the various forces on the system. Second, according to the contact condition of the part with a void at the bottom of the pipe gallery structure, n equations were obtained. This is because, according to this contact condition, the relative displacement between the pipe gallery structure and foundation (the relative displacement includes the settlement value of the foundation at the bar and the displacement of the pipe gallery caused by the load at the bar) is zero. The pipe gallery's bottom void range (x_t) was included in the equation of the displacement caused by the load. In the end, the n equations obtained using the relative displacement contact conditions and the equations obtained using the static equilibrium conditions were combined. Finally, combined with the boundary conditions, a nonlinear system of equations was obtained, and the pipe gallery's bottom void range (x_t) was obtained using MATLAB R2022a programming.

Before a complete void occurred at the bottom of the pipe gallery structure, the relative displacement of the contact part between the pipe gallery structure and foundation was zero; that is, the relative displacement of the pipe gallery structure and foundation at any bar was zero. As shown in Figure 5b, the relative displacement of the pipe gallery structure and foundation at any bar k (Δ_k) is caused by four external loads: Δ_{kpi} denotes the displacement caused by the foundation reaction force in the area where the hanging wall is not a void. Furthermore, Δ_{kq} denotes the displacement along the p_k direction caused by the overlying load at point k in the pipe gallery structure. Δ_{kMC} denotes the displacement along the p_k direction caused by the bending moment M_C at the end of the pipe gallery structure at the k-point of the pipe gallery structure. Finally, Δ_{k1} denotes the displacement at the k-point of the pipe gallery structure caused by Δ_1, the vertical displacement at the end of the pipe gallery structure. Therefore, Equation (2) can be expressed as

$$\Delta_k = \Delta_{kpi} + \Delta_{kM_C} + \Delta_{kq} + \Delta_{k1} = 0. \tag{2}$$

Subsequently, we solved for each term in Equation (2).

① Δ_{kpi} denotes the sum of the settlement value caused by p_{i} at the k-point of the foundation and the deflection caused by p_{i} at the k-point of the pipe gallery, where p_{i} is the interaction force between the pipe gallery and the foundation at any point i.

$$\Delta_{kpi} = \sum_{i=1}^{n} \delta_{ki} p_i, \tag{3}$$

where δ_{ki} is the sum of the settlement value, y_{ki}, generated by the foundation at the k-point and the deflection v_{ki} generated at the k-point of the pipe gallery's hanging wall structure when $p_{\mathrm{i}} = 1$, that is, $\delta_{ki} = y_{\mathrm{ki}} + v_{\mathrm{ki}}$.

According to the subsidence of a semi-infinite plane, the following can be obtained:

$$y_{ki} = \frac{2}{\pi E_0' c} F_{ki}, \tag{4}$$

where $E_0' = \frac{E_0}{1-\mu_0^2}$; E_0 is the elastic modulus of the foundation; and μ_0 is the Poisson coefficient of the foundation. Furthermore, $F_{ki} = c\ln c - c\ln(x-c) - x\ln x + x\ln(x-c)$. For ease of calculation, the lnc and ln(x − c) power series is expanded, and it is assumed that $0 < c \leq 2$, $0 < x - c \leq 2$ to obtain $F_{ki} = -2c(a_k - a_i)$.

$$v_{ki} = \frac{1}{2E_1 I}(a_i + x_t)^2\left(\frac{2}{3}x_t + a_k - \frac{1}{3}a_i\right), \tag{5}$$

where E_1 is the elastic modulus of the pipe gallery, and I is the moment of inertia of the pipe gallery section. The other parameters are shown in Figure 7.

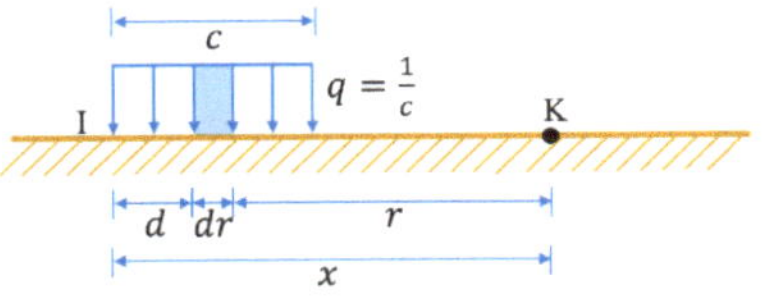

Figure 7. Settlement calculation model.

Solving Equations (4) and (5) yields the following:

$$\delta_{ki} = -\frac{4(k-i-1)}{n\pi E_0'}(l_2 - x_t) + \frac{1}{2E_1 I}(a_i + x_t)^2\left(\frac{2}{3}x_t + a_k - \frac{1}{3}a_i\right). \tag{6}$$

Finally, Equation (6) is input into Equation (3) to obtain Δ_{kpi}.

② Δ_{kMC} denotes the displacement caused by the bending moment M_C at point C to the pipe gallery structure at the k-point in the hanging wall non-void area, which can be calculated using the deflection integral formula of structural mechanics.

$$\Delta_{kMc} = \frac{(x_t + a_k)^2}{2E_1 I} M_C, \tag{7}$$

where $M_C = R\theta_c$; θ_c is the angular displacement between the end parts of the components; R is the joint stiffness of the node; and $R = E_1 I$.

③ Δ_{kq} denotes the displacement of the overlying load at the k-point of the pipe gallery structure in the hanging wall non-void area, which can be obtained using the integral formula of the structural mechanical deflection.

$$\Delta_{kq} = -\frac{q_0}{2E_1 I}\left[\frac{l_2^2}{2}(x_t + a_k)^2 - \frac{l_2}{3}(x_t + a_k)^3 + \frac{1}{12}(x_t + a_k)^4\right]. \tag{8}$$

④ The vertical displacement of point C at the end of the pipe gallery structure was small compared to the length of the hanging wall of the pipe gallery structure; therefore, Δ_{k1}, the vertical displacement of point C at the end of the hanging wall of the pipe gallery structure, caused the downward vertical displacement at the k-point, which can be solved directly by using the geometric method.

$$\Delta_{k1} = \frac{\Delta_1}{l_2}(x_t + a_k). \tag{9}$$

In conclusion, by inputting Equations (3), (7), (8), and (9) into Equation (2), the system of equations containing x_t and the bottom void range can be obtained.

Because the structure of the hanging wall of the pipe gallery satisfies the static equilibrium condition, the following equation can be obtained by considering the moment at point B:

$$x_t p_1 + (x_t + a_2)p_2 + \cdots + (x_t + a_k)p_k + \cdots + (x_t + a_n)p_n - M_q + M_C = M_B = \theta_B E_1 I, \tag{10}$$

where M_q is the bending moment generated by the overlying load at point A at the end of the hanging wall of the pipe gallery structure; $M_q = (q_0/2)l_2^2$; and θ_B is the angular displacement at point B.

In addition, the end of the hanging wall of the pipe gallery structure satisfies the following end-constraint condition:

$$\frac{q_0 l_2^3}{6EI} + \theta_B = \theta_C, \tag{11}$$

$$\frac{q_0 l_2^4}{EI} + \theta_B l_2 = \Delta_1 \leq \Delta_1^u, \tag{12}$$

where Δ_1^u is the maximum end vertical displacement.

To calculate the bottom void range, x_t, of the pipe gallery, first, the different dislocations of the hanging wall (Δ_2) were substituted into Equation (1) to obtain the end displacement (Δ_1) of the hanging wall of the pipe gallery. Second, by inputting Δ_1 into Equation (12), the angle θ_B at the fixed end of the hanging wall of the pipe gallery was obtained. Subsequently, θ_B was input into Equation (11) to obtain the angle θ_C at the end of the hanging wall of the pipe gallery. Finally, a system of equations was constructed using Equations (2) and (10). By solving the system of equations using MATLAB, the x_t value and the bottom void range of the pipe gallery structure were obtained.

4. Model Test for Ground Fissure Dislocation

Using a self-designed device, model tests of orthogonal (90°) and oblique (45°) ground fissure dislocations between the pipe gallery and the ground fissure were conducted. Furthermore, the change law of the surface strain, bottom displacement, and contact pressure between the pipe gallery structure and its soil was obtained, and the bottom void range of the pipe gallery structure calculated using the theory was compared.

4.1. Test Background

The underground pipe gallery structure crossing the ground fissure in Xi'an is considered as the research object. According to the engineering geological survey report of the section, the main soil layers within the buried depth of the pipe gallery are prime fill soil, loess, and paleosol, as shown in Figure 8. Based on the stratigraphic mechanical parameters of other pipe galleries in Xi'an, the loess layer was selected as the representative prototype soil in the model tests, and its mechanical parameters are as follows: c = 33.0 kPa, $\varphi = 19°$, E_0 = 8 MPa, and γ = 18.2 kN/m^3. The prototype pipe gallery was constructed using open-cut cast-in-place construction. The design strength of the pipe gallery is C40; the compressive strength is 26.8 MPa; the tensile strength is 2.39 MPa; and the elastic modulus is 3.45×10^4 MPa. The main aim of this model test was to determine the influence of the ground fissure on the pipe gallery structure in the loess region, focusing on the bottom void range of the pipe gallery structure. After comprehensive consideration, the pipe gallery structure was simplified as a single cabin pipe gallery; its size was determined to be 4.0 m × 4.0 m, and the wall thickness was 0.56 m. The dip angle of the ground fissure was 80°, and the vertical subsidence of the Xi'an ground fissure was predicted to be 0.8 m in one hundred years.

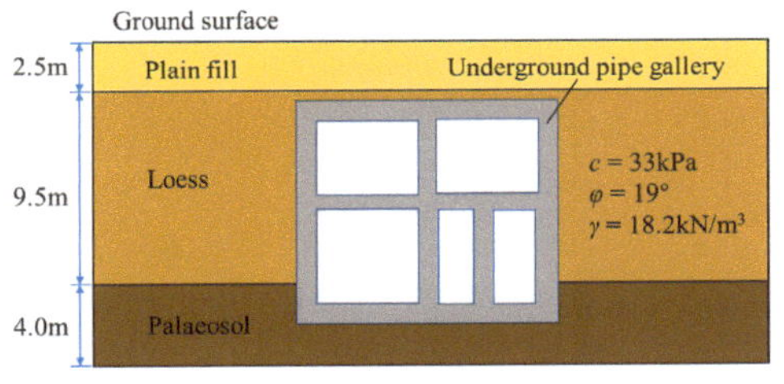

Figure 8. Stratigraphic profile of pipe gallery structure.

The model box used in this test is a cuboid with length, width, and height values of 2 m, 1.6 m, and 1.2 m, respectively. At one end of the box, there is a round hole with a diameter of 30 cm for observing the interior of the pipe gallery, and the other end is made of a removable steel plate spliced together. The front of the box is made of tempered glass, and the top of the model box is open for easy filling of the model soil, as shown in Figure 9a,b. The bottom plate of the model box comprises four plates, A, B, C, and D, which are spliced using latches. Plates A and B are spliced to perform orthogonal (90°) model tests, and plates A and C are spliced to perform oblique (45°) model tests, as shown in Figure 9c,d. A schematic of the model test device is shown in Figure 9e.

4.2. Similarity Ratio of Model Test

To simulate the stress of a comprehensive pipe gallery under the action of ground fissure displacement, similarity ratios between the main physical parameters of the pipe gallery and soil should be consistent to the greatest extent. Considering the size and bearing capacity of the model box, the similarity ratio of the model test was determined as 1:20. The model pipe gallery was prefabricated with deformable acrylic material; its elastic modulus was 2.7×10^3 MPa; and the Poisson's ratio was 0.372 at a test temperature of 14 °C. To make the pipe gallery model convenient and reflect the actual size of the prototype pipe gallery, based on the geometric similarity ratio C_l = 20, a model pipe gallery with a length of 1.9 m, a cross-sectional size of 0.20 m × 0.20 m, and a wall thickness of 6 mm was fabricated. According to the pipe gallery design report, the pipe gallery structure was cast in situ using C40 reinforced concrete, and its elastic modulus E was 32.5 GPa; therefore, the elastic modulus similarity ratio C_E was 1:12.

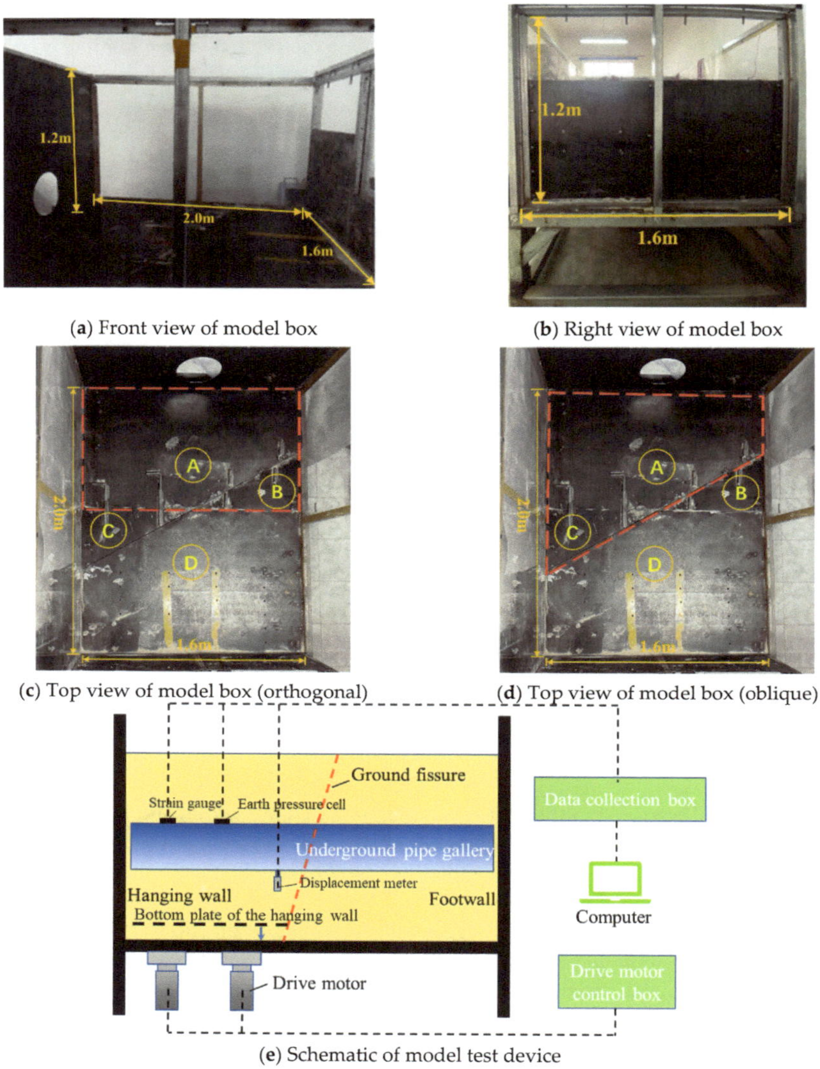

Figure 9. Model test device and schematics.

According to the similarity theorem, the similarity ratios between the model structure and the model soil were designed and are listed in Table 1.

Table 1. The main similarity ratios of the model test.

Type	Physical Parameters	Similar Relationships	Similarity Ratios (Prototype/Model)	
			Pipe Gallery Structure	**Soil**
Geometric features	Length	C_l	20	—
	Moment of inertia	$C_I = C_l^4$	160,000	—
	Displacement	$C_u = C_l$	20	20

Table 1. *Cont.*

Type	Physical Parameters	Similar Relationships	Similarity Ratios (Prototype/Model)	
			Pipe Gallery Structure	Soil
Material characteristics	Elasticity modulus	C_E	12	15
	Stress	$C_\sigma = C_E$	12	—
	Strain	C_ε	1	—
	Bulk density	C_γ	—	1
	Moisture content	C_ω	—	1
	Cohesion	$C_c = C_l$	—	20
	Internal friction angle	C_φ	—	1

4.3. Sensor Layout and Loading Schemes

The model tests must measure the contact pressure, bottom displacement, and surface strain of the pipe gallery under two working conditions. The layout of the sensor used during the tests is shown in Figure 10.

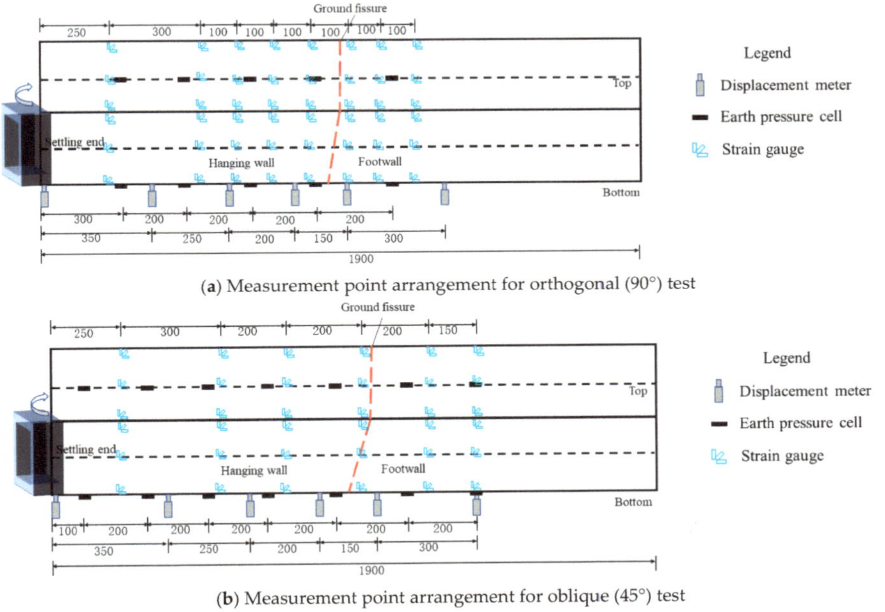

(**a**) Measurement point arrangement for orthogonal (90°) test

(**b**) Measurement point arrangement for oblique (45°) test

Figure 10. Schematic of measurement point arrangements.

The variation trends of the longitudinal stress and strain of the pipe gallery structure with an increase in displacement was monitored by arranging a row of strain gauges on the four sides of the structure. Owing to the symmetry of the orthogonal model test, the strain gauges were arranged only on one face of the side plate of the pipe gallery. Six displacement meters were arranged longitudinally along the central axis of the pipe gallery bottom to measure the hollowing out of the pipe gallery bottom. Micro-earth pressure boxes were laid along the top and bottom axes of the pipe gallery to measure the change in soil pressure during soil settlement.

Based on the maximum predicted number of ground fissures in Xi'an over 100 years, the expected design value of ground fissures in Xi'an within 100 years was set at 800 mm. The activity mode of ground fissures is creep. To simulate the ground fissure settlement reasonably, a graded settlement was adopted. Based on the geometric similarity ratio of 20, the dislocation of the ground fissures in the model test was determined to be 40 mm. In the test, the hanging wall plate was controlled to settle in incremental steps in a sequence of 5 mm, 10 mm, 15 mm, 20 mm, 25 mm, 30 mm, 35 mm, and 40 mm, with a settling speed of 5 mm/d. After each settlement, we allowed it to stand for 24 h, recorded the data every 8 h until the data no longer changed, and then started the next settlement process.

4.4. Test Phenomenon

Figure 11 shows the macroscopic phenomenon of the model soil after the test. Prior to the test, the soil surface in the model box was flat. With the subsidence process of the hanging wall and floor of the model box, the hanging wall soil sank along the ground fissure, resulting in a height difference with the footwall soil. After soil excavation, it was found that owing to the disharmony between the pipe gallery structure and soil deformation, bottom voids appeared under both orthogonal and oblique working conditions. The bottom void ranges of the pipe gallery structure were determined to be 0.6 m and 0.7 m under the orthogonal and oblique conditions, respectively. The maximum void displacement occurred near the ground fissure, and the footwall of the pipe gallery remained in contact with the foundation, as shown in Figure 12.

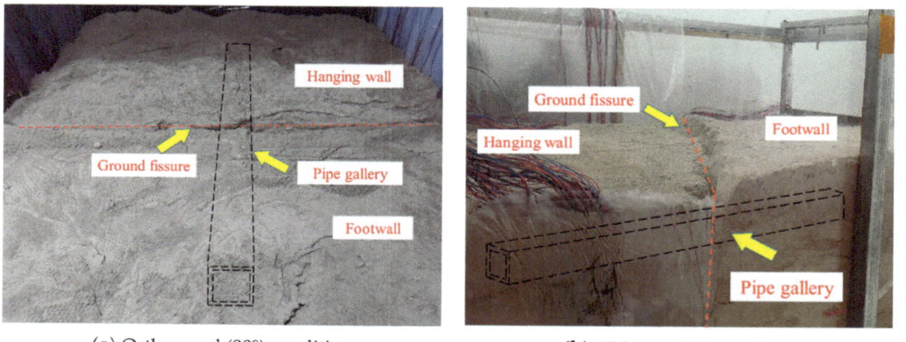

(**a**) Orthogonal (90°) condition (**b**) Oblique (45°) condition

Figure 11. Ground settlement of model test.

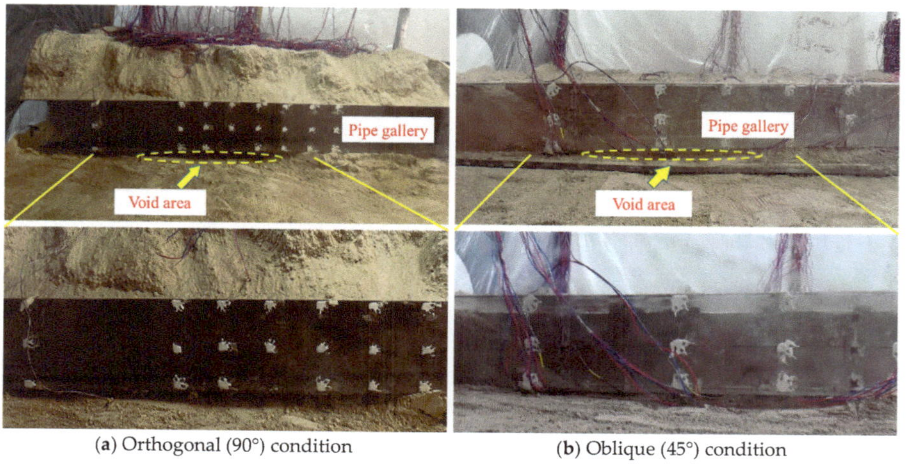

(**a**) Orthogonal (90°) condition (**b**) Oblique (45°) condition

Figure 12. Bottom void phenomenon of pipe gallery.

5. Analysis of Test Results

5.1. Structural Strain of Pipe Gallery

(1) Longitudinal strain analysis

Figures 13 and 14 show the longitudinal strain curves of the top and bottom plates of the pipe gallery structure under different ground fissure displacements. As observed from the data shown in the figure, under the two conditions, the top plate of the pipe gallery was under tension. When the dislocation moment of the hanging wall was between 5 mm and 25 mm (that of the actual prototype was between 10 cm and 50 cm), the strain of the pipe gallery considerably increased. When the dislocation moment of the hanging wall exceeded 25 mm (that of the actual prototype was 50 cm), the strain change in the pipe gallery tended to be stable. The bottom plate of the pipe gallery was under pressure in the footwall area, whereas it was under pressure near the ground fissure in the hanging wall area and under tension far from the ground fissure. The relationship between the strains of the bottom plate and settlement was the same as that of the roof of the pipe gallery. Under the orthogonal condition, the strain of the pipe gallery increased most obviously at a hanging wall position that was 0.175 m (0.875 L, where L is the length of the bottom edge of the pipe gallery) away from the ground fissure. However, under the oblique condition, the obvious strain changes were mostly distributed near the ground fissure and at a footwall position that was 0.375 m (1.875 L) away from the ground fissure.

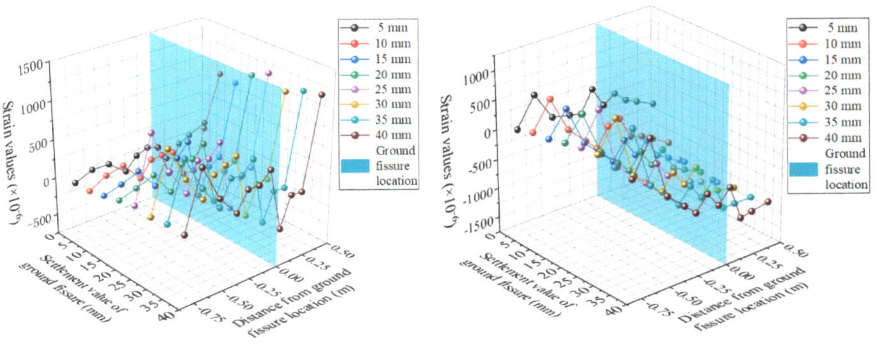

(**a**) Central axis strain of top plate (**b**) Central axis strain of bottom plate

Figure 13. Longitudinal strain curves under the orthogonal (90°) condition.

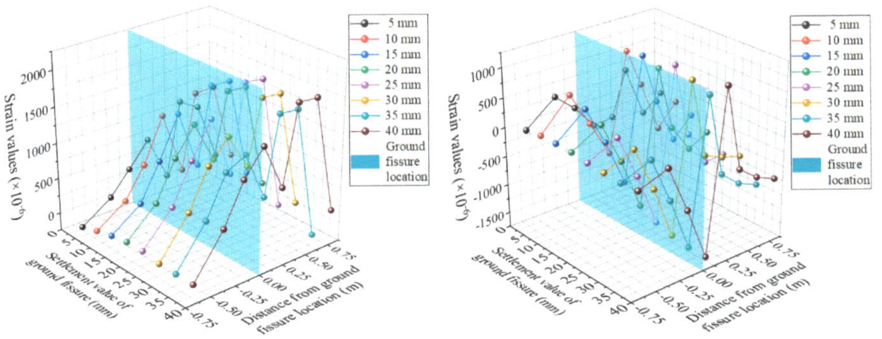

(**a**) Central axis strain of top plate (**b**) Central axis strain of bottom plate

Figure 14. Longitudinal strain curves under the oblique (45°) condition.

Owing to the dislocation of the hanging wall, the stress on the side plate of the pipe gallery became complicated; the footwall of the side plate was under compression, and the hanging wall was under both tension and compression. Because the stresses on the left- and right-side plates were symmetrical under the orthogonal condition, the strain on the side plate under the oblique condition was mainly analyzed here. The longitudinal strain curves along the central axis of the left- and right-side plates of the pipe gallery under the oblique condition are shown in Figure 15. As observed from the figure, the asymmetry of the left- and right-side plates of the pipe gallery is obvious, which indicates that both longitudinal bending and torsional deformation exist in the pipe gallery structure under the oblique condition. By comparing the strains on the top and bottom plates of the pipe gallery structure, it can be observed that the strains on the left- and right-side plates were significantly smaller than those on the top and bottom plates. This indicates that the bottom void of the pipe gallery structure had less influence on the left- and right-side plates than on the top and bottom plates. Therefore, the strengths of the top and bottom plates of the pipe gallery structure should be improved in practical engineering applications.

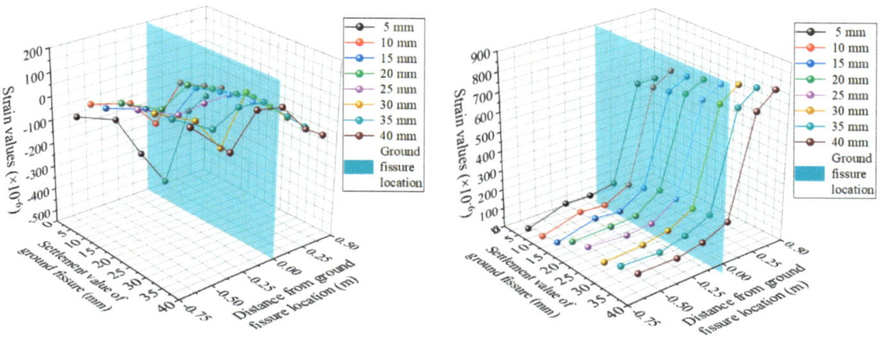

(**a**) Central axis strain of left plate (**b**) Central axis strain of right plate

Figure 15. Strain variation curves of the side plates under the oblique (45°) condition.

(2) Transverse strain analysis

Figures 16 and 17 show the circumferential cross-sectional strain values of the pipe gallery at different positions from the ground fissure when the dislocation of the hanging wall reaches 40 mm. Because the stresses on the left- and right-side plates are symmetrical under the orthogonal condition, only the strain on the left side of the cross section was measured, and the strain on the right side was obtained according to the symmetry of the left side. From all the transverse strains of the pipe gallery structure, under the orthogonal condition, the top plate of the pipe gallery structure was under tension, the hanging wall of the bottom plate was under tension, and some footwall areas were under compression. The strain on the hanging wall was smaller than that on the footwall. Under the oblique condition, the top plate of the pipe gallery was under tension, the footwall was under compression, the bottom plate was under tension, and some hanging wall areas were under compression. Under ground fissure dislocation, the tensile and compressive properties of the side plates of the pipe gallery structure changed significantly; the hanging wall structure was mainly under tension, and the footwall was under compression. Under the oblique condition, the left- and right-side plates of the pipe gallery structure were subjected to torsion, which caused the tensile and compressive properties of the side plates of the pipe gallery structure to show obvious differences in the same section. The stress on the section of the structure was asymmetric, which indicates that the structure was subjected to torsional shear force the under oblique condition.

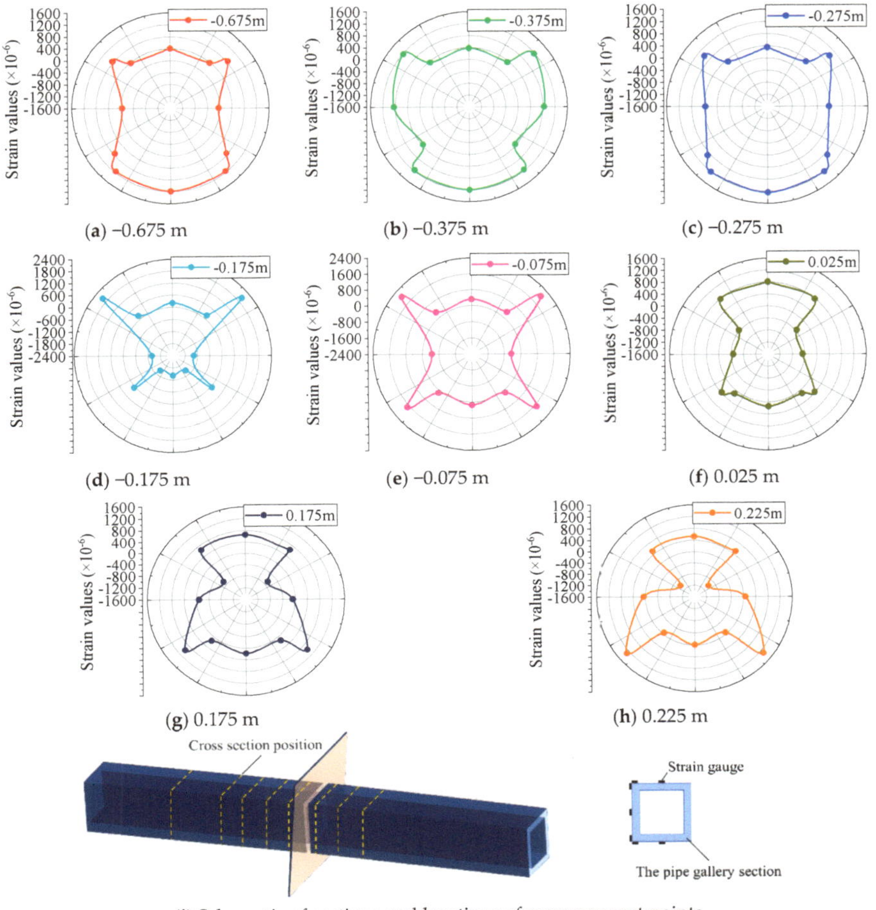

(**a**) −0.675 m (**b**) −0.375 m (**c**) −0.275 m (**d**) −0.175 m (**e**) −0.075 m (**f**) 0.025 m (**g**) 0.175 m (**h**) 0.225 m (**i**) Schematic of sections and locations of measurement points

Figure 16. Transverse strain distributions under the orthogonal (90°) condition.

5.2. Displacement Analysis of Pipe Gallery's Bottom

Figure 18a,b show the relative displacement diagrams between the bottom of the pipe gallery structure and the foundation soil under the orthogonal and oblique conditions, respectively. As observed from the figure, when the hanging wall at the ground fissure started to sink, the pipe gallery structure exhibited a small positive displacement under both conditions. Furthermore, the displacement of the pipe gallery structure at the edge of the ground fissure suddenly changed, indicating that the pipe gallery structure was squeezed by ground fissure dislocation. With an increase in the hanging wall dislocation, the vertical relative displacement between the bottom of the pipe gallery structure and the base near the ground fissures in the hanging wall area gradually increased, which directly reflects the occurrence of the void phenomenon at the bottom of the pipe gallery structure. Simultaneously, the phenomenon of increasing vertical relative displacement gradually extended in the direction of the hanging wall, away from the ground fissures, indicating that the range of the bottom void gradually increased. When the dislocation reached 35 mm, the bottom void was completely developed under both conditions, and it ranged between −0.575 and 0 m (between −2.875 and 0 L). Moreover, the displacement between the bottom and base of the pipe gallery structure did not increase subsequently. The variation law of the relative displacement between the bottom and base of the pipe gallery structure under the oblique condition was the same as that under the orthogonal condition, but the relative displacement at the same position was larger than that under the orthogonal condition.

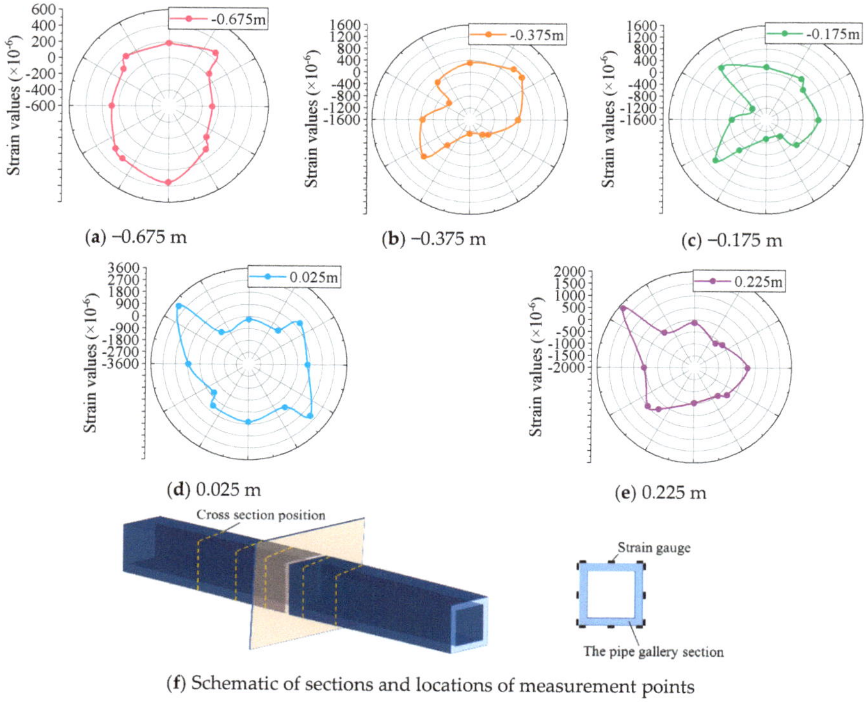

Figure 17. Transverse strain distributions under the oblique (45°) condition.

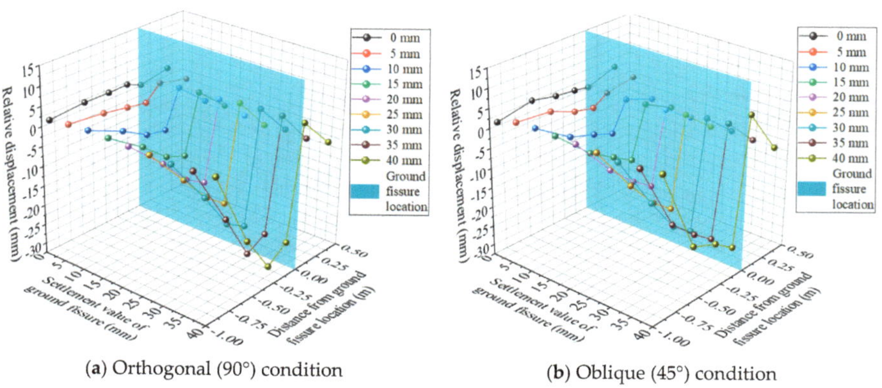

Figure 18. Relative displacement variations between the bottom of the pipe gallery and the foundation.

5.3. The Contact Pressure Between the Pipe Gallery Structure and the Surrounding Soil

Figures 19 and 20 show the variation curves of the longitudinal contact pressure between the bottom plate, top plate, and soil layer of the pipe gallery structure under the orthogonal and oblique conditions, respectively. As observed from the figure, with an increase in the hanging wall dislocation, the longitudinal contact pressure between the bottom plate and the hanging wall decreased, and that between the bottom plate and the footwall increased. In contrast, the longitudinal contact pressure between the top plate of the pipe gallery and the hanging wall increased, and that between the top plate and the footwall decreased. Under the orthogonal condition, when the hanging wall dislocation reached 15 mm, the contact pressure between the hanging wall and the pipe gallery's bottom plate decreased to zero in the range of −0.625 m to 0 m (−3.125–0 L). This indicates that the bottom plate of the hanging wall of the pipe gallery structure and the soil was voided; therefore, void formation under the oblique condition occurred earlier than that under the orthogonal condition. Under the load of the overlying soil, the hanging wall of the pipe gallery dragged the footwall of the pipe gallery, causing it to squeeze the soil layer near the ground fissures. With an increase in settlement, similar to the lever principle, the footwall of the pipe gallery structure was slightly upturned. Consequently, the contact pressure between the footwall and the pipe gallery increased near the ground fissure and decreased away from it. The maximum contact pressure between the structure and the soil was distributed on the top plate of the hanging wall of the pipe gallery structure and the bottom plate of the footwall of the pipe gallery structure near the junction between the ground fissure and the structure.

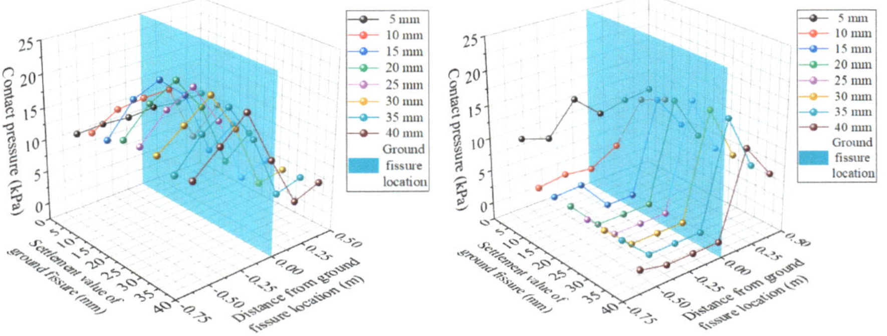

(a) Top plate (b) Bottom plate

Figure 19. Contact pressure curves in the orthogonal (90°) condition.

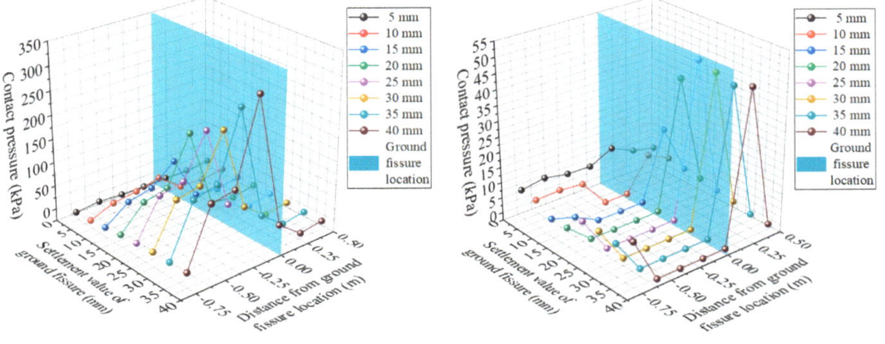

(a) Top plate (b) Bottom plate

Figure 20. Contact pressure curves under the oblique (45°) condition.

5.4. A Discussion of the Void Area

The test results of the three stages of the bottom void formation of the pipe gallery structure were analyzed, and the following observations were made: ① When the dislocation of the hanging wall was in the range of 0–5 mm, after the interaction between the pipe gallery and the foundation, a slight void occurred near the ground fissure, but the void range was very small. The pipe gallery structure was considered to have undergone a simultaneous deformation and critical void stage. ② When the hanging wall dislocation exceeded 5–10 mm, with an increase in the hanging wall dislocation, the void range at the bottom of the pipe gallery structure increased. When the settlement reached 40 mm, the pipe gallery's bottom void range obtained from the model test under the orthogonal condition was in the range of 0.575–0.60 m (2.875–3 L), and that under the oblique condition was in the range of 0.625–0.70 m (3.125–3.5 L). Therefore, in the void development stage of the pipe gallery structure, after comparison, the pipe gallery's bottom void range under the oblique condition in the model test was larger than that under the orthogonal condition in the model test.

The bottom void is an important factor affecting the strain change in the pipe gallery structure. When the hanging wall dislocation was 5 mm, the pipe gallery structure under both orthogonal and oblique conditions was mainly affected by the tensile failure of the top plate. With an increase in the hanging wall dislocation, the bottom plate of the pipe gallery structure was mainly affected by the compressive failure of the footwall. The above results are summarized in Table 2.

Table 2. Bottom void ranges of pipe gallery structure from model tests.

Hanging Wall Dislocations	Void Stages	Bottom Void Range of Pipe Gallery	Damage Range of Pipe Gallery
5 mm	Simultaneous deformation and critical void stages	There is only a slight void between the pipe gallery and the foundation near the ground fissure.	Tensile failure of the top plate occurs.
10 mm 15 mm 20 mm 25 mm 30 mm 35 mm 40 mm	Void development stage	With an increasing hanging wall dislocation, the bottom void of the pipe gallery structure gradually increased. At the end of the test, the bottom void ranges were 2.875–3 L and 3.125–3.5 L under the in orthogonal and oblique conditions, respectively.	In the orthogonal test, fracturing failure occurred when the hanging wall dislocation was 10 mm. In the test under oblique conditions, compressive failure occurred when the hanging wall dislocation was 15 mm, and then the range gradually expanded, mainly in the footwall's bottom plate.

According to the model test results, the bottom hollowed-out corridor structure mainly influences the roof and bottom plate of the pipe corridor structure, where the roof can easily be strained and damaged, and the bottom plate can easily be fractured. Aiming at these specific areas in which it is easy to produce tension and compression failure, the strength of the steel bar and concrete can be locally strengthened within the 6D range of the upper and lower coil corridors. It can be seen from the model experiment that the roof of the pipe corridor is prone to tensile stress and the bottom floor is prone to compressive stress. In view of the stress concentration caused by the cavitation area, the stress concentration of the structure can be eliminated by setting deformation joints at reasonable positions of the structure. The range of setting deformation joints is consistent with that of locally strengthening the pipe corridor structure.

5.5. Verification of Calculation Method

To verify the rationality of the calculation method for the bottom void range of the pipe gallery structure, the calculation results were compared with the pipe gallery model test results. The basic parameters of the model are as follows: the length of the pipe gallery model is 1.9 m, the cross section is 0.20 m × 0.20 m, and the wall thickness is 0.006 m. The length of the hanging wall of the pipe gallery is 0.925 m, the elastic modulus of the model material is 2.47×10^3 MPa, and the moment of inertia is 1.53×10^{-5} m^4. The stratum was made using similar materials with a weight of 18.2 kN/m^3, an elastic modulus of 8 MPa, a Poisson's ratio of 0.25, a maximum settlement of ground fissures of 0.04 m, and an overlying soil depth of 1 m. When calculating, the number of rigid bars was set to 30. The theoretical calculation values of the bottom void range of the pipe gallery were obtained using MATLAB, as listed in Table 3.

Table 3. Bottom void ranges of the pipe gallery structure from theoretical calculations.

Hanging Wall Dislocation Δ_2	Vertical Displacement of Pipe Gallery End Δ_1 (m)	θ_B	θ_C	Calculated Value (m)
0.005	0.0049	0.3757	0.3122	0.1230
0.010	0.0097	0.3705	0.307	0.2328
0.015	0.0143	0.3656	0.3021	0.0811
0.020	0.0184	0.3611	0.2976	0.1536
0.025	0.0221	0.3571	0.2936	0.3876
0.030	0.0254	0.3536	0.2901	0.4591
0.035	0.0281	0.3506	0.2871	0.6737
0.040	0.0304	0.3481	0.2846	0.6561

The calculated values are compared to the test values, as listed in Table 4.

Table 4. A comparison of the calculated values and test values.

Method	Orthogonal Bottom Void Range	Oblique Bottom Void Range
Theoretical calculations	0.6561	0.6561
Model tests	0.6	0.7
$\frac{\text{Calculated value} - \text{Test value}}{\text{Test value}} \times 100\%$	9.35%	−4.39%

From the comparative analysis results, it can be seen that the calculation results of the bottom void range of the pipe gallery are consistent with the model test results, and the maximum error does not exceed 9.35%.

6. Conclusions

In this study, based on the phenomenon of voids at the bottom of a pipe gallery under the action of ground fissure dislocations, a mechanical model was established, and model tests under two working conditions were conducted. The following conclusions were drawn:

1. A calculation method for the void range at the bottom of a pipe gallery structure under the action of ground fissures was established, and the rationality of the method was verified through model tests.

2. When the ground fissure dislocation reached the predicted settlement of one hundred years, the void range at the bottom of the pipe gallery structure was 0.575–0.6 m (2.875–3 times the length of the bottom edge of the pipe gallery section) under the orthogonal condition and 0.575–0.7 m (3.125–3.5 times) under the oblique condition.

3. Based on the model test data, the following observations are made:

① During the process of ground fissure dislocation, the maximum tensile stress on the pipe gallery appears in the upper part of the structure, and the maximum compressive stress appears in the lower part of the structure.

② Longitudinal bending and torsion of the pipe gallery structure occur simultaneously under oblique conditions.

③ The maximum point of contact pressure between the structure and the surrounding soil is distributed at the top of the hanging wall and the bottom of the footwall near the junction of the ground fissure and the structure.

This study provides a theoretical basis for the design optimization of pipe gallery structures in ground fissure areas.

Author Contributions: Methodology, B.H. and H.L.; Validation, B.H.; Formal analysis, B.H., B.D. and H.L.; Investigation, B.D.; Resources, K.D.; Data curation, Y.Z. and K.D.; Writing—original draft, X.Y.; Writing—review & editing, X.Y.; Supervision, Y.Z.; Funding acquisition, B.D. All authors have read and agreed to the published version of the manuscript.

Funding: This work is funded by the National Natural Science Foundation of China (No. 42177155), the Open Fund of State Key Laboratory of Green Building in Western China (LSKF202323), and the Shanxi Provincial Key R&D Program (2023-YBSF-506).

Data Availability Statement: The original contributions presented in the study are included in the article, further inquiries can be directed to the corresponding author.

Conflicts of Interest: Authors Xintao Yu and Baoan Han were employed by the company China Construction Silk Road Construction Investment Co., Ltd. Author Yubo Zhao was employed by the company China Construction No. 6 Bureau Construction Engineering Co., Ltd. The remaining authors declare that the research was conducted in the absence of any commercial or financial relationships that could be construed as a potential conflict of interest.

References

1. Canto-Perello, J.; Curiel-Esparza, J.; Calvo, V. Criticality and threat analysis on utility tunnels for planning security policies of utilities in urban underground Space. *Expert Syst. Appl.* **2013**, *40*, 4707–4714. [CrossRef]
2. Canto-Perello, J.; Curiel-Esparza, J.; Calvo, V. Strategic Decision Support System for Utility Tunnel's Planning Database A 'WOT Method. *Tunn. Undergr. Space Technol.* **2016**, *55*, 146–152. [CrossRef]
3. Qian, Q. Construction of urban underground comprehensive pipe corridor and transformation of urban development mode. *Tunn. Constr.* **2017**, *37*, 647–654.
4. Canto-perello, J.; Curielesparza, J. Assessing governance issues of urban utility tunnels. *Tunn. Undergr. Space Technol.* **2013**, *33*, 82–87. [CrossRef]
5. Huang, Q.; Liang, A.; Men, Y.; Liu, H.; Ma, Y. A full-scale model test on the influence of ground fracture activity on underground water transmission pipeline. *Chin. J. Rock Mech. Eng.* **2016**, *35*, 2968–2977. [CrossRef]
6. Peng, J. *Ground Fissure Disaster in Xi'an*; Science Press: Beijing, China, 2012.
7. Peng, J.; Qiao, J.; Leng, Y.; Wang, F.; Xue, S. Distribution and mechanism of the ground fissures in Wei river basin, the origin of the silk road. *Environ. Earth Sci.* **2016**, *75*, 718. [CrossRef]
8. Peng, J.; Sun, X.; Wang, W.; Sun, G. Characteristics of land subsidence, Earth Fissures and Related Disaster Chain Effects with Respect to Urban Hazards in Xi'an, China. *Environ. Earth Sci.* **2016**, *75*, 1190. [CrossRef]
9. Wang, G.; You, G.; Zhu, J.; Yu, J.; Li, W. Earth Fissures in Su–Xi–Chang Region, Jiangsu, China. *Surv. Geophys.* **2016**, *37*, 1095–1116. [CrossRef]
10. Youssef, A.M.; Sabtan, A.A.; Maerz, N.H.; Zabramawi, Y.A. Ground fissure in Wadi Najran. Kingdom of Saudi Arabia. *Nat. Hazards* **2014**, *71*, 2013–2027. [CrossRef]
11. Ha, D.; Abdoun, T.H.; O'Rourke, M.J.; Symans, M.D.; O'Rourke, T.D.; Palmer, M.C.; Stewart, H.E. Centrifuge Modeling of Earthquake Effects on Buried High-Density Polyethylene (HDPE) Pipelines Crossing Fault Zones. *J. Geotech. Geoenviron. Eng.* **2008**, *134*, 1501–1515. [CrossRef]
12. Michael, O.; Vikram, G.; Tarek, A. Centrifuge modeling of PGD response of buried pipe. *Earthq. Eng. Eng. Vib.* **2005**, *4*, 74–78.

13. Polynikis, V.; Spyros, A.K.; Panos, D. Finite element analysis of buried steel pipelines under strike-slip fault displacements. *Soil Dyn. Earthq. Eng.* **2010**, *30*, 1361–1376. [CrossRef]
14. Prentice, C.S.; Ponti, D.J. Coseismic deformation of the Wrights tunnel during the 1906 San Francisco earthquake: A key to understanding 1906 fault slip and 1989 surface ruptures in the southern Santa Cruz Mountains. *J. Geophys. Res.* **1997**, *102*, 635–648. [CrossRef]
15. Rojhani, M.; Moradi, M. Centrifuge Modeling of Buried Pipelines Response Due to Normal Faulting. In Proceedings of the 14th Asian Regional Conference on Soil Mechanics and Geotechnical Engineering Geotechnics for transportation and infrastructure, Hong Kong, China, 23–27 May 2011.
16. Russo, M.; Germani, G.; Amberg, W. Design and construction of large tunnel through active faults a recent application. In Proceedings of the International Conference of Tunnelling & Underground Space Use, Istanbul, Turkey, 16–18 October 2002.
17. Susumu, Y.; Hiroya, K. Countermeasures of buried steel pipes against surface fault rapture. In Proceedings of the Canada 13th World Conference on Earthquake Engineering, Vancouver, BC, Canada, 1–6 August 2004; p. 916.
18. Vazouras, P.; Dakoulas, P.; Karamanos, S.A. Pipe-soil interaction and pipeline performance under strike-slip fault movements. *Soil Dyn. Earthq. Eng.* **2015**, *72*, 48–65. [CrossRef]
19. Zhou, M.; Moore Ian, D.; Lan, H. Experimental Study of Structural Response of Lined-Corrugated HDPE Pipe Subjected to Normal Fault. *J. Geotech. Geoenviron. Eng.* **2019**, *145*, 04019117. [CrossRef]
20. Liu, C.; Qiao, J.; Yan, T.; Lu, Q.; Zhan, J.; Peng, J. Ground fissure water supply pipeline 45° inclined wear full scale model test research. *J. Civ. Eng.* **2022**, *56*, 1–12. [CrossRef]
21. Deng, B.T.; Li, P.; Li, X.; Tian, J.T.; Zhi, B. Mechanical behavior of underground pipe gallery structure considering ground fissure. *J. Mt. Sci.* **2022**, *19*, 547–562. [CrossRef]
22. Deng, B.T.; Li, P.; Li, X.; Tian, J.T.; Zhi, B. Rationality determination method and mechanical behavior of underground utility tunnels in a ground fissure environment. *Bull. Eng. Geol. Environ.* **2022**, *81*, 50. [CrossRef]
23. Gou, Y.; Huang, Q.; Yang, X.; Wang, L.; Shi, W.; Yang, Z. Model test on deformation and failure mechanism of shield tunnel intersecting ground fissure with large angle. *J. Civ. Eng.* **2022**, *55*, 117–130. [CrossRef]
24. Hu, Z.; Peng, J.; Wang, Q.; Zhu, Q.; Zhao, Z. Experimental study on deformation and failure mechanism of shield tunnel with 60° oblique penetrating crack. *Chin. J. Rock Mech. Eng.* **2010**, *29*, 176–183.
25. Li, J.; Shao, S.; Xiong, T. Model test study of tunnel crossing ground crack in oblique direction. *Rock Soil Mech.* **2010**, *31*, 115–120+126. [CrossRef]
26. Li, K.; Men, Y.; Yan, J.; Li, X. Experimental study on interaction mechanism of subway tunnel crossing ground crack in orthogonal way. *Chin. J. Rock Mech. Eng.* **2011**, *30*, 556–563.
27. Liu, N.; Huang, Q.; Ma, Y.; Bulut, R.; Peng, J.; Fan, W.; Men, Y. Experimental study of a segmented metro tunnel in a ground fissure area. *Soil Dyn. Earthq. Eng.* **2017**, *100*, 410–416. [CrossRef]
28. Peng, J.; He, K.; Tong, X.; Huang, Q.; Liu, C. Failure Mechanism of an Underground Metro Tunnel Intersecting Steep Ground Fissure at Low Angle. *Int. J. Geomech.* **2017**, *17*, E4016006. [CrossRef]
29. Xiong, T.; Shao, S.; Wang, T.; Gao, Z. Model test study of orthogonal ground fracture tunnel in Xi'an metro. *Rock Soil Mech.* **2010**, *31*, 179–186. [CrossRef]
30. Yan, Y.; Huang, Q.; Xie, Y.; Qiu, J.; Han, Y.; Teng, H.; Liu, T. Failure analysis and deformation mechanism of segmented utility tunnels crossing ground fissure zones with Different intersection angles. *Eng. Fail. Anal.* **2022**, *139*, 106456. [CrossRef]
31. Yan, Y.; Huang, Q.; Yang, X. Wang P Research on the deformation and force characteristics of underground utility tunnel crossing ground fissure. *J. Eng. Geol.* **2018**, *26*, 1203–1210. [CrossRef]
32. Peng, J.; Huang, Q.; Hu, Z.; Wang, M.; Li, T.; Men, Y.; Fan, W. A proposed solution to the ground fissure encountered in urban metro construction in Xi'an. *Tunn. Undergr. Space Technol.* **2017**, *61*, 12–25. [CrossRef]
33. Yan, Y.; Qiu, J.; Huang, Q.; Wang, Z.; Xie, Y.; Liu, T. Ground fissures geology in Xi'an and failure mitigation measures for utility tunnel system due to geohazard. *Arab. J. Geosci.* **2021**, *14*, 1207. [CrossRef]
34. Wang, H.; Li, S.; Zhang, Q.; Li, Y.; Guo, X. Development of A New Geomechanical Similar Material. *Chin. J. Rock Mech. Eng.* **2006**, *9*, 1842–1847.
35. Maqsood, Z.; Koseki, J.; Ahsan, M.K.; Shaikh, M.; Kyokawa, H. Experimental study on hardening characteristics and loading rate dependent mechanical behaviour of gypsum mixed sand. *Constr. Build. Mater.* **2020**, *262*, 119992. [CrossRef]
36. Li, S.; Zhou, Y.; Li, L.; Zhang, Q.; Song, S.; Li, J.; Wang, K.; Wang, Q. Development and Application of A New Similar Material For Underground Engineering Fluid-Solid Coupling Model Test. *Chin. J. Rock Mech. Eng.* **2012**, *6*, 1128–1137.
37. Li, S.; Wang, Q.; Li, W.; Li, Z.; Wang, H.; Jiang, B.; Zhang, H. Research on application of flexible uniform pressure loading device to model test. *Rock Soil Mech.* **2014**, *1*, 61–66. [CrossRef]
38. Li, K.; Men, Y.; Liu, Y.; Jiang, R. Model test study on influence of ground fracture profile morphology on metro tunnel deformation. *Rock Soil Mech.* **2011**, *32*, 1691–1697. [CrossRef]

39. Pang, X.; Yan, J. Model test study on oblique crossing of subway tunnel through ground crack. *J. Railw. Sci. Eng.* **2016**, *13*, 1354–1359. [CrossRef]
40. Hu, Z.; Zhang, D.; Zhang, Y.; Wang, S.; Li, F. Experimental study on deformation and failure mechanism of underground integrated pipe corridor structure through active ground cracks. *Chin. J. Rock Mech. Eng.* **2019**, *38*, 2550–2560. [CrossRef]
41. Zhang, D.; Hu, Z.; Lu, G.; Wang, R.; Ren, X. Experimental Study on Deformation Mechanism of a Utility Tunnel in a Ground Fissure Area. *Adv. Mater. Sci. Eng.* **2020**, *2020*, 6758978. [CrossRef]
42. Mei, Y.; Zhang, Y.; Yuan, Y.; Ma, H.; Zhang, L.; Zhao, J. Study on mechanical response and warning standard of underground integrated pipe corridor crossing ground cracks. *J. Saf. Environ.* **2020**, *20*, 1255–1304. [CrossRef]
43. Men, Y.; Zhang, J.; Liu, H.; Han, D.; Wang, H. Discussion on calculation model of Xi'an subway tunnel crossing ground fissure zone. *J. Earth Sci. Environ.* **2011**, *33*, 95–100.
44. Qiao, L.; Liu, J.; Li, S.; Wang, C.; Jiang, Y.; Wang, Z. Underground engineering excavation surface characteristics of the effect of space and application. *Rock Soil Mech.* **2014**, *35*, 481–487. (In Chinese)

Disclaimer/Publisher's Note: The statements, opinions and data contained in all publications are solely those of the individual author(s) and contributor(s) and not of MDPI and/or the editor(s). MDPI and/or the editor(s) disclaim responsibility for any injury to people or property resulting from any ideas, methods, instructions or products referred to in the content.

buildings

MDPI

Article

Study on the Underpinning Technology for Fixed Piers of Concrete Box Girder Bridges on Mountainous Expressways

Honglin Ran [1], Lin Li [1], Yi Wei [1], Penglin Xiao [1] and Hongyun Yang [2,*]

[1] Chongqing Highway Maintenance Engineering (Group) Co., Ltd., Chongqing 404100, China
[2] State Key Laboratory of Mountain Bridge and Tunnel Engineering, School of Civil Engineering, Chongqing Jiaotong University, Chongqing 400074, China
* Correspondence: yanghy@cqjtu.edu.cn

Abstract: To address the challenge of repairing the damage to concrete box girder bridge piers on mountainous highways caused by falling rocks, this paper proposes an active underpinning technique that integrates a "井"-shaped cap system, graded preloading of the foundation, and synchronized beam body correction. The technique utilizes lateral beam preloading (to eliminate the inelastic deformation of the new pile foundation) and longitudinal beam connections (to form overall stiffness). The method involves building temporary and permanent support systems in stages. Through the two-stage temporary support system transition, the removal and in situ reconstruction of the old piers, a smooth transition from the pier–beam consolidation system to the basin-type bearing system is achieved while simultaneously performing precise correction of beam torsion. The structural safety during the construction process was verified through finite element simulations and dynamic monitoring. Monitoring results show that the beam torsion recovery effect is significant (maximum lift of 5.2 mm/settlement of 7.9 mm), and the pier strain (−54.5~−51.3 με) remains within a controllable range. Before the bridge was opened to traffic, vehicle load and impact load tests were conducted. The actual measured strength and vertical stiffness of the main beam structure meet the design requirements, with relative residual deformation less than 20%, indicating that the structure is in good, elastic working condition. The vehicle running and braking dynamic coefficients (μ = 0.058~0.171 and 0.103~0.163) are both lower than the theoretical value of 0.305. The study shows that this technique enables the rapid and safe repair of bridge piers and provides important references for similar engineering projects.

Keywords: bridge engineering; fixed pier; temporary support; active underpinning technique; monitoring and measurement

1. Introduction

Girder bridges are widely used in highway networks due to their simple structure, low cost, and fast construction speed. In western China, the complex geological conditions, along with frequent natural disasters such as earthquakes, mudslides, and mountain floods, lead to significant damage to the bridge substructures under impact loads. Therefore, researching rapid repair techniques for bridge substructures has become a critical issue.

Underpinning technology, as a rapid repair technique that minimally affects the normal functioning of structures, is widely used in old bridge renovation [1], tunnel engineering [2,3] and building construction. To ensure the safety of the superstructure, pile foundation underpinning technology is often used to transfer the overlying load from existing pile foundations, enabling the smooth progress of new construction projects [4]. However, underpinning technology carries high risks during application, and comprehensive monitoring is required to ensure safe and reliable construction [5]. Numerous scholars have conducted studies on the construction techniques and monitoring of bridge underpinning processes. Peng [6] addressed the challenges of deformation control in deep foundation pit projects passing under elevated metro bridges. Based on existing projects, he used numerical simulations and on-site measurements to study the impact of pile foundation underpinning and deep foundation pit construction on elevated metro bridges and rail structures and designed the key parameters. Zhang [7] studied a typical case of metro shield tunnel construction passing through urban interchange bridge piers, using numerical simulations and on-site measurements to analyze the underpinning process. Li [8] conducted a case study on the pile foundation group of the Guangzhou–Shenzhen Expressway bridge crossed by Shenzhen Metro Line 10. He analyzed the stress transfer mechanisms of the pile foundation during the underpinning process and investigated the impact of the shield tunnel on the stability of the pile foundation. Yan [9] created three local node models of underpinning structures with a similarity ratio of 1/1 to study the theory and application of pile foundation underpinning technology, and conducted progressive repeated static loading tests. They studied the shear and antislip properties of the joints and proposed an improved formula for calculating shear capacity. Zheng [10] analyzed and studied the impact of the active underpinning process of shield tunnel pile foundations on the deformation of the lower bridge structure through numerical simulation and on-site monitoring. Qiao [11] employed an active underpinning technique using a spatially continuous lattice underpinning structure. By integrating the monitoring and measurement data, reasonable control indicators were developed to maintain operational safety, successfully addressing the underpinning of highway bridge piers under dynamic loading conditions. Liu et al. [12] studied the feasibility of pile foundation underpinning solutions in specific environments and the efficiency of underground space utilization, achieving effective control of engineering risks. Li [13] used software to perform finite element analysis on both the superstructure and substructure of a bridge to simulate and calculate stress conditions. Passive underpinning technology was employed for construction, with settlement deformation monitoring and comparison conducted throughout the process. Key technical control points for construction were proposed.

Kou [14] analyzed the application of active underpinning technology in subway shield tunnel construction in the Central Plains region, using the example of the Zhengzhou Metro Line 5 shield tunnel passing through bridge foundations. Zhou et al. [15] developed a 3D numerical model for the pile foundation underpinning and tunnel construction to ensure the viaduct's structural safety. The model's accuracy was verified by comparing settlement calculations with field monitoring data. Shan et al. [16] and Igba U.T. et al. [17] proposed an optimized pile foundation underpinning scheme based on practical engineering applications. Tian [18] employed a scaffold-free bridge underpinning technique combined with monitoring measures, successfully completing the underpinning of an operational bridge. Based on the research findings of the aforementioned scholars, a case study was conducted where the concrete box girder fixed pier system was rapidly replaced in situ with a pot-bearing support system. This study aims to validate the impact of the pile-pier support system conversion process on the entire bridge, with the goal of facilitating the adoption of this method in similar engineering projects. In light of this, this study investigates active underpinning technology for bridge piers based on the underpinning construction of a concrete box girder bridge on an operational expressway in western China.

2. Engineering Background

2.1. Project Introduction

The study focuses on a dual-column pier box girder bridge divided into left and right segments. The left segment's superstructure consists of a single-span 6 × 20 m reinforced concrete box girder. GPZ(I)5.0DX bearings are installed at piers 1 and 5, while piers 2, 3, and 4 are rigidly connected to the box girder. The right segment consists of a single-span 5 × 20 m reinforced concrete box girder. GPZ(II)5.0DX bearings are installed at piers 1 and 4, piers 2 and 3 are rigidly connected, and GPZ(II)2.5DX bearings are placed at the abutments. The substructure consists of dual-column piers with pile foundations, and the abutments adopt gravity-type U-shaped abutments with expanded foundations. The bridge was completed and opened to traffic in 2010. The design load is classified as Highway-I level; the design speed is 80 km/h; the bridge maintenance and inspection level is II, with a load classification of a Highway-I level.

2.2. Description of Damage

On 21 May 2024, a rockfall from the hillside struck the second fixed pier (Piers 28 and 29) of the right segment, completely destroying Pier 29, exposing the reinforcement on the right side of Pier 28, causing cracks at the pier top, and binders to develop through cracks, with potential pile foundation damage predicted. It was deemed difficult to directly reinforce the original piers to restore their load-bearing capacity; although the main beams showed no significant damage, the pier column failure caused slight torsion in the main beam, as shown in Figure 1. In light of this, this study adopts an in situ active underpinning scheme to urgently repair the damaged piers.

Figure 1. Disease conditions of pier columns 28 and 29 of Pier 2.

The most recent regular inspection (2022) assessed the bridge's technical condition as Category 2. After the damage to the piers, the bridge's technical condition is now rated as Category 5, indicating a hazardous state. The piers are severely damaged, with main reinforcement yielding and bending, diagonal cracks on the piers, and cracks exceeding the allowable width, which compromises the bridge's future safe operation. Concrete near the bottom of the main beam, close to the top of Pier 29 on the right side, has been torn and partially spalled. No significant cracks were found near the bottom of the main beam around Piers 28 and 29, and no noticeable displacement of the main beam was observed.

2.3. Post-Damage Load-Bearing Capacity Check

According to the design standards "General Specification for Highway Bridges and Culverts" (JTG D60-2004) and "Design Code for Highway Reinforced Concrete and Prestressed Concrete Bridges" (JTG D62-2004) [19,20], a finite element analysis was conducted on the bridge based on its reinforced concrete structure; the model consists of 210 beam elements and 193 nodes and the bridge calculation model is shown in Figure 2. The bridge's load-bearing capacity, crack width, and deflection all meet the requirements of the specifications; however, the safety reserve factor for the load-bearing capacity of Pier 2-2 is relatively low.

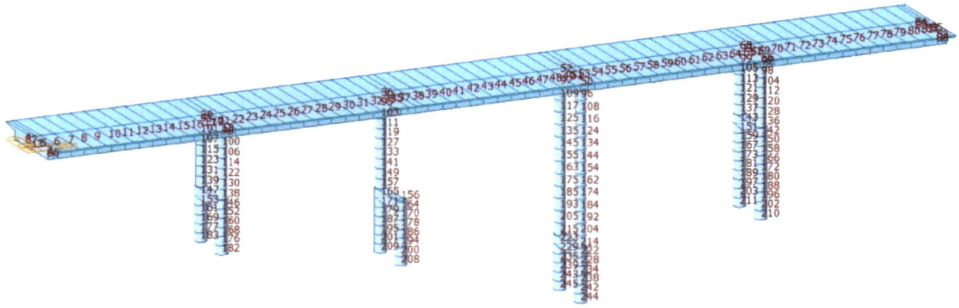

Figure 2. Post-damage bridge calculation model.

As shown in Figure 3, in the ultimate limit state of load-bearing capacity, the minimum safety factor for bending capacity in the positive bending moment zone at the mid-span of the main beam is 1.11, while the minimum safety factor for bending capacity in the negative bending moment zone at the pier top is 1.45. The bending capacity meets the requirements of the specifications.

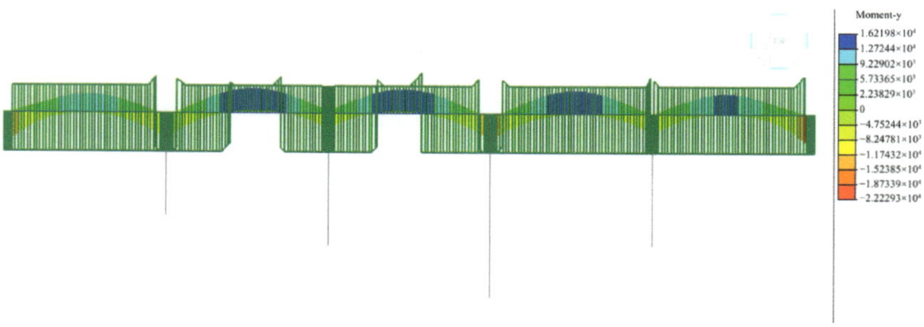

Figure 3. Envelope diagram for positive section bending check of the main beam during normal service stage.

As shown in Figure 4, in the ultimate limit state of load-bearing capacity, the minimum safety factor for the shear capacity of the main beam is 1.20, and the shear capacity meets the requirements of the specifications.

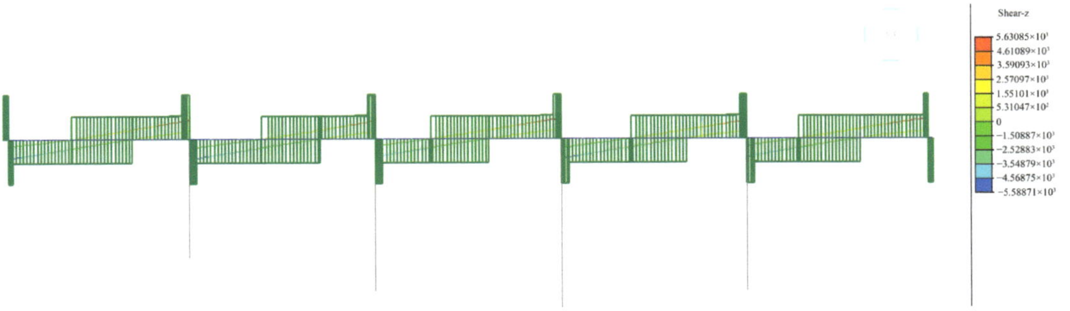

Figure 4. Envelope diagram for inclined section shear check of the main beam during normal service stage.

As shown in Table 1, under the basic combination, except for Pier 29, which is completely damaged, the safety reserve factors for the control sections of the remaining piers range from 1.4 to 3.69. The load-bearing capacity of the bridge piers meets the specifications; however, the safety reserve factor for the load-bearing capacity of the damaged Pier 28 is relatively low.

Table 1. Strength calculation results of the bridge piers under a basic combination of loads.

Verification Position	Axial Force (KN)	Resistance (KN)	Safety Factor
Pier 26	5553.3	13,393.9	2.41
Pier 27	3631.7	13,393.9	3.69
Pier 28	7631.4	10,715.1	1.40
Pier 30	6157.1	13,393.9	2.18
Pier 31	3964.9	13,393.9	3.38
Pier 32	4266.7	13,393.9	3.14
Pier 33	3990.9	13,393.9	3.36

As shown in Figure 5, the crack verification of the reinforced concrete components is carried out considering the short-term effect combination and the influence of long-term effects, and the crack resistance meets the requirements. The deflection value calculated based on the short-term effect combination stiffness is multiplied by the long-term growth factor of 1.45 for this bridge. The deflection of the main beam is 17.86 mm, which is less than the calculated span of L/600 (33.33 mm), and the deflection meets the specification requirements.

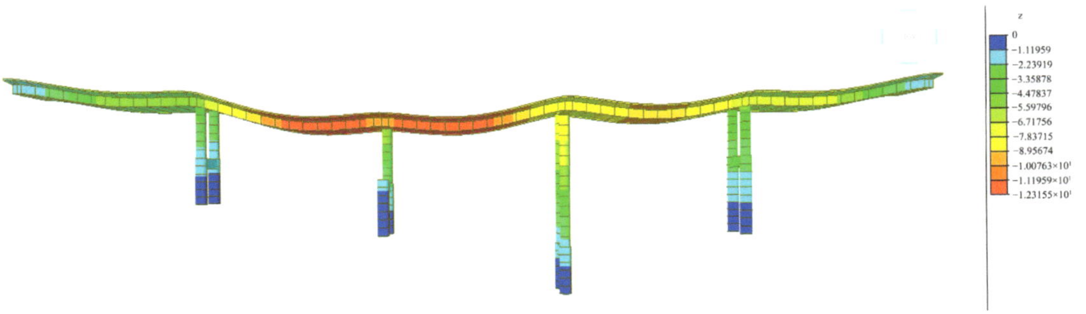

Figure 5. Cloud diagram of the maximum vertical displacement of the main beam under the short-term effect combination (mm).

3. Pier Underpinning Plan

3.1. Design Principles of the Plan

First, emergency support is applied to the main beam to prevent secondary damage caused by changes in the structural stress state. Then, new pile foundations and caps are constructed. The newly designed underpinning structure requires strong anti-settlement capacity and excellent structural performance. The newly built piers align with the original piers to maintain the span arrangement of the original bridge. Considering that the reinforcement at the original pier–beam junction has yielded, the original connection system is modified to a GPZ (II)-5.0-GD bearing support system. During the underpinning process, temporary steel supports and jacks are used to correct the torsion of the beam, and the reaction force is transferred to the newly built pile foundations and caps. The preloading process can eliminate some of the deformations of the newly constructed piles and caps, with the deformation being compensated for by the elongation of the jacks.

3.2. Pier Underpinning Structure Design

3.2.1. Selection of the Underpinning Method

The method of load transfer for underpinning is divided into passive underpinning and active underpinning. The comparison of their characteristics is shown in Table 2.

Table 2. Comparison of the characteristics of different underpinning methods.

Underpinning Structure Name	Load Transfer Method	Deformation Control	Applicable Scenarios
Passive underpinning	Direct transfer	Structural deformation is difficult to control.	Scenarios with small tonnage and structures that are not very strict about deformation.
Active underpinning	Staged transfer	Partial deformation of the underpinning structure is eliminated through preloading, allowing for control of the settlement of the superstructure.	Scenarios with large tonnage and structures that have strict requirements for deformation.

(1) Comparison of the applicability of active underpinning and passive underpinning

As shown in Table 2, active underpinning eliminates deformation through staged preloading, making it suitable for large tonnage and high-precision scenarios, while passive underpinning has weaker deformation control capability.

(2) Potential risks and countermeasures

Differential settlement risk: The difference in stiffness between the new and old pile foundations may lead to long-term uneven settlement. To address this, the new foundation adopts a pile group design, and concrete beams are used to connect and form a '井' (well) shaped cap to optimize the load distribution, thereby reducing the differential settlement between the new and old pile foundations.

Material degradation risk: Concrete creep and steel corrosion may reduce the durability of the underpinning structure. Countermeasures include upgrading the pile foundation concrete from C25 to C30, upgrading the cap and pier concrete from C35 to C40, and adding rust inhibitors.

Construction disturbance risk: The removal of temporary supports may induce secondary deformation. This risk is mitigated by dynamically adjusting the lifting force through real-time monitoring, ensuring that displacement fluctuations remain below the design value.

3.2.2. Emergency Temporary Support

After the grouting treatment of the concrete cracks in the binder of Pier 2, it was used as the emergency temporary support foundation. Steel pipe columns of 600 × 14 mm were used, with sway bracing and diagonal braces made of 28-channel steel. At the top of the steel pipe columns, I-beam 50 steel pad beams were arranged, and jacks were placed on the pad beams. The temporary support is shown in Figure 6. The lifting capacity of the jack at the top of the steel pipe column for the third temporary support is 20 tons, and the remaining jacks only need to be tightened.

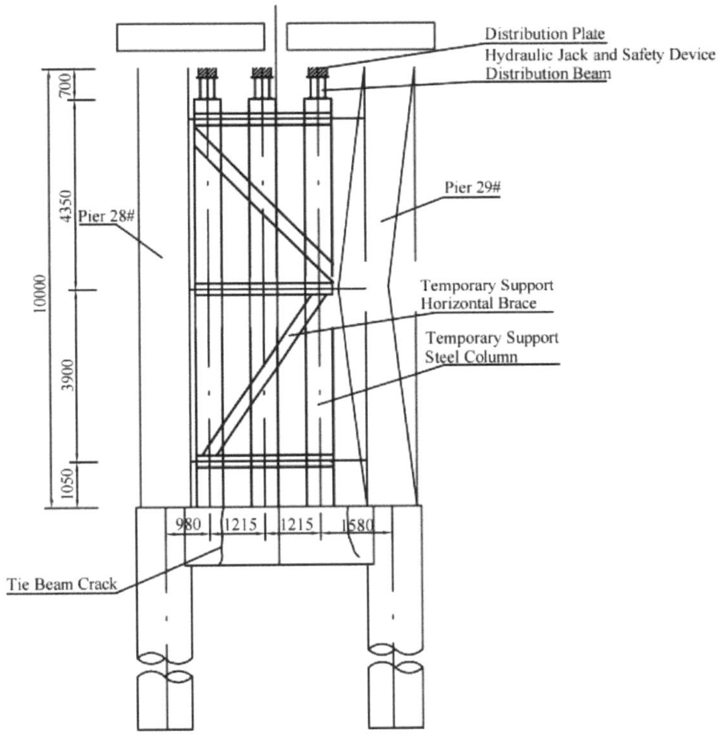

Figure 6. Emergency temporary support.

3.2.3. Underpinning Support System

As shown in Figure 7, four new piles were added on both sides of the existing pile foundation in the longitudinal direction of the bridge, along with the installation of bearing platforms. The new piles have a diameter of 1.3 m, with a lateral center-to-center spacing of 5 m and a pile length of 12 m; consistent with the original design, the pile foundation concrete grade is upgraded from C25 to C30. The longitudinal spacing from the center of the new piles to the original piles is 3.15 m. The bearing platform is a frame structure with dimensions of 8.6 × 9.5 × 2 m. Steel pipe supports were installed on the newly constructed bearing platform, using 600 × 14 mm pipes, with three supports evenly distributed on each side. On top of the steel pipe supports, I-beam 50 steel distributive beams were arranged, with jacks positioned on the distributive beams for lifting purposes. Between the steel pipe supports, 351 × 16 mm steel pipes and 28-channel steel were used as horizontal and diagonal braces, ensuring the structural stability of the entire support system. The pier was damaged by external impact, leading to damage to the original transverse tie beams and in situ pile foundations. The extent of the pile foundation damage cannot be accurately assessed, so the bearing capacity of the original pile foundations is not considered in the new foundation system. New pile foundations of the same diameter are constructed in the longitudinal direction of the bridge, centered around the existing pile foundation. Initially, these new piles are connected with concrete tie beams in the transverse direction to form a temporary foundation for the active replacement system. After the beam is lifted, the damaged pier and tie beams are removed. The pile foundations are connected with concrete tie beams in the longitudinal direction, thus forming a footing in shape resembling the Chinese character "井", which serves as the permanent foundation for the newly constructed piers, the newly built piers align with the original piers, and the concrete grade is upgraded from C35 to C40.

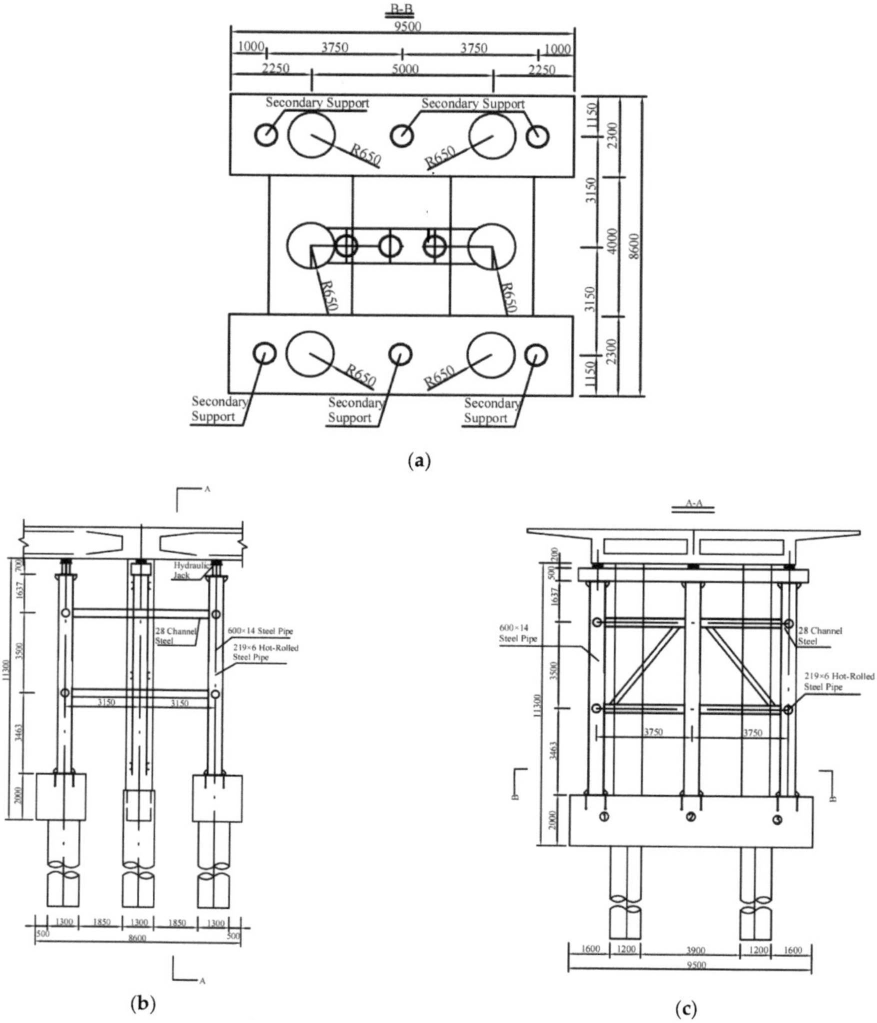

Figure 7. Schematic diagram of the underpinning support system. (**a**) Layout plan of the new pile foundation and bearing platform. (**b**) Elevation view of the steel pipe support system. (**c**) Section view of the steel pipe support system.

3.3. Construction Process Flow

The construction process flow is shown in Figure 8.

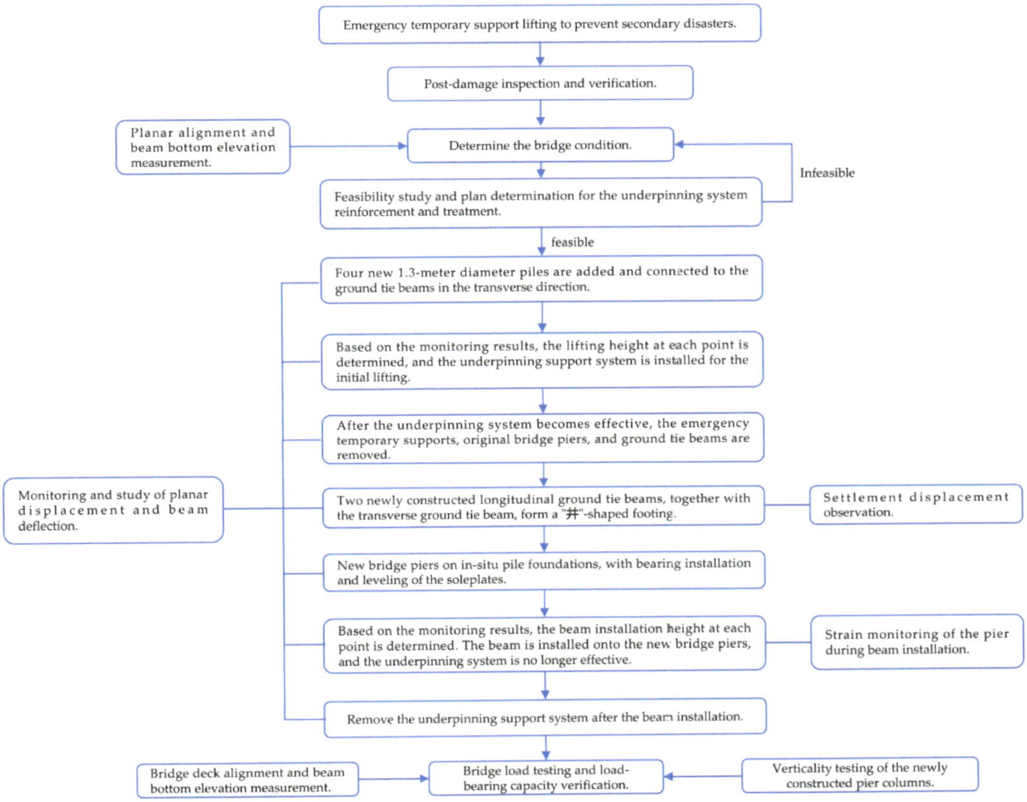

Figure 8. Construction process flow.

3.4. Pier Underpinning Structural Verification

The design does not consider the original bridge pile foundation's load; all external forces are borne by the newly installed underpinning components. According to the 'Seismic Ground Motion Parameter Zoning Map of China' (GB 18306-2015) and the 'Seismic Design Code for Highway Bridges' (JTG/T 2231-01-2020) [21,22], the seismic fortification category of this bridge is Category B, with a seismic fortification intensity of VII, a peak ground acceleration of 0.10 g, and a seismic measure level of 3. The seismic importance factor for E1 seismic action is 0.5 with a recurrence period of 100 years, and for E2 seismic action, the seismic importance factor is 1.7 with a recurrence period of 2000 years. The seismic design method selected is Category 1.

The calculation model consists of 247 nodes and 229 elements. The major components of the bridge are modeled using beam elements. The finite element model of the bridge adopts common linear and isotropic assumptions. The model boundaries are defined by the types and directions of constraints, with elastic connections, while also considering the effects of local overall temperature and temperature gradients as specified in the standards. The primary materials used for the main beam, piers, and caps are C40 concrete, while the newly constructed Pile Foundation 2 uses C30 concrete. The load and load combination calculations are as follows:

(1) Load calculations: Class I highway.
(2) Temperature load: Overall temperature variation, ±20 °C.
(3) Load combinations: The load combinations are based on the "General Specifications for Highway Bridge and Culvert Design" (JTG D60-2015) [23].

Based on the design calculation parameters, we used finite element software to analyze the load-bearing capacity of the bridge after underpinning. The simulation results serve as a reference for the design phase, and during the implementation process, real-time monitoring data is used to guide construction. The finite element calculation model is shown in Figure 9, and the calculation results are shown in Figures 10 and 11.

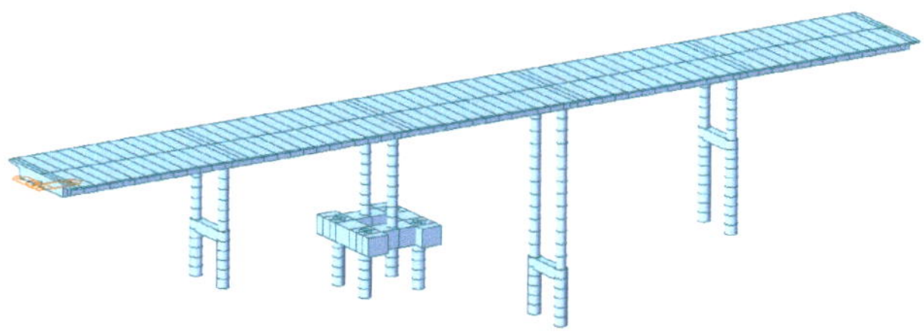

Figure 9. The finite element calculation model.

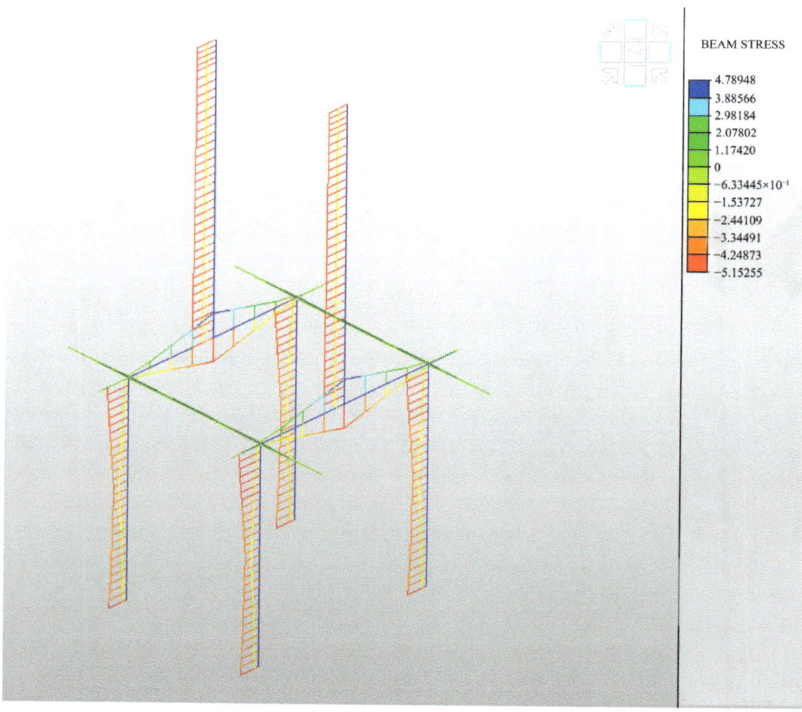

Figure 10. Stress diagram of the underpinning structure under basic load combination.

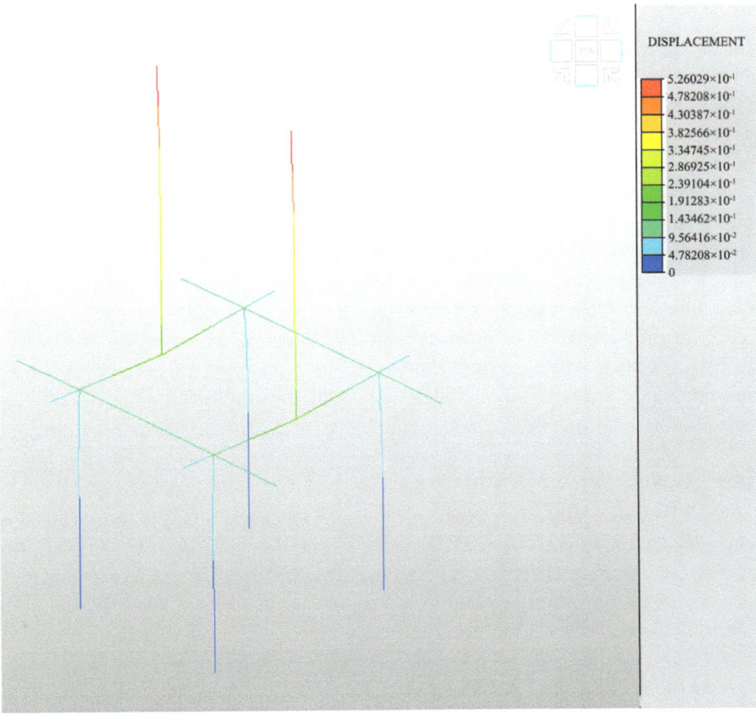

Figure 11. Maximum displacement diagram of the underpinning structure under moving load.

According to the "Code for Design of Highway Reinforced Concrete and Prestressed Concrete Bridges and Culverts" (JTG 3362-2018) [24], the bearing capacity of the eccentrically compressed Pier 2 and the bearing capacity of the newly added bearing platform for Pier 2 are checked. The calculation results are shown in Tables 3 and 4.

Table 3. Calculation table for the reinforced Pier 2.

Verification Location	Design Load		Safety Factor
	Axial Force (KN)	Bending Moment (KN·m)	
Pier Base	4069.64	1169.79	1.99

Table 4. Calculation table for the reinforced Pier 2 bearing platform.

Verification Location	Design Load (KN)	Bearing Capacity (KN·m)	Safety Factor
Transverse Bridge Direction (During Process)	1486	5397.8	3.3
Longitudinal Bridge Direction (After Reinforcement)	6996	14,800	1.92

The verification results show that the bearing capacity of Pier 2 and the newly added pier cap after reinforcement meet the code requirements.

3.5. Jacking Force and Control Displacement Calculation

According to the calculation results, the total load on the supporting steel pipes under dead load is 510 tons, and the initial lifting force of the secondary support jacks is controlled at 50% of the dead load, which is 42 tons. After the original pier columns are dismantled, secondary lifting is performed to reach the design elevation. Based on the testing results, the original Pier 28 is used as the reference point, and the vertical displacement controls the lifting elevation of the Pier 29 beam. By comparing the design slope, girder bottom elevation, and bridge deck alignment across multiple survey lines, the degree of girder torsion and the adjusted elevation are calculated. The controlled displacement for lifting the beam is set to 13 mm (calculated based on the beam bottom elevation, where the beam experiences a twisting deflection of approximately −8 mm at the damaged pier, requiring an uplift of 8 mm for recovery; a reserved height of 5 mm is provided for the bearing installation space, which will decrease after the beam is installed). The jacking force is adjusted as per actual conditions but must not exceed 120 tons. If the force exceeds 120 tons, construction must be paused, and the causes analyzed and resolved before resuming.

4. On-Site Construction Monitoring

To monitor and study the structural conditions of damaged bridges throughout the construction process, ensuring construction safety and quality control, structural monitoring of the bridge is essential. This includes tracking changes in structural performance and working conditions under various incidental loads such as temperature fluctuations, wind, rain, and construction angles during the construction period. Through dynamic monitoring and research on bridge deck profile measurement, lateral displacement monitoring, beam deflection monitoring, settlement displacement observation, strain monitoring of the pier during beam installation, and verticality measurements, the safety and reliability of the bridge construction are ensured. This provides reliable data and a scientific basis for accurate analysis, evaluation, prediction, and management of the project.

4.1. Monitoring Categories and Frequency

The monitoring categories and frequencies are shown in Table 5.

Table 5. Monitoring categories and frequencies.

Monitoring Category	Monitoring Frequency
Bridge deck profile	Once before steel pipe support installation and once after completion.
Planar displacement	Once daily before completion.
Beam deflection	Once daily before completion, real-time monitoring during beam placement.
Settlement displacement observation	Once daily from pier cap pouring to completion.
Beam placement pier strain monitoring	During beam placement, incremental loading strain gauges are arranged on two newly built piers.
Verticality measurement	Once after completion of the newly built piers.

(1) Bridge Deck Profile

The bridge deck profile is measured using a DiNi03 precision-level instrument. A total of four measurement lines are set up, with 84 measurement points, as shown in Figure 12.

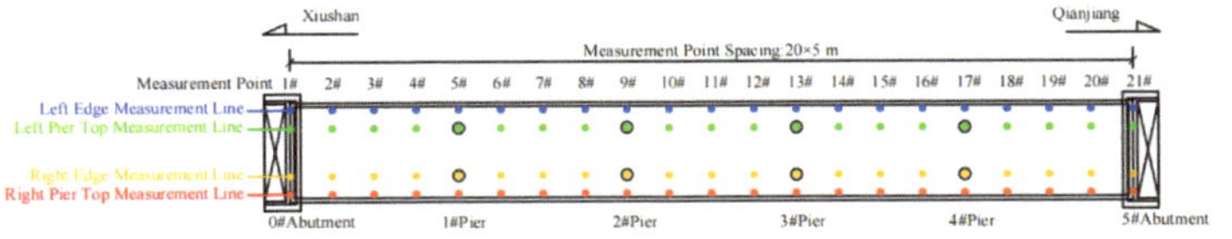

Figure 12. Schematic diagram of bridge deck profile measurement point layout.

(2) Planar Displacement

Planar displacement of the bridge is measured using a GM-101 total station. A total of five measurement points are set up, as shown in Figure 13.

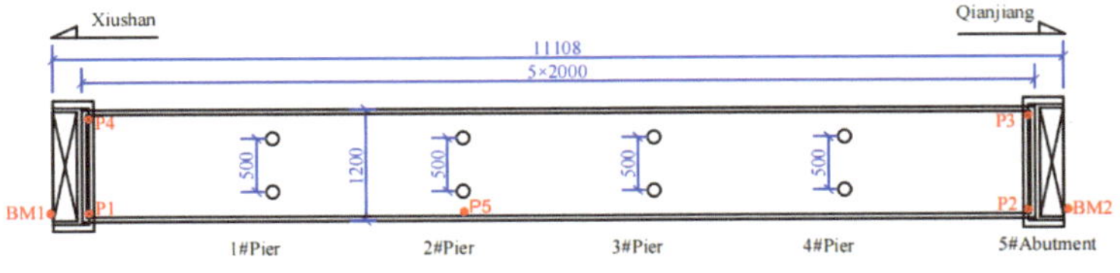

Figure 13. Schematic diagram of planar displacement measurement point layout.

(3) Bridge Deflection

Bridge deflection is measured using a TS60I0.5″R1000 total station. A total of 11 measurement points are set up on the beam, as shown in Figure 14.

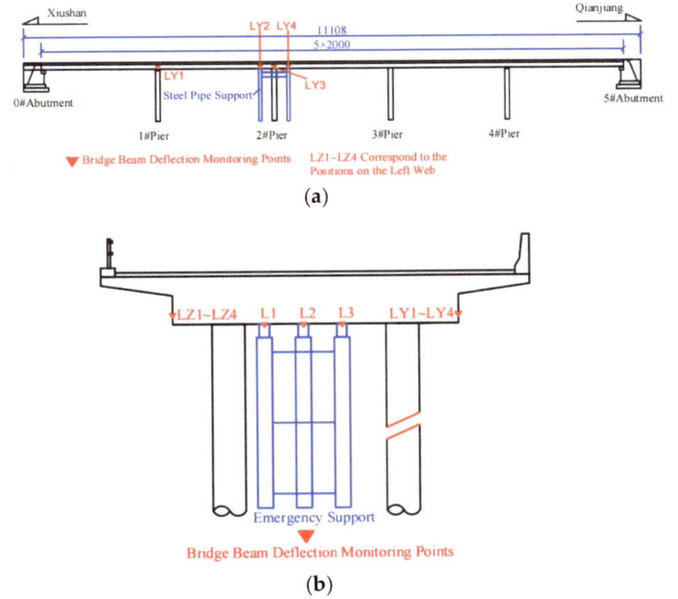

Figure 14. Schematic diagram of bridge beam deflection measurement points. (**a**) Elevation view and (**b**) cross-section diagram.

(4) Settlement Displacement

Settlement displacement of the pier caps is measured using a TS60I0.5″R1000 total station. A total of five measurement points are set up on the piers and pier caps, as shown in Figure 15.

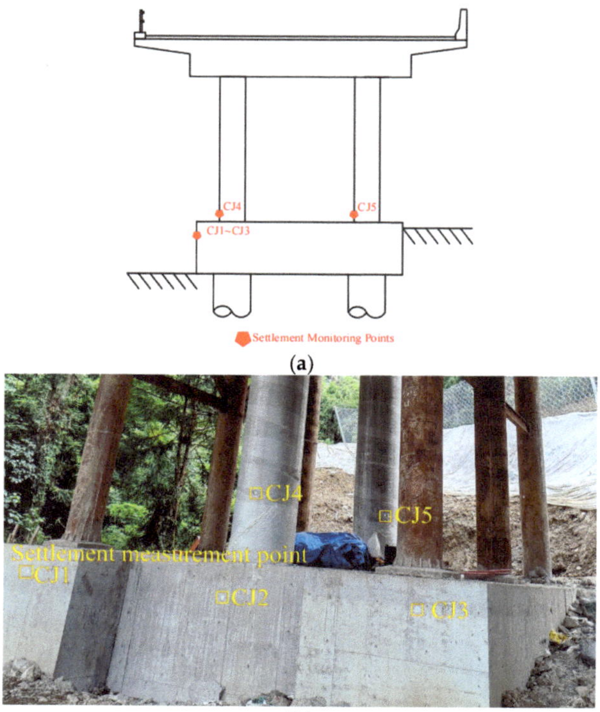

(a)

(b)

Figure 15. Schematic diagram of pier and abutment settlement displacement monitoring points.

(5) Pier Strain Monitoring

During beam placement, the stress (strain) of the piers is monitored using the ZiBee + DH3819 wireless static strain testing and analysis system. A total of eight measurement points are set up on the newly constructed piers, as shown in Figure 16.

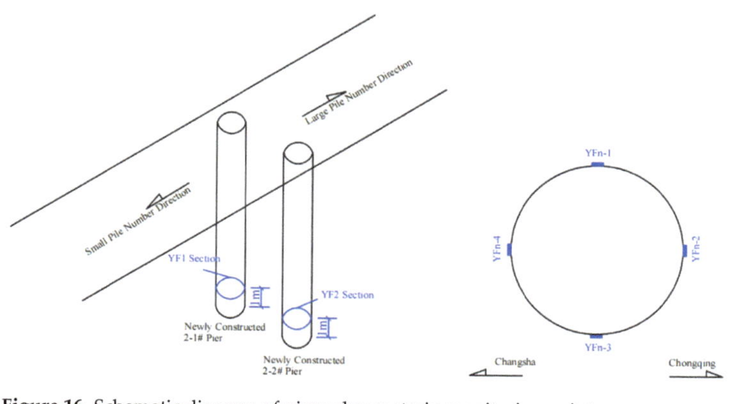

Figure 16. Schematic diagram of pier column strain monitoring points.

(6) Pier Verticality Measurement

The verticality of the newly constructed piers is measured using a TS60I0.5″R1000 total station. Verticality is measured in four directions for each of the two piers.

4.2. Analysis of Monitoring and Measurement Results

(1) Bridge Deck Profile

As shown in Table 6 and Figure 17, the initial measurement data of the points between Piers 1 and 3 were compared with the data obtained after completion. At measurement point 9 on Pier 2 (the section at the damaged location), the bridge deck elevation at the right side of the guardrail edge increased by 5.2 mm; the right pier top increased by 3.3 mm, the left pier top decreased by 5.0 mm, and the left side of the guardrail edge decreased by 7.9 mm. The slight torsion of the beam supported by the steel pipe was largely restored after the beam was installed. After analysis, it was found that these changes are all less than the uplift control displacement of 8 mm.

Table 6. Elevation measurement changes.

Measurement Point	Measurement Line				
	Right Edge	Right Pier Top	Left Edge	Left Pier Top	
5	1.8	1.4	0.6	0.0	Pier 1
6	1.2	2.4	−1.5	−2.4	/
7	2.0	3.1	−2.5	−4.8	/
8	4.3	1.8	−4.2	−8.1	/
9	5.2	3.3	−5.0	−7.9	Pier 2
10	4.5	3.4	−4.3	−6.8	/
11	3.3	3.4	−1.7	−5.2	/
12	2.5	3.1	−0.3	−1.2	/
13	1.8	3.0	1.1	0.6	Pier 3

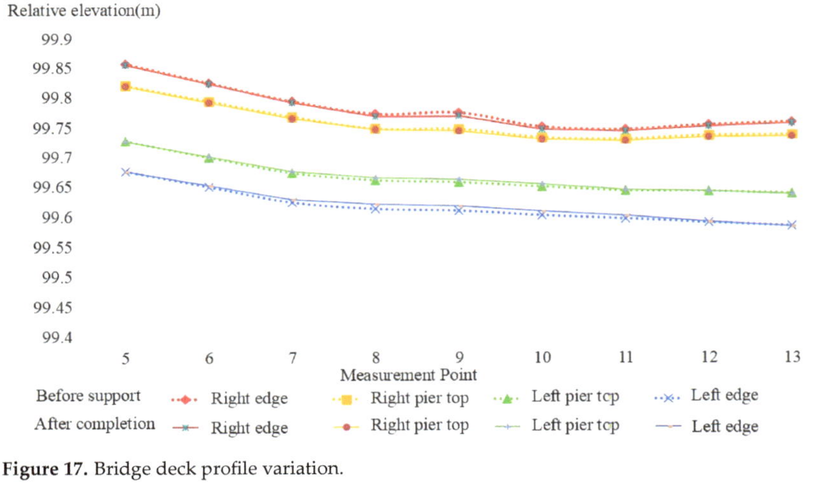

Figure 17. Bridge deck profile variation.

(2) Planar Displacement

The lateral displacement of the bridge deck ranged from −6.7 mm to 7.6 mm in the longitudinal direction from −7.4 mm to 4.6 mm in the transverse direction. During the monitoring period, no significant abnormal changes in lateral displacement were observed, which aligns with the theoretical research results.

(3) Beam Deflection

As shown in Table 7 and Figure 18, the main measurement points deflection variation stage diagram, the deflection changes during the beam lifting, beam installation, and after the installation are normal, and the adjustment of the beam bottom elevation meets the design requirements with no significant abnormalities. The beam deflection underwent the following main changes: 1. At the damaged pier, the beam experienced downward deflection on the right side, with slight twisting; 2. After the temporary support, the beam was lifted, causing a slight upward curvature on the right side and slight twisting; 3. After the second support was completed at the pile foundation, the beam returned to a slight torsion state, and the beam at the tops of the left and right piers was adjusted to a final beam installation elevation of 4 mm to 5 mm higher; 4. After the beam was installed onto the new pier supports, the beam bottom elevation reached the design position. The entire process is consistent with the theoretical research results.

Table 7. Beam deflection changes at key stages.

Stage	State	LZ2	LZ3	LZ4	LY2	LY3	LY4
After initial temporary support	Slight twisting	0	0	0	0	0	0
After secondary steel pipe support	Slight twisting	2.1	2.6	2.9	14.9	14.9	15
Adjustment before beam placement	Restored to a non-twisted state	−3.7	−3	−2.3	9.3	10	9.6
After beam placement	Restored to a non-twisted state	−6.4	−4.7	−4.2	5	5.1	5.2

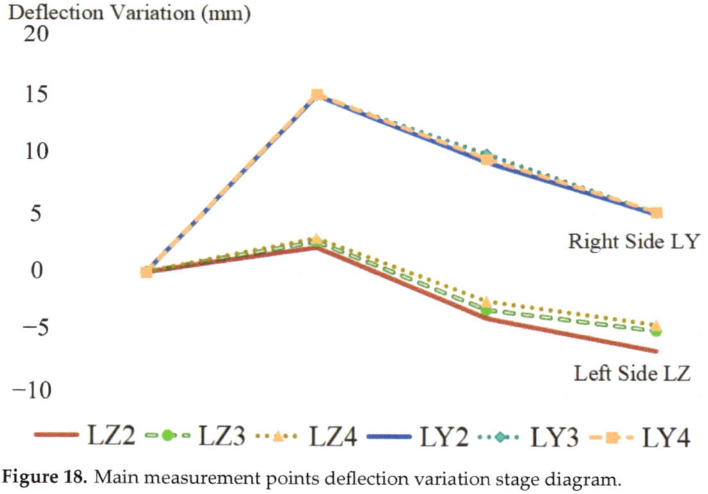

Figure 18. Main measurement points deflection variation stage diagram.

(4) Pier and Pier Cap Settlement

The settlement changes at each measurement point on the piers and pier caps ranged from −0.9 mm to 0.9 mm. No significant settlement was observed during the monitoring period.

(5) Pier Stress and Strain

As shown in Tables 8 and 9, the strain values of the newly constructed piers during beam placement ranged from −54.5 με to −51.3 με, with calibration factors between 0.87 and 0.92. The stress and strain of the piers showed no significant abnormalities, and the piers performed normally under the dead load.

Table 8. Pier column stress–strain calculation table for support reaction forces during beam placement.

Operating Condition	Pier Number	Support Reaction (KN)	Pier Column Cross-Sectional Area (m^2)	Stress (MPa)	Strain (με)
Dead Load (Beam Placement)	Column 2-1 Column 2-2	2556.1	1.327	−1.926	−59.3

Table 9. Pier column stress–strain monitoring table for support reaction forces during beam placement.

Operating Conditions	Pier Number	Measurement Point	Strain (με)				Calibration Factor
			Level 1 (55%)	Level 2 (80%)	Level 3 (100%)	Theoretical Value	
Dead Load (Beam Placement)	Pier 2-1	YF1-1	−20.3	−40.9	−51.3	−59.3	0.87
		YF1-2	−22.4	−40.4	−52.6	−59.3	0.89
		YF1-3	−23.4	−40.6	−54.8	−59.3	0.92
		YF1-4	−21.7	−40.3	−51.8	−59.3	0.87
	Pier 2-2	YF2-1	−25.4	−42.1	−53.3	−59.3	0.90
		YF2-2	−25.3	−39.7	−52.6	−59.3	0.89
		YF2-3	−24.2	−41.1	−52.8	−59.3	0.89
		YF2-4	−24.8	−40.9	−54.5	−59.3	0.92

(6) Pier Verticality

The newly constructed piers exhibited longitudinal inclination values of 1.0 mm to 8.0 mm and transverse inclination values of 2.0 mm to 5.0 mm. The verticality was within a range of 0.1‰ to 1.0‰, meeting the specified code requirements.

(7) Measurement error correction

The measurement uncertainty mainly includes the accuracy of the strain testing system, the accuracy of displacement measurements at various points, and the influences of environmental wind speed and temperature. In this measurement, high-precision equipment was used, with a strain testing system accuracy of 1 με and displacement measurement equipment accuracy of 0.5″ and 0.001 mm. The measurements were conducted in an environment with wind speeds of 1.5 m/s and wind force below level 1. The strain test is compensated for temperature effects using strain gauge temperature compensation sheets, and the displacement measurement results are corrected for temperature variations through finite element analysis. However, it should be noted that the surface temperature of the beam changes much faster than the temperature of the bridge concrete, so when performing correction analysis during periods of rapid temperature change, the range of change should be appropriately reduced.

5. Bridge Load Testing

To study and evaluate the safety and performance of the structure, a load test is conducted on the bridge before reopening to traffic. Factors such as impact load, construction disturbances, and changes in the original support system are considered. By measuring the deformation and internal forces of the bridge structure under static load test conditions, the actual performance of the reinforced bridge structure is verified. This allows for the evaluation and study of the actual working state of the bridge after reinforcement and its compliance with the design expectations. Additionally, dynamic load tests are conducted to understand the inherent dynamic characteristics of the bridge after reinforcement, clarifying its dynamic performance under operational loads. The load test is shown in Figure 19.

Figure 19. Photograph of the load-testing vehicle during the loading process.

5.1. *Vertical Static Load Test*

5.1.1. Test Contents

The test content includes the following four aspects: stress testing of the test section, vertical deflection and stress testing of Pier 2, vertical deflection testing at the mid-span section of the test span, and crack testing of the beam body.

5.1.2. Theoretical Calculation

The bridge load test selected the following sections for evaluation: A-A: Maximum positive bending moment section at mid-span of the 1st span. B-B: Maximum negative bending moment section at the top of Pier 1. C-C: Maximum positive bending moment section at mid-span of the second span. D-D: Maximum negative bending moment section at the top of Pier 2. E-E: Maximum positive bending moment section at mid-span of the third span. The specific locations of the sections for the static load test are shown in Figure 20.

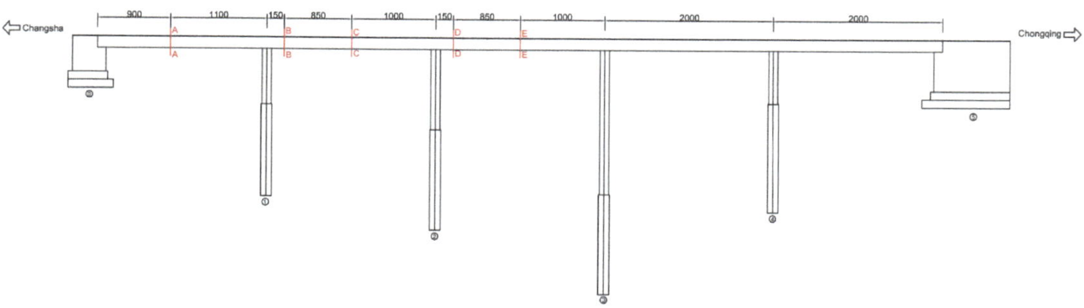

Figure 20. Schematic diagram of the static load test section location.

The test control load uses the standard value of vehicle load (considering the impact factor) and follows the highway class I load specified in the "General Specifications for Highway Bridge and Culvert Design" (JTG D60-2004) [19]. The vehicle load considers both central and eccentric loading as the most unfavorable conditions. The loading parameters are shown in Figure 21. Under the action of moving vehicle loads, the internal force diagrams and deformation envelope diagrams of the bridge structure are shown in Figures 22 and 23. The calculated bending moment, test bending moment, and corresponding load efficiency coefficients for static loading conditions of the test span are shown in Table 10.

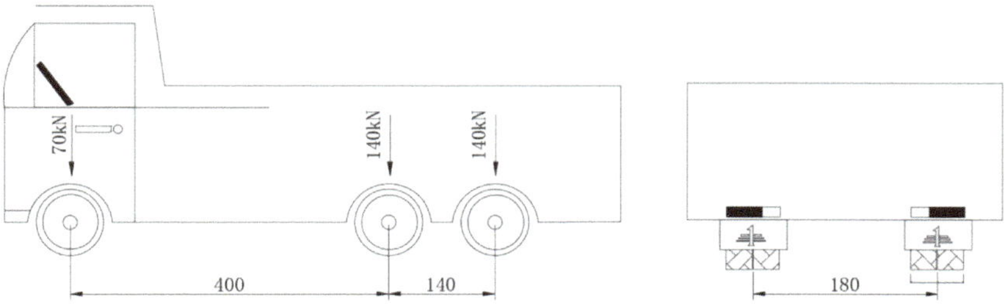

Figure 21. Load vehicle parameter diagram (cm).

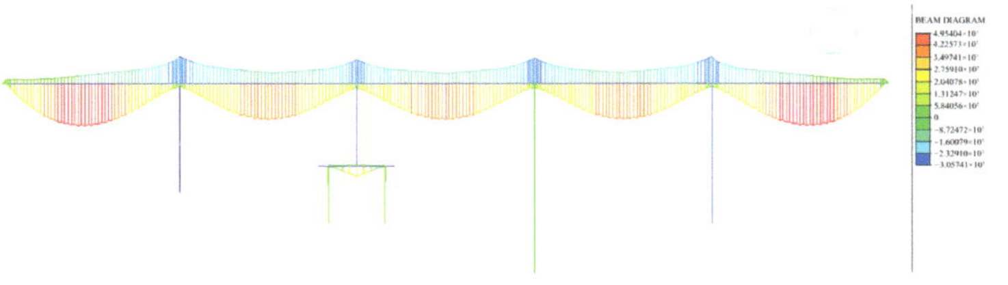

Figure 22. Moment envelope diagram under moving vehicle load (KN·m).

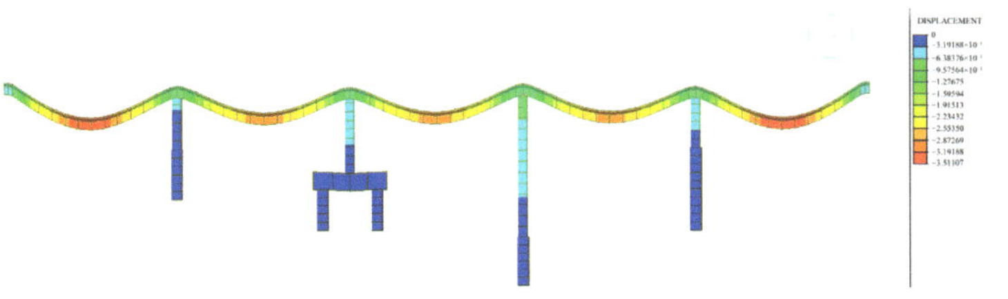

Figure 23. Maximum vertical deformation diagram under moving vehicle load (mm).

Table 10. Calculated bending moment, experimental bending moment, and corresponding load efficiency coefficients for static loading conditions of the test span.

Test Section	Calculated Value (KN·m)	Test Value (KN·m)	Efficiency Coefficient	Loading Method (Transverse × Longitudinal)
Span 1 Midspan Maximum Positive Bending Moment (A-A), Central Load	4936.9	4845.0	0.98	3 × 2 = 6 vehicles
Span 1 Midspan Maximum Positive Bending Moment (A-A), Eccentric Load	4936.9	4845.0	0.98	3 × 2 = 6 vehicles
Pier 1 Top Maximum Negative Bending Moment (B-B), Central Load	−3057.4	−2906.6	0.95	3 × 2 = 6 vehicles
Pier 1 Top Maximum Negative Bending Moment (B-B), Eccentric Load	−3057.4	−2906.6	0.95	3 × 2 = 6 vehicles

Table 10. *Cont.*

Test Section	Calculated Value (KN·m)	Test Value (KN·m)	Efficiency Coefficient	Loading Method (Transverse × Longitudinal)
Span 2 Midspan Maximum Positive Bending Moment (C-C), Central Load	4220.4	4163.7	0.99	3 × 2 = 6 vehicles
Span 2 Midspan Maximum Positive Bending Moment (C-C), Eccentric Load	4220.4	4163.7	0.99	3 × 2 = 6 vehicles
Pier 2 Top Maximum Negative Bending Moment (D-D), Central Load	−2718.4	−2615.8	0.96	3 × 2 = 6 vehicles
Pier 2 Top Maximum Negative Bending Moment (D-D), Eccentric Load	−2718.4	−2615.8	0.96	3 × 2 = 6 vehicles
Span 3 Midspan Maximum Positive Bending Moment (E-E), Central Load	4234.0	4049.7	0.96	3 × 2 = 6 vehicles
Span 3 Midspan Maximum Positive Bending Moment (E-E), Eccentric Load	4234.0	4049.7	0.96	3 × 2 = 6 vehicles

To obtain the continuous curve of the relationship between test load and displacement for the structure and to prevent unexpected structural damage, static test loads are applied in three loading levels and one unloading level for different test items. The loading method involves incrementally increasing the load to the maximum level in a single step-by-step process, followed by direct unloading to zero load.

5.1.3. Test Results

The load efficiency coefficient ranged from 0.95 to 0.99, meeting the basic load requirements specified in the "Specifications for Highway Bridge Load Test" (JTG/T J21-01-2015) [25].

(1) Strain Test Results

The calibration coefficients for the measured strain in the control sections of the main beam during the static load test ranged from 0.70 to 0.83. The measured strain values were all smaller than the calculated strain values, indicating that the structural strength meets the design requirements. The measured relative residual strain ranged from 6.1% to 17.2%, all below 20%, which demonstrates that the structure is in good elastic working condition and complies with the relevant requirements of the "Specifications for Highway Bridge Load Test" (JTG/T J21-01-2015).

(2) Deflection Test Results

The calibration coefficients for the measured deflection in the control sections of the main beam during the static load test ranged from 0.75 to 0.85. The measured deflection values were all smaller than the calculated deflection values, indicating that the vertical stiffness of the structure meets the design requirements. The relative residual displacement after unloading ranged from 0% to 16.1%, all below 20%, indicating that the structure is in good elastic working condition and satisfies the requirements of the "Specifications for Highway Bridge Load Test" (JTG/T J21-01-2015).

(3) Crack Test Results

No significant cracks were observed in the test bridge segment during the loading process.

5.2. Dynamic Load Test

5.2.1. Test Contents

(1) Pulsation Test

Under conditions with no traffic load on the bridge deck and no irregular vibration sources near the bridge, the pulsation test measures the slight vibration response of the bridge span structure caused by random loads such as wind loads and ground pulsations at the bridge site.

(2) Driving Test

The driving test, conducted under unobstructed conditions, primarily measures the dynamic load response and impact factor at the midspan section (E-E) of the central span under driving speeds of 10, 20, 40, and 60 km/h on a bridge deck in good condition. The test focuses on measuring dynamic strain.

(3) Braking Test

The braking test primarily measures the dynamic load response and impact factor at the midspan section (E-E) of the central span under emergency braking conditions at driving speeds of 10, 20, and 30 km/h. The test content for the corresponding operating conditions is shown in Table 11.

Table 11. Dynamic load test condition table.

Operating Conditions	Test Section	Operating Conditions Description	Test Content
1	Mid-span Section	Pulsation Test	Frequency
2	Mid-span Section	Driving Test at 10 km/h	Frequency, Dynamic Strain
3	Mid-span Section	Driving Test at 20 km/h	Frequency, Dynamic Strain
4	Mid-span Section	Driving Test at 40 km/h	Frequency, Dynamic Strain
5	Mid-span Section	Driving Test at 60 km/h	Frequency, Dynamic Strain
6	Mid-span Section	Braking Test at 10 km/h	Frequency, Dynamic Strain
7	Mid-span Section	Braking Test at 20 km/h	Frequency, Dynamic Strain
8	Mid-span Section	Braking Test at 30 km/h	Frequency, Dynamic Strain

5.2.2. Test Results

(1) Dynamic Characteristics Results

As is shown in Table 12, the overall stiffness of the inspected bridge span structure is normal.

Table 12. Dynamic characteristics results of Longdonggou Bridge.

Serial Number	Measured Fundamental Frequency (Hz)	Calculated Fundamental Frequency (Hz)	Measured Damping Ratio (%)	Measured/ Calculated Frequency Ratio
1	7.666	6.136	1.450	1.249

(2) Driving Test Results

As is shown in Table 13, the theoretical fundamental frequency is 6.136 Hz. According to the General Specifications for Highway Bridge and Culvert Design (JTG D60-2015) [23], the theoretical value of the impact factor is 0.305. The measured dynamic stress impact factor (μ) ranges from 0.058 to 0.171, which is lower than the theoretical value of 0.305.

Table 13. Driving test dynamic response results.

Load Case	10 km/h Driving	20 km/h Driving	40 km/h Driving	60 km/h Driving
Midspan Section Dynamic Strain (με)	10	11	11	11
Impact Factor (μ)	0.058	0.072	0.146	0.171

(3) Braking Test Results

As is shown in Table 14, the theoretical fundamental frequency is 6.136 Hz, and the theoretical impact coefficient value, calculated according to the "General Code for Design of Highway Bridges and Culverts" (JTG D60-2015), is 0.305. The measured dynamic stress impact coefficient (μ) ranges from 0.103 to 0.163, which is lower than the theoretical value of 0.305.

Table 14. Dynamic response detection results.

Condition	10 km/h Braking	20 km/h Braking	30 km/h Braking
Dynamic Strain at Mid-span Section (με)	7	10	11
Impact Coefficient (μ)	0.103	0.119	0.163

6. Conclusions

This study systematically investigates the in situ active underpinning technique for the repair of damaged concrete box girder bridge piers on a highway in the western region. The safety and effectiveness of the technique are validated through numerical simulations, dynamic monitoring, and load testing. The main conclusions are as follows:

(1) Technological Innovation and Structural Performance Validation

This study introduces an innovative active underpinning technique that combines a "井"-shaped cap system, graded preloading of the foundation, and synchronized beam body correction. Through a two-stage temporary support system transition (Figure 7), a smooth transition from the pier–beam consolidation system to the basin-type bearing system is achieved, with precise control of beam torsion displacement (maximum uplift of 5.2 mm/settlement of 7.9 mm). Finite element simulations confirm that after underpinning, the safety factor at key cross-sections is significantly higher than the standard limit (e.g., the bending safety factor for Pier 2 is 1.99, and the safety factor for the newly added cap in the bridge direction is 1.92). The bearing performance meets the Class I highway load requirements. It is recommended that a long-term seismic monitoring system be installed for the bridge to accumulate seismic response data and provide a basis for seismic design and maintenance in similar engineering projects.

(2) Construction Monitoring and Load Test Results

Dynamic monitoring throughout the construction process indicates that the strain values of the new piers (−54.5~−51.3 με) and calibration coefficients (0.87~0.92) are within reasonable ranges. The residual deflection deformation of the main beam is less than 16.1%, and the vertical displacement, lateral deviation, and pier verticality meet the regulatory requirements. Load testing conducted before the bridge is opened to traffic confirmed that the repaired bridge exhibits excellent static and dynamic performance: the static load efficiency coefficient ranges from 0.95 to 0.99, and the relative residual strain (6.1%~17.2%) and deflection (0%~16.1%) are both below 20%, indicating that the structure is in an elastic working state. The measured damping ratio is 1.450, and the ratio of the measured to theoretical fundamental frequency is 1.249, confirming that the overall stiffness of the bridge span structure is normal. The dynamic load impact coefficients (driving μ = 0.058~0.171, braking μ = 0.103~0.163) are significantly lower than the theoretical value of 0.305. It is recommended that overloaded vehicles be strictly controlled and that daily maintenance and inspection of the bridge be enhanced.

(3) Long-Term Performance and Durability Study

After repair, the bridge's technical condition grade is restored to Class 2, and regular monitoring shows that the structure remains stable. To address long-term risks, it is recommended to extend the monitoring period in stages (monitoring differential settlement and fatigue damage every 7 days/month/quarter within 3 years). Additionally, non-destructive durability assessments should be included in annual inspections (such as ultrasonic crack extension detection and electrochemical potential method to assess rebar corrosion rate) to control creep and material degradation. Referring to Ding et al. (2024) [3] on the long-term settlement of shield tunnel pile underpinning, future studies could integrate BIM models to predict the evolution of creep.

(4) Future Research Directions and Engineering Applications

This technique provides an efficient and safe solution for the rapid repair of bridges in mountainous areas and has significant potential for wider application. Future work should focus on further optimizing the support system design, exploring concrete-filled steel tube composite structures, and incorporating smart monitoring technologies (such as 5G real-time transmission and AI-based deformation early warning) to enhance the level of construction automation in complex environments. Furthermore, it is recommended that validation be conducted under extreme conditions (such as simulating 1.3 times Class I highway heavy-load traffic and performing E2 earthquake time-history analysis) to improve the rigor of the technical system.

Author Contributions: Conceptualization, H.R. and H.Y.; methodology, H.R.; software, H.R.; validation, L.L. and P.X.; formal analysis, H.R.; investigation, L.L.; resources, P.X.; data curation, L.L.; writing—original draft preparation, H.R.; writing—review and editing, H.Y.; visualization, Y.W.; supervision, Y.W.; project administration, H.Y.; funding acquisition, H.Y. All authors have read and agreed to the published version of the manuscript.

Funding: The work described was supported by the National Natural Science Foundation of China (No.52078089, No.52274176, and No.52078090), the Guangdong Province Key Areas R & D Program (2022B0101070001), the Chongqing Elite Innovation and Entrepreneurship Leading Talent Project (CQYC20220302517), the Natural Science Foundation of Chongqing, China(cstc2021jcyj-msxmX1075), the Chongqing Natural Science Foundation Innovation and Development Joint Fund

(CSTB2022NSCQ-LZX0079), and the Chongqing Municipal Education Commission "Shuangcheng Economic Circle Construction in Chengdu-Chongqing Area" Science and Technology Innovation Project (KJCX2020031).

Informed Consent Statement: Informed consent was obtained from all subjects involved in the study.

Data Availability Statement: All data, models, or codes that support the findings of this study are available from the corresponding author upon reasonable request.

Conflicts of Interest: Authors Honglin Ran, Lin Li, Yi Wei and Penglin Xiao were employed by the company Chongqing Highway Maintenance Engineering (Group) Co., Ltd. The remaining author declares that the research was conducted in the absence of any commercial or financial relationships that could be construed as a potential conflict of interest.

References

1. Yan, L.; Gou, X.; Guo, Z.; Zhang, X.; Jiang, Y.; Ran, X.; Chen, G.; Yue, K. Experimental and numerical investigation on pile foundation underpinning structure system in urban overpass. *Materials* **2023**, *16*, 6576. [CrossRef] [PubMed]
2. Wang, N.; Jiang, Y.; Geng, D.; Huang, Z.; Ding, H. Numerical investigation of the combined influence of shield tunneling and pile cutting on underpinning piles. *Front. Earth Sci.* **2022**, *10*, 896634.
3. Ding, W.; Cao, K.; Wen, Y.; Du, Y.; Shang, C.; Li, Z.; Huang, X. Analysis of the Influence of Pile Underpinning of the Pile Group Under the Viaduct of the Overlapping Shield Tunnel on the Surrounding Environment. *Int. J. Civ. Eng.* **2024**, *22*, 1807–1833.
4. Ma, W.J.; Wang, X.; Wang, B.L.; Wang, B.L.; Wang, X.L. Application of Passive Pile Foundation Underpinning Technology on Lanzhou Rail Transit Line 1. *Urban Mass Transit.* **2020**, *23*, 51–55.
5. Park, Y.H.; Kim, J.P.; Cho, K.H. Stability analysis of subway box structure supported by modified underpinning method. *Tunn. Undergr. Space Tech.* **2015**, *50*, 199–208.
6. Peng, H.; Meng, B.; Tan, S. Study on Deformation Control of Road-Deep Foundation Pit Passing under Elevated Subway Bridge. *Appl. Sci.* **2024**, *14*, 6357. [CrossRef]
7. Zhang, C.; Zhao, Y.; Zhang, Z.; Zhu, B. Case study of underground shield tunnels in interchange piles foundation underpinning construction. *Appl. Sci.* **2021**, *11*, 1611. [CrossRef]
8. Li, Z.; Chen, Z.Q.; Wang, L.; Zeng, Z.K.; Gu, D.M. Numerical simulation and analysis of the pile underpinning technology used in shield tunnel crossings on bridge pile foundations. *Undergr. Space* **2021**, *6*, 396–408.
9. Yan, L.; Wang, G.; Chen, M.; Yue, K.; Li, Q. Experimental and application study on underpinning engineering of bridge pile foundation. *Adv. Civ. Eng.* **2018**, *2018*, 5758325.
10. Zheng, F.; Jiang, Y.; Wang, N.; Geng, D.; Xu, C. Investigation on the influence of active underpinning process on bridge substructures during shield tunnelling: Numerical simulation and field monitoring. *Buildings* **2023**, *13*, 241. [CrossRef]
11. Qiao, C.Q. Key Construction Technology of Continuous Beam Jacking-underpinning Pier Column Foundation. *Build. Constr.* **2020**, *42*, 1699–1701.
12. Liu, C.J.; Liu, Y.X. Application of various pile foundation underpinning technologies in Shenzhen metro line 9. *Mod. Urban Transit.* **2020**, *12*, 79–83.
13. Li, P.F. Analysis on Pile Foundation Passive Underpinning of Viaduct on Urban Expressway. *Transp. Sci. Technol.* **2022**, *4*, 56–61.
14. Kou, W. Application of Active Underpinning Technique in Construction of Zhengzhou Metro Shield Tunnel Passing Under Bridge Pile Foundation. *Rail. Eng.* **2018**, *058*, 87–90.
15. Zhou, Y.; Liu, Y.; Chen, Q.; Ou, X.; Li, Y. Case study of an underpinning pile foundation for an interval tunnel crossing an existing bridge. *Appl. Sci.* **2022**, *12*, 12566. [CrossRef]
16. Shan, H.F.; Xia, T.D.; Yu, F.; Tao, H.B.; He, S.H. Influence of underpinning pile drilling construction on the bearing behavior of existing loaded foundation piles: Case study. *Adv. Civ. Eng.* **2020**, *2020*, 9568279. [CrossRef]
17. Igba, U.T.; Igba, P.O.; Adekunle, A.A.; Labiran, J.O.; Oyebisi, S.O.; Cosmas, C.A. Strengthening and underpinning of a sinking two storey building in Lagos State Nigeria. *Eng. Fail. Anal.* **2023**, *150*, 107334. [CrossRef]
18. Tian, W.W. Summary of key points of bridge underpinning construction process control in operation. *Highway* **2022**, *67*, 275–278.
19. *JTG D60-2004*; General Code for Design Highway Bridges and Culverts. China Communication Press: Beijing, China, 2004.
20. *JTG D62-2004*; Code for Design of Highway Reinforced Concrete and Prestressed Concrete Bridges and Culverts. China Communication Press: Beijing, China, 2004.
21. *GB 18306-2015*; Seismic Ground Motion Parameter Zonation Map of China. China Standards Press: Beijing, China, 2016.
22. *JTG/T 2231-01-2020*; Specifications for Seismic Design of Highway Bridges. China Communication Press: Beijing, China, 2020.

23. *JTG D60-2015*; General Specifications for Design of Highway Bridges and Culverts. China Communication Press: Beijing, China, 2015.
24. *JTG 3362-2018*; Specifications for Design of Highway Reinforced Concrete and Prestressed Concrete Bridges and Cul-verts. China Communication Press: Beijing, China, 2018.
25. *JTG/T J21-01-2015*; Load Test Methods for Highway bridges. China Communication Press: Beijing, China, 2015.

Disclaimer/Publisher's Note: The statements, opinions and data contained in all publications are solely those of the individual author(s) and contributor(s) and not of MDPI and/or the editor(s). MDPI and/or the editor(s) disclaim responsibility for any injury to people or property resulting from any ideas, methods, instructions or products referred to in the content.

buildings

MDPI

Article

Dynamic Prediction Method for Ground Settlement of Reclaimed Airports Based on Grey System Theory

Ke Ma [1,2,4], He Weng [2], Zhaojun Luo [3], Saeed Sarajpoor [4] and Yumin Chen [5,*]

[1] School of Civil Engineering, Chongqing University, Chongqing 400045, China; marco1991@126.com
[2] Key Laboratory of Ministry of Education for Geomechanics and Embankment Engineering, Hohai University, Nanjing 210098, China; 17826026885@163.com
[3] Jiangsu SiWeiEr Architectural Design Institute Co., Ltd., Changzhou 213332, China; learnay8698@163.com
[4] Institute for Smart City of Chongqing University in Liyang, Chongqing University, Changzhou 213332, China; sarajpoor@163.com
[5] School of Civil Engineering, Suzhou University of Science and Technology, Suzhou 215011, China
* Correspondence: ymch@usts.edu.cn

Abstract: Settlement issues at airports pose a significant threat to operational safety, particularly in coastal regions, where land reclamation introduces unique challenges. The complexities of marine foundations, the difficulties in investigating reclaimed land, and the heightened risks of excessive settlement require timely and accurate monitoring and prediction to effectively identify risks and minimize unnecessary maintenance costs. To address these challenges, this study introduces a dynamic prediction model based on grey system theory, enhanced by a variable-size sliding window mechanism that continuously integrates the latest monitoring data. Validation using datasets from Kansai International Airport and Xiamen Xiang'an International Airport demonstrates that the model improves prediction accuracy by over 20% compared to existing models. Additionally, an exponential forecasting mechanism for long-term settlement prediction is developed and verified with data from Pudong International Airport. The proposed model demonstrates robust predictive capabilities across both long-term and short-term forecasting scenarios.

Keywords: dynamic settlement prediction; variable-size sliding windows; grey system

1. Introduction

With rapid urbanization, economic development, and the growing exploration and utilization of marine resources, coastal regions face increasing land scarcity due to high population density and extensive infrastructure projects such as ports and airports. Land reclamation through dredging and filling has emerged as a widely adopted solution to alleviate this issue. However, such marine reclamation projects, especially the construction of reclaimed airports, often involve significant geological challenges and engineering difficulties.

Lessons learned from incidents such as the large-scale settlement of Kansai Airport in Japan [1], the major landslide at Nice Airport in France [2], the sinking of the UAE's "World Islands", and the severe damage to protective structures in the Maldives caused by the Indian Ocean tsunami [3] demonstrate that artificial island construction projects face severe threats, including post-construction settlement, poor overall stability, and erosion by dynamic marine environments. During island reef development, airports—characterized by their massive scale, critical functions, high maintenance requirements, and difficult repair processes—play a pivotal role. Their safety constitutes a key factor determining the normal operation of island reef facilities.

Marine foundations are complex and challenging to investigate thoroughly, and the fill materials used in reclamation are often derived from seabed sediments or transported from other regions. These materials are prone to settlement, particularly during the initial stages of construction and after project completion. The resulting excessive settlement poses substantial risks to the structural integrity and operational safety of marine airports. Thus, timely monitoring and accurate settlement predictions are essential to maintain airport infrastructure stability and ensure safe, uninterrupted operations.

At present, settlement prediction methods can generally be divided into three main categories. The first category comprises methods based on consolidation theory, which use soil properties, compaction density, and filling rates to construct geometric models, often analyzed via numerical simulations using finite element methods (FEM) or finite difference methods (FDM) [4–9]. The second category includes function-fitting models, which rely on measured settlement data to generate predictive curves, such as the hyperbolic method [10], Poisson curve model [11,12], logistic curve model [13], and other curve-fitting techniques. The third category leverages intelligent machine-learning algorithms, such as ant colony optimization [14], support vector machines [15], random forests [16], and neural networks [17–19], to enhance prediction accuracy and adaptability.

Each of these three prediction methods has its own limitations. Prediction methods based on FEM or FDM for numerical calculations are highly dependent on mesh size. As the number of meshes increases, the computational complexity also rises, and achieving a globally convergent solution may not always be possible. Function model fitting methods often require manual parameter tuning, and inappropriate parameter choices can result in poor fitting performance. Furthermore, these parameters need to be adjusted as application scenarios change, which limits the general applicability of the method. Machine-learning-based prediction methods, while adaptable, typically require a large amount of high-dimensional data for effective modeling. In practical engineering projects, however, it is often challenging to obtain sufficient valid data, which can compromise prediction accuracy.

In recent years, grey system theory has been increasingly applied to settlement prediction. Positioned between the function fitting and machine-learning approaches, the grey model focuses on uncovering the internal relationships between data. It does not require manual parameter adjustment and is particularly well suited for linear predictions based on time-series data. Introduced by Deng in 1982 [20], grey system theory effectively predicts data evolution using small, incomplete, and less reliable data sets. As it employs differential equations to explore the underlying patterns in the data, it requires minimal information for modeling, achieves high accuracy, and avoids the need to account for specific distribution patterns or trends. Grey models have been successfully used in various predictive applications, including population growth [21], grain output [22], GDP [23], and landslide prediction [24,25]. In the field of landslide prediction, several advanced methods based on grey models have also been developed.

In terms of settlement prediction based on grey system theory, Jin et al. used the grey model Grey Model(1,1) (GM(1,1)) to predict the settlement of road embankments [26]. Wang et al. proposed a multivariable settlement prediction model, the multivariable grey model (1,n) (MGM(1,n)), which has better prediction accuracy compared to the traditional GM(1,1) model [27]. Zhang et al. combined the grey model with a fractional order, introducing a fractional-non-attribute GM(1,1) model to enhance settlement prediction accuracy by extending the cumulative order range of the model [28]. Zhang et al. proposed a composite prediction model based on the optimized discrete grey model and back-propagation neural network for better prediction of pit settlement [29].

However, these prediction methods based on grey system theory are often static, meaning they are only suitable for short-term predictions and lack the mechanisms for dynamic prediction. This study uses the GM(1,1) grey model with sliding window technology to propose a dynamic prediction model that can effectively use time series data of historical settlements for settlement prediction.

The novelty of this study is that the proposed model leverages grey system theory to predict airport settlement while employing a sliding window mechanism to continuously incorporate the latest monitoring data. When new data become available, the model is updated using this information, and the sliding window size is dynamically adjusted to refine the predictions. This approach effectively decomposes a long-term nonlinear problem into multiple short-term linear subproblems, thereby enhancing prediction accuracy. Comparative experiments were conducted using historical settlement data from Kansai International Airport and Xiamen Xiang'an International Airport—two offshore reclaimed airports—to validate the model's performance. Furthermore, the model's capability for long-term settlement prediction was verified by applying an exponential forecasting mechanism to historical data from Pudong International Airport, which is constructed on a soft alluvial deposit. The experimental results demonstrate that the proposed model achieves an improvement in prediction accuracy of over 20% compared to existing models.

2. Method

This section introduces the dynamic grey prediction model, as illustrated in the flow chart in Figure 1. The process begins with the collection of real-time geotechnical data using sensors and other equipment. Based on the current sliding window size, the data required for preprocessing are determined. The collected engineering data are then transformed into a format suitable for application within the grey model. The preprocessed data are analyzed using the grey model for prediction, and the sliding window size is adjusted as necessary to optimize performance. Finally, a decision is made on whether to continue collecting real-time data and performing subsequent predictions.

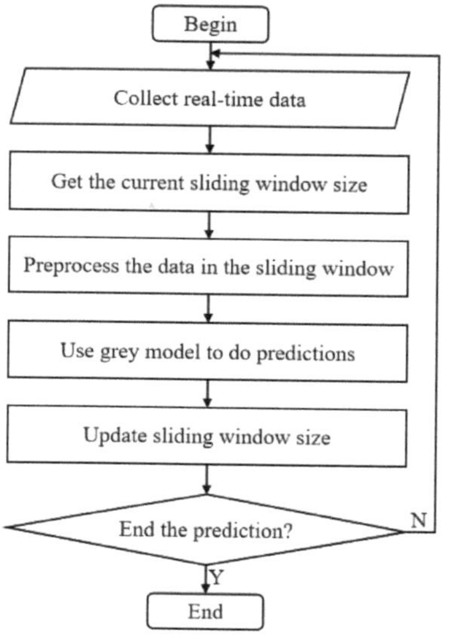

Figure 1. Flow chart of dynamic grey model.

Sections 2.1 and 2.2 describe the traditional grey model and sliding window technology, Section 2.3 discusses the data preprocessing methods, and Section 2.4 describes the methodology for selecting the sliding window size.

2.1. Grey Model

The preprocessed equidistant sequence is defined as $X^{(0)} = (x^{(0)}(1), x^{(0)}(2), \ldots, x^{(0)}(n))$, and the 1st-order accumulated generating operation sequence of $X^{(0)}$ is defined as $X^{(1)} = (x^{(1)}(1), x^{(1)}(2), \ldots, x^{(1)}(n))$, where $x^{(1)}(k) = \sum_{i=1}^{k} x^{(0)}, k = 1, 2, \ldots, n$.

Additionally, let $Z^{(1)} = (z^{(1)}(2), z^{(1)}(3), \ldots z^{(1)}(n))$, where $z^{(1)}(k) = (x^{(1)}(k) - x^{(1)}(k-1))/2$. Then, let

$$x^{(0)}(k) + az^{(1)}(k) = b \tag{1}$$

be the mean form of the GM(1,1) model, which is essentially a difference equation, where $-a$ is the development coefficient, and b is the grey action quantity. In this equation, $-a$ functions as the development coefficient, capturing the trend in the settlement rate, while b serves as the grey input, reflecting the impact of external environmental factors on the foundation settlement process. Specifically, a relatively large value of b suggests that the settlement may be significantly influenced by the intrinsic properties of the soil, resulting in a slower stabilization of the settlement rate; it may also indicate that the foundation settlement is under continuous loading, or that notable fluctuations in groundwater levels are accelerating soil consolidation, among other possibilities.

The parameter vector $\hat{a} = [a, b]^T$ can be estimated using the least squares method.

$$\hat{a} = (B^T B)^{-1} B^T Y \tag{2}$$

where

$$Y = \begin{bmatrix} x^{(0)}(2) \\ x^{(0)}(3) \\ \cdots \\ x^{(0)}(n) \end{bmatrix}, \quad B = \begin{bmatrix} -z^{(1)}(2) & 1 \\ -z^{(1)}(3) & 1 \\ \cdots & \cdots \\ -z^{(1)}(n) & 1 \end{bmatrix} \tag{3}$$

Here, the predictive value of the k-th element in the sequence of the settlement $\hat{x}^{(1)}(k)$ can be calculated based on a, b, and the first element $x^{(0)}(1)$ in the sequence.

$$\hat{x}^{(1)}(k) = (x^{(0)}(1) - \frac{b}{a}) \times (\frac{1}{1+a})^{k-1} + \frac{b}{a}, k = 1, 2, 3, \ldots \tag{4}$$

The cumulative reduction restoration formula of the GM(1,1) model is:

$$\hat{x}^{(0)}(k) = a^{(1)}\hat{x}^{(1)}(k) = \hat{x}^{(1)}(k) - \hat{x}^{(1)}(k-1),\ k = 2, 3, \ldots \tag{5}$$

After processing the monitoring data to equidistant intervals and establishing the corresponding GM(1,1) model, the development coefficient $-a$ and the grey action quantity b can be computed. Then, based on Equations (4) and (5), $\hat{x}^{(0)}(n+1)$, $\hat{x}^{(0)}(n+2)$, $\hat{x}^{(0)}(n+3)\ldots$ can be computed. Finally, after using the corresponding k, the corresponding predicted values can be calculated.

2.2. Sliding Window

In practical engineering, long-term system changes are typically nonlinear, but over shorter periods, these changes can often be approximated as linear. This section utilizes a sliding window approach to select the most recent valid data, thereby improving prediction accuracy.

Figure 2 illustrates an example of a sliding window with a length of 5. Initially, the data points within the window are $\{0.24, 0.248, 0.25, 0.246, 0.269\}$. When a new data point (0.262) is acquired, the window updates by removing the oldest data point (0.24) and adding the new data point (0.262) at the end. If the first, third, and fifth data points in the sliding window are selected as input data, the initial input is $\{0.24, 0.25, 0.269\}$. After incorporating the new data point (0.262), the input data change to $\{0.248, 0.246, 0.262\}$, resulting in two distinct input datasets.

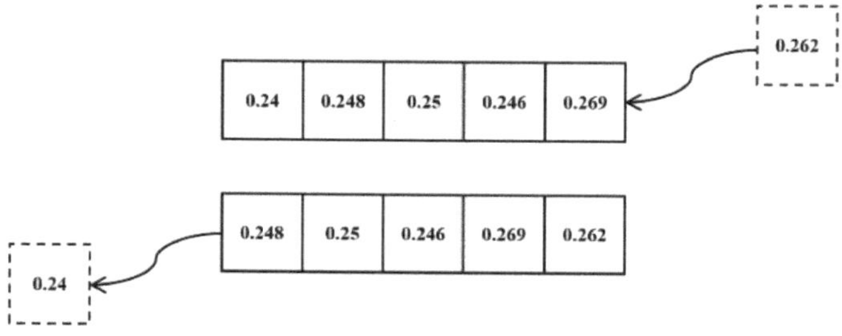

Figure 2. An example of a sliding window (size: 5).

The application of sliding window technology in geotechnical engineering enables the dynamic replacement of outdated data with real-time monitoring data. This process enhances the accuracy and adaptability of prediction models. However, using a fixed window size can introduce challenges. A window that is too small may not contain sufficient data, leading to potential errors due to inadequate information. Conversely, a window that is too large may include excessive redundant data, and outdated information could compromise the model's prediction accuracy and real-time performance.

This study examines the prediction accuracy of existing models under different window sizes and proposes a method for dynamically adjusting the sliding window size in real time (see Section 2.4).

2.3. Preprocessing Methods

Due to factors such as sensor malfunctions, ensuring that data acquisition intervals meet the equidistant requirement of grey models can be challenging in settlement applications. Therefore, preprocessing of the acquired data is necessary. In this study, the mean value interpolation method was utilized for data preprocessing, although other interpolation methods are also viable.

Suppose the time at which data need to be acquired is denoted as t. Let t_1 and t_2 be the two nearest monitoring time points before and after t, respectively, with $t_1 < t < t_2$. The measured values at these time points are denoted as $y(t_1)$ and $y(t_2)$, respectively. The processed monitoring value $y(t)$ at time t can be computed as follows using mean value interpolation:

$$y(t) = y(t_1) + \frac{t - t_1}{t_2 - t_1} \times (y(t_2) - y(t_1)) \tag{6}$$

Based on different application scenarios, this study proposes two mechanisms for settlement prediction:

(1) Equidistant mechanism: After preprocessing, the data sequence provided to the grey system maintains uniform time intervals. For example, in settlement prediction, if the monitoring interval is set to one month, the grey system also uses data collected at one-month intervals. This approach allows the model to predict settlement for subsequent months using the available data. Data collection by the equidistant grey model is straightforward, as it only requires specifying the time interval for data acquisition. However, its predictive horizon is relatively short. For instance, a model using the equidistant mechanism may leverage one year of settlement data to predict settlement for the next three months, but it may not be suitable for longer-term predictions.

(2) Exponential increment mechanism: In this approach, the data sequence fed into the grey system has exponentially increasing time intervals after preprocessing. For example, in settlement prediction, the minimum monitoring interval is m days, and the exponential increment factor is a^x. The monitoring times then become m days, $m \times a^x$ days, $m \times a^{2x}$ days, $m \times a^{3x}$ days, and so on. These intervals appear "equidistant" on a logarithmic scale and serve as inputs for the grey model to make predictions. This mechanism is particularly suitable for long-term predictions, such as using data collected over one year to forecast settlement over a decade. However, it requires more extensive data collection, especially during the initial phase, as multiple sets of data with shorter monitoring intervals are needed at the outset.

Figure 3 shows a flow chart of prediction when using the exponential increment mechanism (the size of the sliding windows is 5). Assuming "today" is the 100th day of data acquisition, the input data sequence corresponds to days $\{10^{1.2}, 10^{1.4}, 10^{1.6}, 10^{1.8}, 100\}$ $(x^{(0)}(1), x^{(0)}(2), x^{(0)}(3), x^{(0)}(4), x^{(0)}(5))$, and the data for 10^x days are calculated using Formula (6). Subsequently, this input sequence is used for modeling, yielding the predicted result for days $\{10^{2.2}, 10^{2.4}, 10^{2.6}, \ldots\}$ $(\hat{x}^{(1)}(6), \hat{x}^{(1)}(7), \hat{x}^{(1)}(8), \ldots)$. If predictions for one year later (day 465) and two years later (day 830) are required, the corresponding results can be calculated based on the predicted data for days $10^{2.6}$, $10^{2.8}$, and $10^{3.0}$ $(\hat{x}^{(1)}(8), \hat{x}^{(1)}(9), \hat{x}^{(1)}(10))$ using Formula (6).

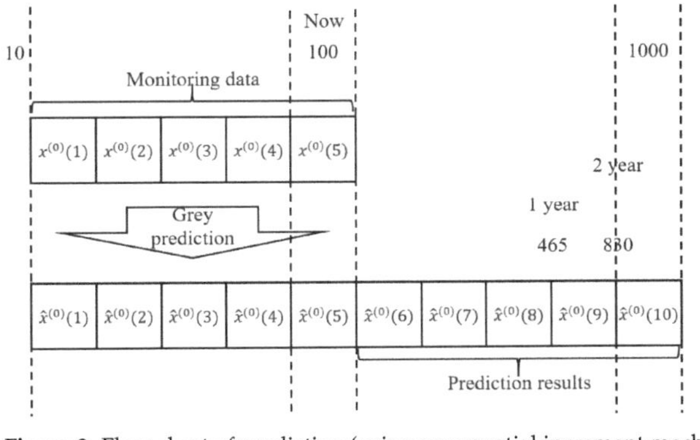

Figure 3. Flow chart of prediction (using exponential increment mechanism).

2.4. Choosing Size of Sliding Window

This study employs real-time monitoring data to validate the suitability of the chosen sliding window size. The process involves computing the prediction accuracy for various sliding window sizes and adopting the optimal size as the subsequent prediction criterion.

Given that the current sliding window size s is l, the candidate window sizes are set to $(l-1)$, l, and $(l+1)$, where the potential sliding window size fluctuates by up to 1 unit from the current sliding window size.

We assume that the value of the k-th data point in the data sequence is $y(k)$, and the last data point obtained in the sequence is the n-th data point with a value of $y(n)$. Subsequently, a GM(1,1) model is established using the most recent s data points, ranging from $(n-4-s)$ to $(n-5)$, as inputs to forecast the values at positions $(n-2)$, $(n-1)$, and n within the sequence. The predicted values are denoted as $y'(n-2)$, $y'(n-1)$, and $y'(n)$, respectively. The weighted absolute difference between the predicted values and the true measured values is used as the evaluation metric, denoted as GAP:

$$GAP(s=l) = \beta_1|y'(n-2) - y(n-2)| + \beta_2|y'(n-1) - y(n-1)| + \beta_3|y'(n) - y(n)| \quad (7)$$

where $\beta_1 + \beta_2 + \beta_3 = 1$, and β_i $(i = 1,2,3)$ is the weight for each predicted position. In general, it can be set as $\beta_1 = \beta_2 = \beta_3$. If long-term prediction is more important in the application, it can be set as $\beta_1 < \beta_2 < \beta_3$, whereas if short-term prediction is more important, it can be set as $\beta_1 > \beta_2 > \beta_3$.

If $GAP(s = l-1) = min\{GAP(s = l-1), GAP(s = l), GAP(s = l+1)\}$, then the sliding window size for the next stage is set to $l-1$.

If $GAP(s = l) = min\{GAP(s = l-1), GAP(s = l), GAP(s = l+1)\}$, then the sliding window size for the next stage is set to l.

Otherwise, the sliding window size for the next stage is set to $l+1$.

It is important to ensure that the sliding window size s is appropriately chosen to provide sufficient information for establishing the grey model. In geotechnical engineering applications, it is recommended that s be greater than 2 to avoid insufficient data. At the same time, to minimize excessive data redundancy, s should not exceed 9.

3. Results

In this section, comparative experiments are conducted based on three different datasets for settlement prediction to validate the predictive accuracy of the dynamic grey model proposed in this study.

3.1. Settlement Prediction Experiment on Kansai International Airport Dataset

This section utilizes publicly available settlement data from Kansai International Airport (http://www.kansai-airports.co.jp/efforts/our-tech/kix/sink/sink3/sink3_e.html, accessed on 31 August 2024) to verify the prediction accuracy of the proposed dynamic grey model, where the settlement of each monitoring point is recorded every year. The monitoring data from the A-8 and B-5 monitoring point at Kansai International Airport from 2003 to 2022 are used as inputs for comparison experiments between the traditional grey prediction model, the grey prediction model with a fixed-size sliding window, and the dynamic grey prediction model proposed in this study. For the proposed model, the equidistant mechanism is used, and the interval of the time series is one year.

Figure 4a shows the comparative experimental results of the settlement predictions for Kansai International Airport A-8 monitoring point. The results show that the traditional grey model can effectively predict settlement for the first 4 years, but after 2012, the accuracy is low, with a prediction deviation of more than 2 m. The proposed model can continuously and effectively predict settlement, with a prediction deviation of less than 0.1 m for the last few years. Moreover, the proposed model can use existing data to predict settlement after 2 years, making it more accurate than the traditional grey model.

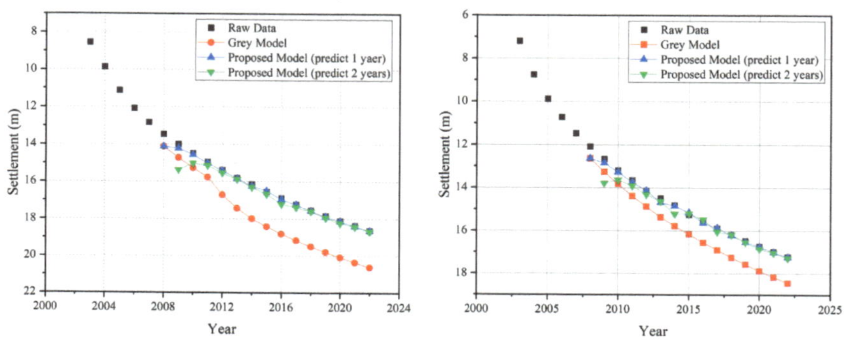

(a) A-8 monitoring point **(b)** B-5 monitoring point

Figure 4. Experimental results of settlement predictions at Kansai International Airport.

Figure 5a,b shows QQ plots of the actual and predicted values using the three models and for different years at Kansai International Airport. The figure illustrates the one-year prediction results obtained using the proposed model. The data points closely align with the y = x line, demonstrating strong agreement between the predicted values and the actual monitoring data.

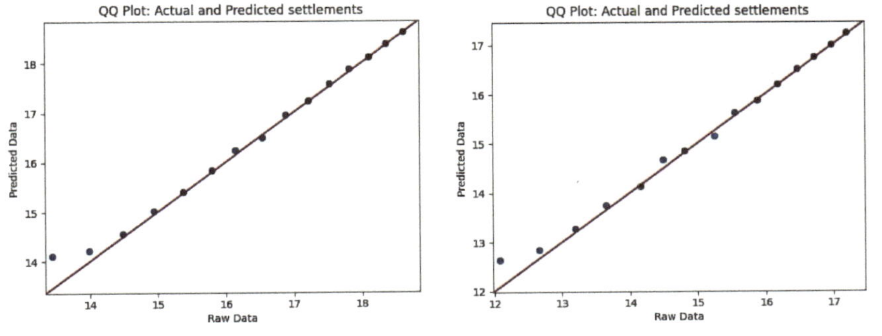

(a) A-8 monitoring point **(b)** B-5 monitoring point

Figure 5. Experimental results of QQ plots at Kansai International Airport.

Figure 6a and Table 1 present the settlement data for the A-8 monitoring point at Kansai International Airport from 2003 to 2022, along with the deviation results obtained using the three prediction models. For the traditional grey model, only the next position of the existing data sequence is predicted (e.g., using the settlement data sequence from 2003 to 2012 to predict settlement in 2013, or comparing the predicted result with the monitored value in 2013 to obtain the prediction error). In the other two prediction models, the next one to three positions of the existing data sequence can be predicted effectively.

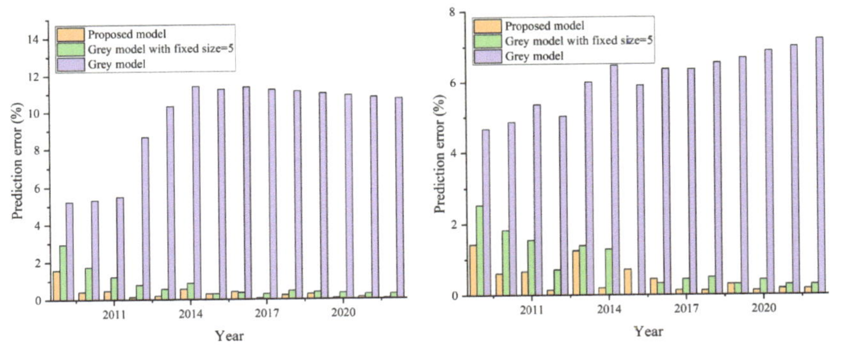

(a) A-8 monitoring point **(b)** B-5 monitoring point

Figure 6. Experimental results of prediction errors at Kansai International Airport.

Table 1 shows that the soil settlement prediction accuracy using the traditional grey model is poor, with errors exceeding 10%. Based on the traditional grey model, adding a fixed-size sliding window mechanism (with a size of 5) can effectively predict the next one to three positions of the existing data sequence. Using the proposed model, the prediction accuracy can be further improved based on the fixed-size sliding window grey model. According to the experimental results, the proposed model maintains prediction deviation below 0.5% when forecasting the next position in the data sequence. For the next two to three positions, the prediction deviation remains below 2% in most cases, except for certain instances (e.g., predicting settlement in 2014 using data from 2003 to 2011). Among the three models, the proposed model consistently demonstrates the highest prediction accuracy. Additionally, the t-test results for the last 14 years (with a degree of freedom of 13) indicate that the proposed model outperforms both the traditional grey model and the grey model with a fixed-size sliding window. The absolute value of t for the proposed model is less than 3, significantly lower than those of the other two models, confirming its superior performance.

Table 1. Experimental results of Kansai International Airport A-8 monitoring point.

	Measured Value (m)	Proposed Model		Grey Model	Fixed-Size Window	
		Next 1	Next 2	Next 1	Next 1	Next 2
2003	8.55	-	-	-	-	-
2004	9.87	-	-	-	-	-
2005	11.13	-	-	-	-	-
2006	12.08	-	-	-	-	-
2007	12.82	-	-	-	-	-
2008	13.45	4.91%	-	4.91%	4.91%	-
2009	13.99	1.57%	9.79%	5.22%	2.93%	9.79%
2010	14.49	0.41%	3.52%	5.31%	1.73%	5.73%
2011	14.94	0.47%	1.27%	5.49%	1.20%	3.61%
2012	15.38	0.13%	1.04%	8.65%	0.78%	2.41%
2013	15.80	0.19%	0.51%	10.32%	0.57%	1.58%
2014	16.14	0.56%	0.99%	11.40%	0.87%	1.61%
2015	16.54	−0.30%	0.79%	11.25%	0.30%	1.27%
2016	16.88	0.41%	1.90%	11.37%	0.36%	0.89%
2017	17.22	0.06%	0.87%	11.21%	0.29%	0.75%
2018	17.53	0.23%	0.29%	11.12%	0.46%	0.74%
2019	17.82	0.28%	0.62%	11.00%	0.39%	0.95%
2020	18.10	0.06%	0.61%	10.88%	0.33%	0.77%
2021	18.36	0.11%	0.27%	10.78%	0.27%	0.71%
2022	18.61	0.05%	0.32%	10.69%	0.27%	0.59%
RSME	-	0.1851	0.4145	1.6395	0.2615	0.4857
t-value (DF = 13)	-	−2.8494	−2.6659	−12.3249	−3.8979	−3.5789

Figure 4b shows the comparative experimental results of settlement predictions for the Kansai International Airport B-5 monitoring points. The settlement variation shows that the prediction deviation of the traditional grey model increases with time, with deviation of more than 1.2 m in 2022. The proposed model can continuously and effectively predict settlement, with a prediction deviation of less than 0.1 m for the last few years. Moreover, the proposed model can use existing data to predict settlement after 2 years, making it more accurate than the traditional grey model.

Figure 6b and Table 2 present the settlement data for monitoring point B-5 at Kansai International Airport from 2003 to 2022, and the deviation results of the predictions made using the three different models are presented. Table 2 shows that the traditional grey model exhibits high prediction deviations for soil settlement. After adding the fixed-size sliding window mechanism, predictions for the next one to three positions in the existing data sequence can be effectively made. The experimental results demonstrate that the proposed model achieves a prediction error of less than 1% when predicting the next position in the data sequence. For predictions of the next two to three positions, the error remains below 3% in most cases, with only a few exceptions. Among the three models, the proposed model consistently delivers the highest prediction accuracy. Furthermore, the *t*-test results for the last 14 years (with a degree of freedom of 13) confirm that the proposed model outperforms both the traditional grey model and the grey model with a fixed-size sliding window. The absolute value of t for the proposed model is less than 2.5, significantly lower than those of the other two models, underscoring its superior performance.

Table 2. Experimental results of Kansai International Airport B-5 monitoring point.

	Measured Value (m)	Proposed Model		Grey Model	Fixed-Size Window	
		Next 1	Next 2	Next 1	Next 1	Next 2
2003	7.20	-	-	-	-	-
2004	8.75	-	-	-	-	-
2005	9.89	-	-	-	-	-
2006	10.72	-	-	-	-	-
2007	11.46	-	-	-	-	-
2008	12.08	4.55%	-	4.55%	4.55%	-
2009	12.66	1.42%	8.85%	4.66%	2.53%	8.85%
2010	13.19	0.61%	3.34%	4.85%	1.82%	5.16%
2011	13.65	0.66%	1.83%	5.35%	1.54%	3.88%
2012	14.15	−0.14%	1.13%	5.02%	0.71%	2.61%
2013	14.49	1.24%	0.90%	6.00%	1.38%	2.48%
2014	14.81	0.20%	2.70%	6.48%	1.28%	2.90%
2015	15.25	−0.72%	−0.39%	5.90%	0.00%	1.44%
2016	15.55	0.45%	−0.51%	6.37%	0.32%	0.77%
2017	15.88	−0.13%	0.94%	6.36%	0.44%	0.69%
2018	16.18	0.12%	−0.06%	6.55%	0.49%	0.99%
2019	16.46	0.30%	0.43%	6.68%	0.30%	1.09%
2020	16.72	0.12%	0.72%	6.88%	0.42%	0.72%
2021	16.97	0.18%	0.35%	7.01%	0.29%	0.82%
2022	17.20	0.17%	0.47%	7.21%	0.29%	0.70%
RSME	-	0.1644	0.3561	0.9379	0.2031	0.4335
t-value (DF=13)	-	−2.2205	−2.4865	−16.9147	−4.778	−4.3224

3.2. Settlement Prediction Experiment on Data from Xiamen Xiang'an International Airport

This section uses the settlement data of Xiamen Xiang'an International Airport obtained in ref. [30], which include the deformation data of six monitoring points at Xiamen Xiang'an International Airport. The monitoring data from 2016 to 2020 are used as inputs for comparison experiments between the dynamic grey prediction model proposed in this study, the grey prediction model with a fixed-size sliding window, and the traditional grey prediction model. For the proposed model, the equidistant mechanism is used, and the interval of the time series is 0.25 years.

Figure 7a–e shows the comparative experimental results of settlement predictions for the different monitoring points. The experimental results demonstrate that the proposed prediction model more effectively captures the variations in airport settlement and delivers more accurate predictions. Except for a few specific time periods (e.g., around June 2017 for monitoring points P1 and P2, and around January 2018 for monitoring point P3), the settlement prediction error of the proposed model remains below 2 mm. While the grey model with a fixed window performs slightly less effectively, its results are acceptable. In contrast, the traditional grey model produces significant errors, with these inaccuracies compounding over time.

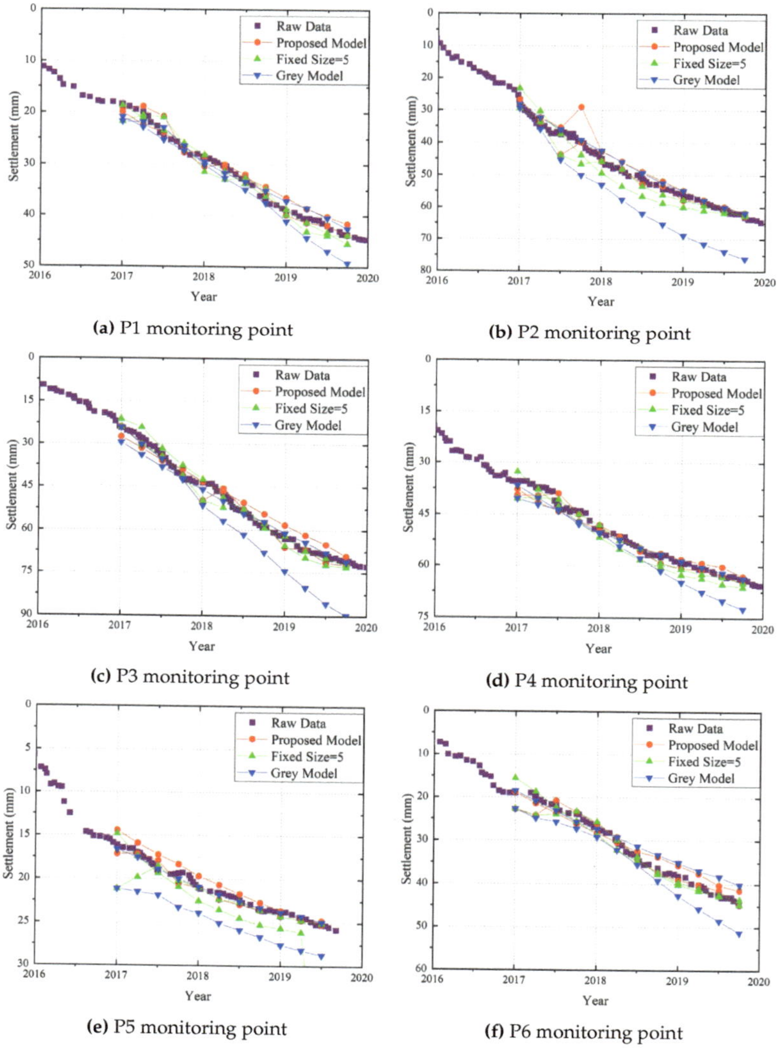

(a) P1 monitoring point
(b) P2 monitoring point
(c) P3 monitoring point
(d) P4 monitoring point
(e) P5 monitoring point
(f) P6 monitoring point

Figure 7. Experimental results of settlement predictions at Xiamen Xiang'an Airport.

Figure 8a–e show QQ plots of the actual and predicted values obtained using the three models and in different years at Xiamen Xiang'an International Airport. The figure illustrates the one-year prediction results obtained using the proposed model. The data points closely align with the y = x line, demonstrating strong agreement between the predicted values and the actual monitoring data.

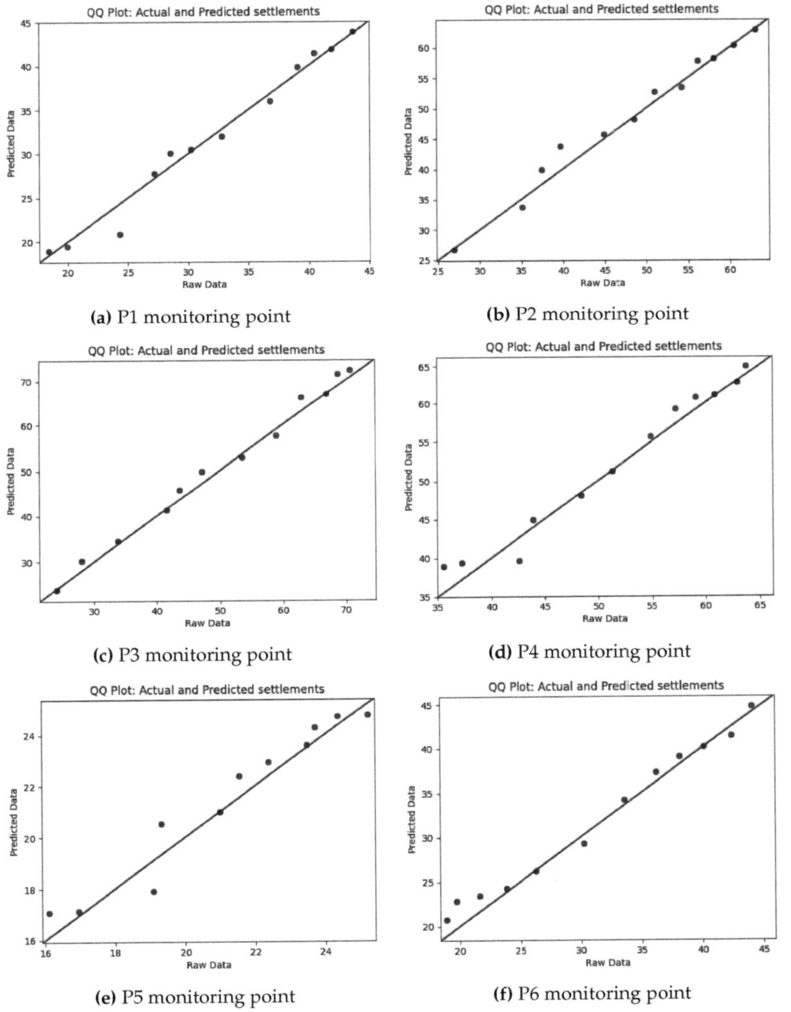

(a) P1 monitoring point **(b)** P2 monitoring point

(c) P3 monitoring point **(d)** P4 monitoring point

(e) P5 monitoring point **(f)** P6 monitoring point

Figure 8. Experimental results of QQ plots at Xiamen Xiang'an Airport.

Figure 9a presents the root-mean-square error (RMSE) of the proposed model, grey model with a fixed size, and the fitting methods proposed in ref. [30]. The experimental results demonstrate that the proposed prediction model consistently achieves the lowest root-mean-square error (RMSE) across all monitoring points. Furthermore, excluding a few specific time periods (e.g., around June 2017 for monitoring points P1 and P2, and around January 2018 for monitoring point P3), the RMSE of the proposed model is even lower. The RMSE of the grey model with a fixed window is also within an acceptable range, roughly aligning with the hyperbolic fitting model referenced in the literature.

In contrast, the prediction errors of the traditional grey model are significantly larger and deemed unacceptable. Overall, the proposed prediction model exhibits outstanding prediction accuracy.

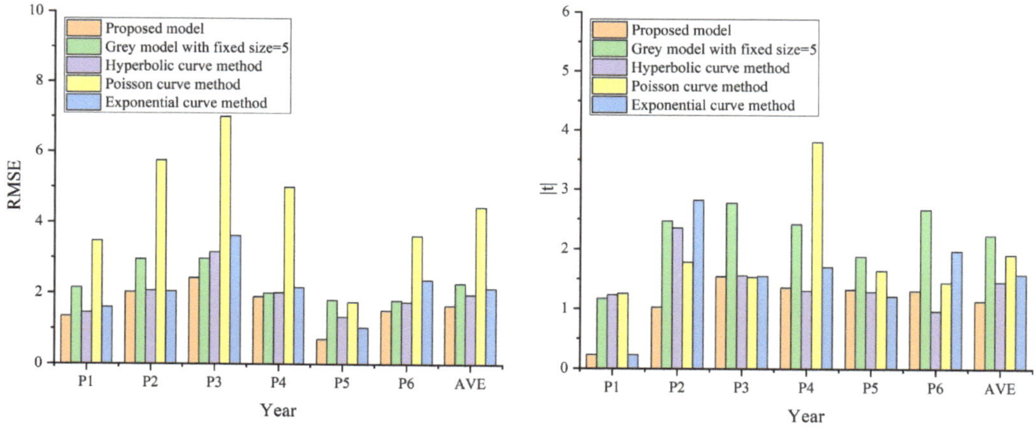

(a) RSME results **(b)** *t*-test results

Figure 9. Experimental results of settlement predictions at Xiamen Xiang'an International Airport.

Figure 9b presents the *t*-test results, showing that, in most cases, the proposed model performs either the best or second best. Additionally, the average absolute *t*-value for the proposed model is less than 1.5, outperforming all three fitting methods described in ref. [30] and the other two grey models. This confirms that the proposed model delivers superior performance.

3.3. Settlement Prediction Experiment on Data from Shanghai Pudong International Airport

In this section, the monitoring data from the P215 and P230 monitoring points at Shanghai Pudong International Airport from 1998 to 2010 are used as inputs for the predictions of the dynamic grey model proposed in this study. Shanghai Pudong International Airport is not an offshore reclaimed airport, but it has a typical soft soil foundation. Therefore, settlement prediction for this airport is equally important. In the prediction of settlement at Pudong International Airport, the exponential increment mechanism is employed. During the experiment, settlement data from each year are used to predict the settlement conditions for the following 0.5 years, 1 year, and 1.5 years. For example, data from March 2000 to March 2001 are used to predict settlement from April to September 2001, April 2001 to March 2002, and April 2001 to September 2002.

Figure 10a,b shows the settlement prediction results for monitoring points P215 and P230. The experimental results show that the proposed prediction model effectively forecasts airport settlement in most cases. As shown in the figures, the prediction performance for the period from 2001 to 2003 is less accurate compared to that for other periods. This is attributed to the lack of settlement monitoring data during this time, requiring older data to be used for the simulation, which diminishes the prediction accuracy. When monitoring data are recorded consistently, the proposed model efficiently leverages a limited amount of historical data (from the last year) to predict long-term settlement conditions over the next 0.5 to 1.5 years.

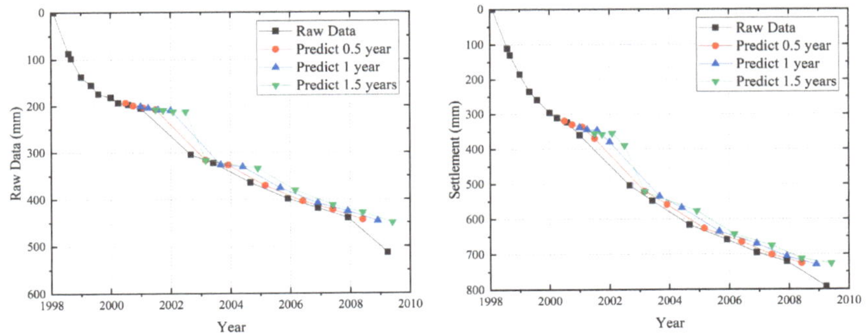

(a) P215 monitoring point **(b)** P230 monitoring point

Figure 10. Experimental results of settlement predictions at Shanghai Pudong International Airport.

4. Discussion

4.1. Performance and Comparisons Between Different Models

This study presents a dynamic grey model based on the GM(1,1) grey model and sliding window technology. The model is designed to utilize a small amount of recent settlement data to perform accurate predictions. Additionally, the sliding window size can be flexibly adjusted to improve prediction accuracy. As long as some monitoring data are available, the model maintains robustness and accuracy, making it applicable across various scenarios.

The model also incorporates two preprocessing mechanisms tailored to different prediction scenarios. The equidistant mechanism is ideal for short-term settlement predictions, such as using one year of settlement data to forecast settlement over the next three months. Conversely, the exponential increment mechanism is better suited for long-term predictions, such as leveraging one year of data to estimate settlement over a decade. While the equidistant mechanism typically delivers higher accuracy, the two mechanisms can be used simultaneously to complement each other, enhancing the model's adaptability to diverse prediction scenarios.

Table 3 compares the proposed model with models based on consolidation theory, function fitting, machine learning, and other grey system-based approaches. Unlike other grey system-based models, the proposed model incorporates sliding window technology, enabling it to utilize the most recent monitoring data for prediction. This makes it a dynamic prediction model.

Table 3. Comparisons between different models.

	Dynamic	Requirements for Data	Demand for Resources
Proposed model	Yes	Small size, easy to obtain	Low
Models based on consolidation theory	No	Small size, easy to obtain	Moderate
Models based on function fitting	No	Small size, easy to obtain	Low
Models based on machine learning	Yes/no	Big size, difficult to obtain	High
Other grey system-based models	No	Small size, easy to obtain	Low

While some machine-learning algorithms are also dynamic, their prediction models often require large amounts of high-dimensional data for effective modeling. However, in practical single engineering projects, obtaining sufficient valid data can be challenging, which negatively impacts prediction accuracy. The other models lack a mechanism for dynamic prediction.

Compared to these alternatives, the proposed prediction model demonstrates superior performance.

4.2. Future Directions

Further research could focus on the following areas:

(1) Expansion to different grey models and data preprocessing methods: The dynamic grey model proposed in this study is based on the traditional GM(1,1) grey model and employs a simple mean-difference method for data preprocessing. Depending on the application requirements, other grey models, such as GM(1,n) or GM(2,1), could also be utilized. This study considered predicting settlement based on previous settlement monitoring data. I have checked. However, if additional multidimensional data—parameters that collectively influence settlement—were available, the GM(1,n) model would likely be more effective. Furthermore, if rapid detection of settlement acceleration is required, the GM(2,1) model would be more appropriate. By combining sliding window techniques with various data preprocessing approaches, dynamic models tailored to a wider range of scenarios could be developed. In that case, the dynamic prediction framework demonstrates significantly enhanced applicability across diverse operational scenarios [31,32].

The GM(1,n) model enhances predictive interpretability in complex scenarios through multivariate expansion, though this requires increased data acquisition costs (collecting multi-dimensional parameters) and enables the incorporation of geotechnical variables (e.g., soil permeability) and environmental factors (e.g., groundwater fluctuations) as auxiliary inputs—a capability absent in the univariate GM(1,1) framework. In contrast, the GM(2,1) model strengthens nonlinear adaptability via second-order dynamic modeling, but demands extended monitoring sequences (typically more than 10 data points), achieving superior performance in long-term nonlinear settlement prediction where conventional first-order models exhibit progressive error accumulation.

(2) Enhancing early warning systems: The proposed model demonstrates strong predictive performance and can also be effectively applied to early warning systems. If significant deviations between monitoring data and predicted settlement values persist over a certain period, an early warning can be issued.

For instance, the proposed prediction model can be seamlessly integrated into airport infrastructure monitoring and maintenance systems. By continuously comparing real-time monitoring data with predicted settlement values, the model can promptly detect significant deviations, assess potential risks, and issue timely warnings. For example, if the monitored settlement exceeds the predicted value by X cm and this discrepancy continues for N consecutive days (X and N should be calculated depending on factors such as soil type, foundation conditions, and external loads), the model can trigger an early warning to highlight potential issues.

(3) Integration of physical significance into dynamic grey models: The dynamic grey prediction model proposed in this study is purely a mathematical model driven by monitoring data, without requiring an analysis of the physical mechanisms behind settlement. Future research could achieve more accurate predictions by integrating data-driven methods with the physical principles underlying settlement behavior. For instance, physical mechanisms such as soil consolidation, secondary settlement, and creep deformation can be used to constrain the prediction range of the model proposed in this study.

(4) Application to other settlement predictions: The settlement prediction model proposed in this study is a versatile dynamic forecasting model that can be used for predictions as long as continuous monitoring data are available. It is believed that this model can be applied to other settlement prediction scenarios, particularly in settlement predictions for building construction in other land reclamation projects.

5. Conclusions

Based on grey system theory and sliding window techniques, this study proposes a dynamic model for airport settlement prediction. The model evaluates predictions using the most recent data and refines its forecasting capability by dynamically adjusting the sliding window size. In addition, two mechanisms—an equidistant mechanism and an exponential mechanism—are integrated into the model to address short-term and long-term settlement prediction scenarios, respectively. For short-term predictions, comparative experiments using historical settlement data from Kansai International Airport and Xiamen Xiang'an International Airport—two offshore reclaimed airports—demonstrated an improvement in accuracy of over 20% compared to existing models. In the long-term scenario, effective predictions were obtained using historical data from Pudong International Airport, which is constructed on a soft alluvial deposit. The proposed forecasting model is purely mathematical, eliminating the need to account for numerous physical parameters, and is straightforward to implement. However, the model relies solely on a single dimension of real-time monitoring data (i.e., historical settlement data); although it provides an interpolation method to estimate data at specific time points, it still requires high-quality monitoring data and a sufficient acquisition frequency. Poor data quality or prolonged gaps in data collection may significantly compromise the model's predictive accuracy. Incorporated multidimensional data monitoring can be used to mitigate potential issues arising from data gaps.

Author Contributions: Conceptualization, K.M. and H.W.; methodology, K.M. and S.S.; software, K.M.; validation, K.M. and Z.L.; formal analysis, K.M.; investigation, K.M.; resources, Z.L.; data curation, K.M.; writing—original draft preparation, K.M.; writing—review and editing, S.S.; visualization, K.M.; supervision, Y.C.; project administration, Y.C.; funding acquisition, Y.C. and K.M. All authors have read and agreed to the published version of the manuscript.

Funding: This research was funded by the National Natural Science Foundation of China under Grant 52179101, and by Key Laboratory of the Ministry of Education for Geomechanics and Embankment Engineering, Hohai University, under Grant 2024002.

Data Availability Statement: The dataset used in this study is available on request from the authors.

Conflicts of Interest: Author Zhaojun Luo was employed by Jiangsu SiWeiEr Architectural Design Institute Co., Ltd. The remaining authors declare that the research was conducted in the absence of any commercial or financial relationships that could be construed as a potential conflict of interest.

Abbreviations

FEM	finite element method
FDM	finite difference method
GM(1,1)	grey model (1,1)
MGM(1,n)	multivariable grey model (1,n)
$-a$	development coefficient
b	grey action quantity
$\hat{a}$	$[a,b]^T$
$x^{(0)}(k)$	the monitoring value of the k-th element in the sequence of settlement
$X^{(0)}$	the monitoring value sequence of settlement
$X^{(1)}$	the 1st-order accumulated generating operation sequence of $X^{(0)}$
$\hat{x}^{(0)}(k)$	the predictive value of the k-th element in the sequence of settlement
RMSE	root-mean-square error
QQ Plot	quantile-quantile Plot

References

1. Mesri, G.; Funk, J. Settlement of the Kansai international airport islands. *J. Geotech. Geoenviron. Eng.* **2015**, *141*, 04014102.
2. Dan, G.; Sultan, N.; Savoye, B. The 1979 Nice harbour catastrophe revisited: Trigger mechanism inferred from geotechnical measurements and numerical modelling. *Mar. Geol.* **2007**, *245*, 40–64.
3. Fujima, K.; Shigihara, Y.; Tomita, T.; Honda, K.; Nobuoka, H.; Hanzawa, M.; Fujii, H.; Ohtani, H.; Orishimo, S.; Tatsumi, M.; et al. Survey results of the Indian Ocean tsunami in the Maldives. *Coast. Eng. J.* **2006**, *48*, 81–97.
4. Shi, X.S.; Zhao, J. Practical estimation of compression behavior of clayey/silty sands using equivalent void-ratio concept. *J. Geotech. Geoenviron. Eng.* **2020**, *146*, 04020046.
5. Nikakhtar, L.; Zare, S.; Nasirabad, H.M.; Ferdosi, B. Application of ANN-PSO algorithm based on FDM numerical modelling for back analysis of EPB TBM tunneling parameters. *Eur. J. Environ. Civ. Eng.* **2022**, *26*, 3169–3186. [CrossRef]
6. Liu, C.; Wang, Z.; Liu, H.; Cui, J.; Huang, X.; Ma, L.; Zheng, S. Prediction of surface settlement caused by synchronous grouting during shield tunneling in coarse-grained soils: A combined FEM and machine learning approach. *Undergr. Space* **2024**, *16*, 206–223. [CrossRef]
7. Feng, W.Q.; Yin, J.H. A new simplified Hypothesis B method for calculating consolidation settlements of double soil layers exhibiting creep. *Int. J. Numer. Anal. Methods Geomech.* **2017**, *41*, 899–917.
8. Nour, A.; Slimani, A.; Laouami, N. Foundation settlement statistics via finite element analysis. *Comput. Geotech.* **2002**, *29*, 641–672.
9. Sexton, B.G.; McCabe, B.A.; Castro, J. Appraising stone column settlement prediction methods using finite element analyses. *Acta Geotech.* **2014**, *9*, 993–1011.
10. Al-Shamrani, M.A. Applicability of the rectangular hyperbolic method to settlement predictions of sabkha soils. *Geotech. Geol. Eng.* **2004**, *22*, 563–587.
11. Fan, H.; Chen, Z.; Shen, J.; Cheng, J.; Chen, D.; Jiao, P. Buckling of steel tanks under measured settlement based on Poisson curve prediction model. *Thin-Walled Struct.* **2016**, *106*, 284–293. [CrossRef]
12. Zhu, L.; Xing, X.; Zhu, Y.; Peng, W.; Yuan, Z.; Xia, Q. An advanced time-series InSAR approach based on poisson curve for soft clay highway deformation monitoring. *IEEE J. Sel. Top. Appl. Earth Obs. Remote Sens.* **2021**, *14*, 7682–7698.
13. Wang, B.; Wang, X.; Ma, X. Study on optimal combination settlement prediction model based on logistic curve and Gompertz curve. *Stavební Obz.-Civ. Eng. J.* **2020**, *29*, 347–357.
14. Pulket, T.; Arditi, D. Construction litigation prediction system using ant colony optimization. *Constr. Manag. Econ.* **2009**, *27*, 241–251.
15. Song, Z.; Liu, S.; Jiang, M.; Yao, S. Research on the Settlement Prediction Model of Foundation Pit Based on the Improved PSO-SVM Model. *Sci. Program.* **2022**, *2022*, 1921378.
16. Yang, P.; Yong, W.; Li, C.; Peng, K.; Wei, W.; Qiu, Y.; Zhou, J. Hybrid random forest-based models for earth pressure balance tunneling-induced ground settlement prediction. *Appl. Sci.* **2023**, *13*, 2574. [CrossRef]
17. Xie, X.; Pan, C. Safety Prediction of Deep Foundation Pit Based on Neural Network and Entropy Fuzzy Evaluation. *E3S Web Conf.* **2021**, *233*, 03001.
18. Cao, Y.; Zhou, X.; Yan, K. Deep learning neural network model for tunnel ground surface settlement prediction based on sensor data. *Math. Probl. Eng.* **2021**, *2021*, 9488892.
19. Xie, S.L.; Hu, A.; Wang, M.; Xiao, Z.R.; Li, T.; Wang, C. 1DCNN-based prediction methods for subsequent settlement of subgrade with limited monitoring data. *Eur. J. Environ. Civ. Eng.* **2025**, *29*, 759–784.
20. Deng, J.L. Control problems of grey systems. *Syst. Control Lett.* **1982**, *1*, 288–294.
21. Tong, M.; Yan, Z.; Chao, L. Research on a grey prediction model of population growth based on a logistic approach. *Discret. Dyn. Nat. Soc.* **2020**, *2020*, 2416840. [CrossRef]
22. Li, B.; He, C.; Hu, L.; Li, Y. Dynamical analysis on influencing factors of grain production in Henan province based on grey systems theory. *Grey Syst. Theory Appl.* **2012**, *2*, 45–53.
23. Liu, C.; Xie, W.; Lao, T.; Yao, Y.t.; Zhang, J. Application of a novel grey forecasting model with time power term to predict China's GDP. *Grey Syst. Theory Appl.* **2021**, *11*, 343–357. [CrossRef]
24. Zhang, W.; Xiao, R.; Shi, B.; Zhu, H.h.; Sun, Y.j. Forecasting slope deformation field using correlated grey model updated with time correction factor and background value optimization. *Eng. Geol.* **2019**, *260*, 105215. [CrossRef]
25. Wu, L.; Li, S.; Huang, R.; Xu, Q. A new grey prediction model and its application to predicting landslide displacement. *Appl. Soft Comput.* **2020**, *95*, 106543.
26. Jin, S.J.; Zhang, D.S.; Shu, Z.; Zhao, Y. Grey model theory used in prediction of subgrade settlement. *Appl. Mech. Mater.* **2012**, *105*, 1576–1579. [CrossRef]
27. Wang, Y.; Yang, G. Prediction of composite foundation settlement based on multi-variable gray model. *Appl. Mech. Mater.* **2014**, *580*, 669–673. [CrossRef]
28. Zhang, J.; Qin, Y.; Zhang, X.; Che, G.; Sun, X.; Duo, H. Application of non-equidistant GM(1,1) model based on the fractional-order accumulation in building settlement monitoring. *J. Intell. Fuzzy Syst.* **2022**, *42*, 1559–1573. [CrossRef]

29. Zhang, C.; Li, J.z.; He, Y. Application of optimized grey discrete Verhulst–BP neural network model in settlement prediction of foundation pit. *Environ. Earth Sci.* **2019**, *78*, 441. [CrossRef]
30. Xiong, Z.; Deng, K.; Feng, G.; Miao, L.; Li, K.; He, C.; He, Y. Settlement prediction of reclaimed coastal airports with InSAR observation: A case study of the Xiamen Xiang'an International Airport, China. *Remote Sens.* **2022**, *14*, 3081. [CrossRef]
31. Zeng, J.J.; Feng, P.; Dai, J.G.; Zhuge, Y. Development and behavior of novel FRP-UHPC tubular members. *Eng. Struct.* **2022**, *266*, 114540.
32. Douglas, I.; Lawson, N. Airport construction: Materials use and geomorphic change. *J. Air Transp. Manag.* **2003**, *9*, 177–185.

Disclaimer/Publisher's Note: The statements, opinions and data contained in all publications are solely those of the individual author(s) and contributor(s) and not of MDPI and/or the editor(s). MDPI and/or the editor(s) disclaim responsibility for any injury to people or property resulting from any ideas, methods, instructions or products referred to in the content.

buildings

MDPI

Article

Study on the Causes of Cracking in Concrete Components of a High-Pile Beam Plate Wharf

Chao Yang [1,2], Pengjuan He [2], Shaohua Wang [1,2,3,*], Jiao Wang [2] and Zuoxiang Zhu [2]

1 Hubei Key Laboratory of Disaster Prevention and Mitigation, China Three Gorges University, Yichang 443002, China; yangchao0615@ctgu.edu.cn

2 College of Civil Engineering & Architecture, China Three Gorges University, Yichang 443000, China; 202208140021017@ctgu.edu.cn (P.H.); 202308140011018@ctgu.edu.cn (J.W.); 202308140011006@ctgu.edu.cn (Z.Z.)

3 College of Architectural Engineering, Hubei Three Gorges Polytechnic, Yichang 443000, China

* Correspondence: 202208140011004@ctgu.edu.cn; Tel.: +86-1354585-6720

Abstract: The high-pile beam slab structure is a commonly employed design for riverbank wharves; however, the wharf structure may incur damage due to various factors during long-term operation, resulting in potential safety concerns. To illustrate this, an investigation was conducted on a high-pile beam slab wharf, which included on-site examination, testing, and large-scale three-dimensional numerical simulation. The effects of gravity, ship impact, earthquake, lateral impact, water, and crane change were considered to explore the causes of cracking in the wharf concrete components. The results indicated that crane modification significantly augmented loads, precipitating notable deformation (92% increase in maximum vertical displacement), and the maximum tensile stress exceeded concrete tensile strength. The inadequate thickness of the steel reinforcement protective layer caused concrete carbonation, steel exposure, and corrosion, reducing structural capacity. The presence of defects in the pile foundation has been shown to result in high stress concentrations, which can lead to deformation and damage. There was a 58% increase in vertical displacement in the concrete components above the affected area compared to intact piles. Based on analysis of the results, appropriate measures for strengthening and correction have been proposed to ensure the safety and durability of the wharf. A comprehensive multifactor evaluation and 3D simulation of the actual dimensions are recommended to ensure the safety of wharf structures.

Keywords: high-pile beam plate structure; cracks; cause analysis; treatment measures; 3D numerical simulation

1. Introduction

High-pile wharves are extensively used in coastal areas of China and the Yangtze River due to their advantages of small structural displacement, small amount of sand and gravel, and suitability for deep water and soft soil foundation conditions [1]. However, the long-term operation of these wharves can cause damage to concrete components, which can affect the safety and durability of the wharves [2].

The load on a wharf structure is categorized into three main types: horizontal load, which includes ship berthing load and wave load; vertical load, which includes gravity of self-weight, stacked goods, and upper machinery; and environmental load, which includes seismic load and corrosion. Under these loads, the thickness of the concrete protective layer decreases, leading to concrete carbonation, corrosion of exposed steel bars, and the development and expansion of cracks in concrete members [3–7]. Numerous experimental and analytical studies have been carried out on the deformation and failure of high-piled wharf structures under various loads. Zheng and Zhang [6] conducted a model test to evaluate the damage caused by horizontal impact loads on bent structures. The study compared the extent of damage caused by different loads and concluded that ship collisions are the main cause of wharf damage. Xie et al. [8] established formulas for single-pile and pile-supported platform force–displacement for determining the horizontal load-bearing capacity of a high-pile wharf using numerical simulation. Zhang et al. [9] developed a three-dimensional finite element model of a high-piled beam-slab wharf and simulated its dynamic response to horizontal loads. This showed that the frame structure had the weakest bearing capacity at the joints of the members. Yuan [10] developed a model for a beamless high-piled wharf to analyze the effect of post-stack loading on the deformation of a pile–soil system. Zheng et al. [11] proposed a damage identification method based on stiffness, natural vibration period, and experimental acceleration data, which can effectively improve the health monitoring of high-piled wharves. Zhu et al. [12] proposed a method to extract the effective component of dynamic response under wave excitation. The robustness and sensitivity of the new damage index were verified by finite element simulation and an experimental model of a high-piled wharf, which is expected to assist the daily health monitoring of pile foundations of high-piled wharves. Xu et al. [13] used FLAC 3D 6.0 version to carry out dynamic calculations under ship impact load, and effectively identified locations of damage to a high-piled wharf. You et al. [14] investigated the 3D bearing characteristics of a high-piled beam slab wharf under an upper mechanical load. Xiao and Li [15] proposed models for chloride penetration, crack initiation, and crack propagation in reinforced concrete (RC) structures under different stages of marine environments. Li et al. [16] investigated the stress characteristics of a coastal port wharf structure in a special environment with high salt and high humidity, and proposed a health inspection index for the wharf structure. Li et al. [17] established a three-dimensional finite element model of an integral high-piled wharf structure with pile–soil interaction, and revealed the failure mechanism of lateral bearing deformation and the degradation mode of lateral bearing performance under the combined action of chloride attack and ship berthing impact. Coelho and Araújo [18] demonstrated that in instances of intricate geometric configurations, the finite element model is capable of accurately predicting the structural behavior of the beam–column connection. Furthermore, the finite element model can accurately simulate and analyze the structural performance of reinforced concrete components, which demonstrates its good applicability [19–22].

The existing literature has mainly studied the mechanical behavior of wharf structures under single loads. Numerical simulations and laboratory tests have been conducted using small models, such as single piles and bent piles. However, for high-pile beam plate wharves, the foundation, high piles, and superstructure interact and influence various external loads, resulting in a more complex stress and deformation response. This paper presents a study on the causes and failure mechanisms of cracks in concrete components of a high-pile beam slab wharf with a large number of cracks during operation. The study was conducted through on-site investigation, detection, and large-scale 3D numerical simulation. The effects of various loads, including gravity load, ship impact load, seismic load, lateral impact load, water load, and crane load, were considered.

2. Characteristics of Cracks in Concrete Components

2.1. Project Overview

Zigui Port is located in Yinxingtuo Village, Maoping Town, Zigui County, Yichang City, about 8.0 km upstream of the Three Gorges Dam on the Yangtze River. It functions as an emergency and long-term transportation hub for the Three Gorges Reservoir area. The project for protecting the reservoir bank covers a length of approximately 1262.00 m and a width of 130.00 m. It is divided into two areas: a RO–RO terminal area and a miscellaneous terminal area. Figure 1 illustrates that the wharf comprises a wharf platform and four approach bridges. This is a supporting project of Zigui Three Gorges Overturned Dam Logistics Industrial Park. The wharf platform measures 315.00 m in length and 25.00 m in width, with a top elevation of 176.28 m and a bottom elevation of 150.28 m. It is divided into three 3000 t-class miscellaneous berths to meet the import and export requirements of 1.40 million tons of groceries per year.

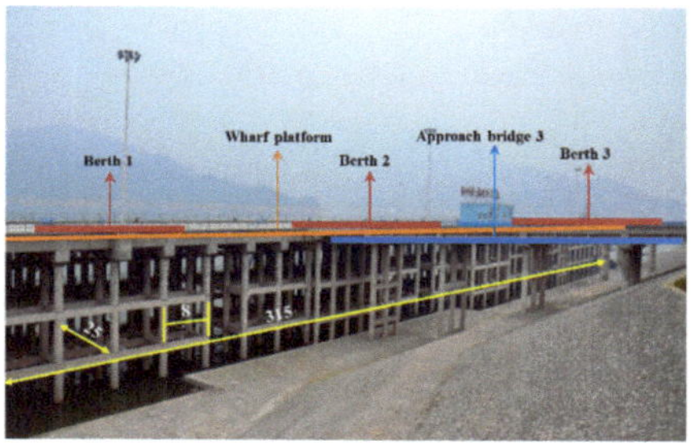

Figure 1. Wharf frame panorama (unit: m).

Figure 2 illustrates a stratigraphic sequence composed of silty gravel, sand, and granitic materials exhibiting varying degrees of weathering. The basal unit consists of slightly weathered granite with a thickness ranging from approximately 1.0 m to 5.0 m, which is conformably overlain by moderately weathered granite. Overlying this unit is a 4.1–11.2 m-thick layer of strongly weathered granite, followed by the uppermost stratum of completely weathered granite with a vertical extent of 0.5–4.3 m. Additionally, localized lens-shaped deposits of silty gravelly sand are present near the terrestrial margin, occupying a sub-horizontal zone with a maximum thickness of 3.0 m.

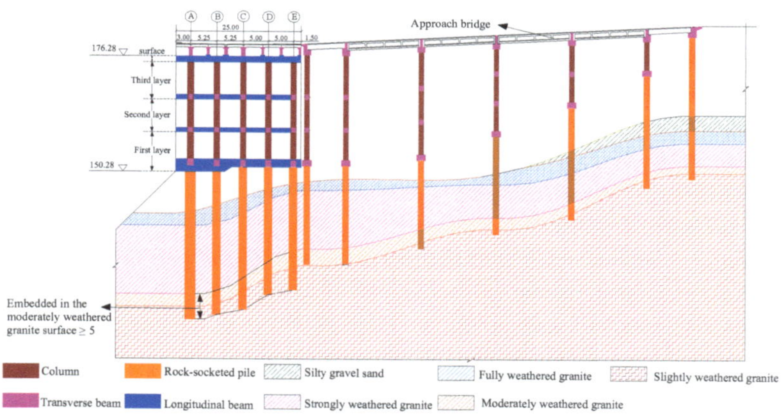

Figure 2. Wharf longitudinal section diagram (unit: m).

The foundation for the rock-socketed pile was designed to take advantage of the high strength and rigidity of granite. The bearing layer is made of moderately and slightly weathered granite, and the pile must be embedded in the weathered rock surface of the granite, at a depth of no less than 5.00 m. The piles are arranged in five columns (A–E) and 40 rows (1–40). Figure 2 illustrates that the spacing between columns A and B and between columns B and C is 5.25 m and the spacing between columns C and D and between columns D and E is 5.00 m. The spacing of each row is 8.00 m. The diameter of the piles in columns B to E was designed to be 1600.00 mm. To consider the effects of ship berthing and wave loads during operation, the diameter of the pile foundation for column A near the river side was increased slightly to 2200.00 mm. The final design of the rock-socketed pile length ranges from 23.00 m and 43.00 m due to the variation in the thickness of the moderately weathered granite.

The structure above was a frame type consisting of cast-in-place columns, longitudinal and transverse foundation connecting beams, cast-in-place crossbeams, prefabricated longitudinal beams, prefabricated panels, cast-in-place surface layers, and wearing layers. The framework comprised three layers (excluding the surface layer). The distribution of columns was consistent with the pile foundation, with one column arranged at the top of each rock-socketed pile. The diameter of columns in columns B to E was 1200 mm. To ensure high impact strength on the river side, the diameter of column A had been increased to 1400 mm. The longitudinal beams measured 2400 mm × 1600 mm, while the transverse beams measured 1200 mm × 1600 mm. The cast-in-place crossbeam measured 800 mm × 1000 mm, and the prefabricated longitudinal beam measured 1400 mm × 1000 mm. The prefabricated panel was 1000 mm thick.

2.2. Characteristics of Cracks

The on-site investigation, carried out after three years of operation, revealed that multiple cracks had appeared in the concrete surface layer, columns, and beams of the wharf (Figure 3).

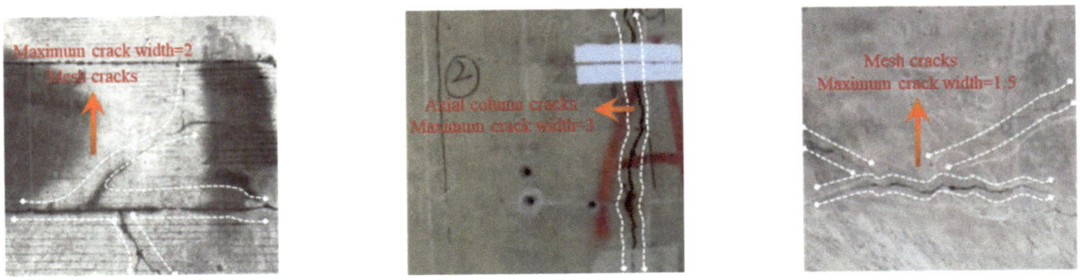

(a) (b) (c)

Figure 3. On-site cracks (unit: mm): (**a**) surface cracks; (**b**) column crack; (**c**) beam crack.

Some components were severely damaged, affecting the normal use of the wharf. The cracks in the surface layer were primarily in a network pattern, with the longest crack measuring 21.80 m and the largest network crack measuring 1.50 m × 7.60 m. The column exhibited numerous axial cracks, some of which combined with circumferential cracks to form mesh cracks. In severe cases, the steel bars were corroded and the hoops were exposed. The maximum crack width in the column was 12 mm, with a depth ranging from 111 mm to 209 mm. The connecting beams had cracks with a maximum width of 2.38 mm and a depth ranging from 139 mm to 144 mm. Additionally, some of the top surfaces of the connecting beams were covered with mesh cracks. The connecting beam developed multiple symmetric cracks on both the upstream and downstream sides, with a maximum width of 12 mm and a depth of 125 mm to 186 mm. Some of these cracks extended to the top surface of the foundation beam, forming a network of cracks. The connecting beams are severely damaged and show signs of exposed reinforcement corrosion.

Li et al. [23] concluded that concrete cracks with a width of less than 0.20 mm generally do not require specialized control measures. For cracks between 0.20 and 3.00 mm, significant deformation occurs, and repair and reinforcement measures are necessary. If the crack width exceeds 3 mm and the degree of deformation is significant, complex repair and reinforcement measures are required. The concrete components are classified into four categories based on crack width: Class A represents components without cracks, Class B represents components with minor cracks that have a width of less than 0.3 mm, Class C represents components with obvious cracks that have a width between 0.3 and 3 mm, and Class D represents components with severe cracks that have a width greater than 3 mm.

Figure 4a–c display the distribution of cracks in columns and connecting beams. The column cracks were mainly concentrated downstream between rows 21 to 40. Out of the 102 columns in this interval, 82 columns had cracks, accounting for 80.4%. The most severe D-class cracks were all present on the downstream side. On the upstream side, out of the 102 columns in rows 1 to 40, 21 columns had cracks, accounting for 20.6%. Moreover, the cracks in the columns were mainly distributed vertically, with over 80% of them appearing on the first layer. The most severe D-class cracks were also concentrated on the first layer. Furthermore, D-class cracks were found to be distributed in all four directions of the column. It is worth noting that all cracks were on the water-facing side.

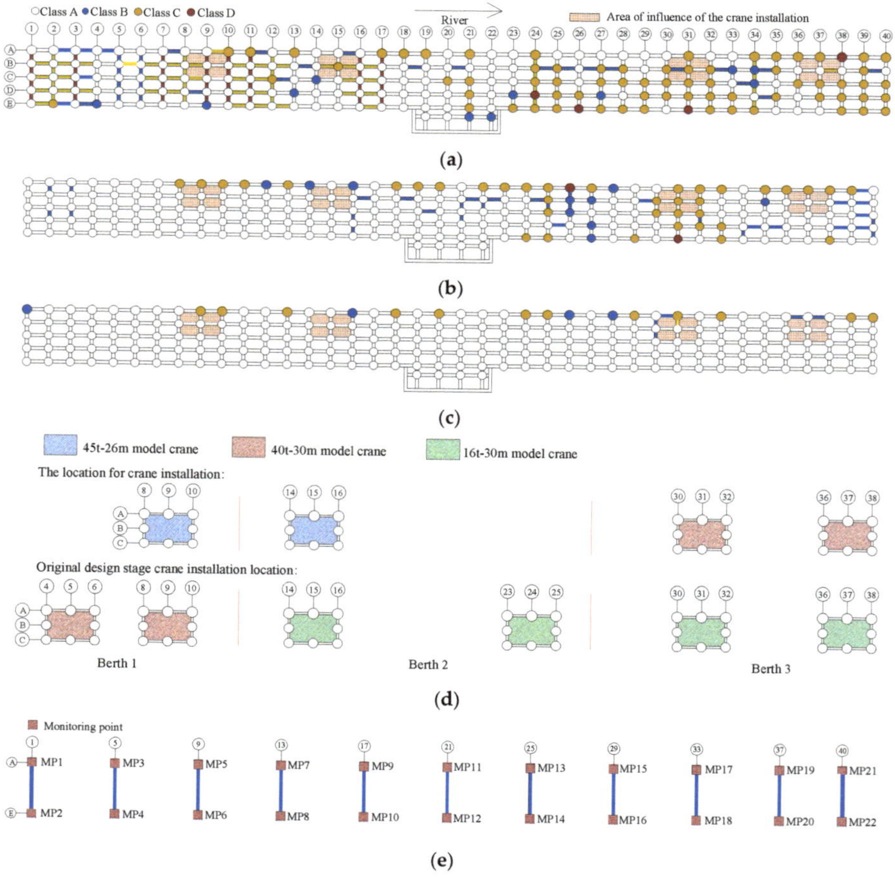

Figure 4. Distribution of cracks in the columns and tie beams and arrangement of the crane: (**a**) first layer; (**b**) second layer; (**c**) third layer; (**d**) surface; (**e**) arrangement of the displacement monitoring project.

The cracks in the connecting beams were mainly concentrated in the first layer, with relatively few cracks in the second and third layers. Furthermore, the connecting beams with C-class and D-class cracks were primarily located in the first layer and concentrated in rows 1 to 20. In the first layer of connecting beams, 20% of the total was found to be cracked D-class beams, while 12.5% and 5% were cracked C-class and B-class beams, respectively.

The cracks on the columns were mostly located on the river side and were concentrated on the downstream side (rows 21 to 40) and the first layer. The cracks in the connecting beams were concentrated in rows 1 to 20 of the first layer. The wharf structure was severely damaged, with reduced bearing capacity and durability that no longer meets the design service life requirements. The maximum width of cracks in the wharf and approach bridge was 12 mm, which poses serious peeling problems and does not meet the requirements for use.

3. Cause Investigation

3.1. Potential Causes

Possible causes of cracking in concrete components were investigated on-site. (1) Operational errors were made by waiting vessels during three months of operation at the wharf due to the influence of strong winds. These errors caused impacts on the structures of rows 1 to 5 of the wharf, resulting in the peeling and chipping of some concrete component protective layers. (2) Two blasting operations were carried out near the wharf during its operation. Earthwork blasting was conducted at the Xiongjialing site behind the wharf, with the closest distance to the wharf being 50 m. Underwater blasting construction was also carried out to excavate the harbor pool, with a minimum distance of 100 m from the wharf. (3) A 4.5 magnitude earthquake occurred in Zigui County (31.03° N, 110.47° E), where the wharf is located, with a depth of 7 km, after two years of operation at the wharf. Local residents reported the earthquake as strong. The horizontal loads generated by earthquakes may have an impact on the wharf structure. (4) The elevation of the bedrock varies significantly, with the moderately weathered bedrock layer ranging from 77.6 to 195.4 m and the slightly weathered bedrock layer ranging from 75.4 to 192.9 m. This uneven distribution of the bearing rock layer may result in differential deformation of the pile foundation. (5) Due to functional requirements, the initially designed lifting equipment had undergone changes. Figure 4d illustrates the layout before and after the change of lifting equipment. The lifting equipment of Berth 1 had been altered from two 40 t–30 m gantry cranes to one 45 t–26 m shore container crane. Similarly, the lifting equipment of Berth 2 had been altered from two 16 t–30 m gantry cranes to one 45 t–26 m shore container crane. The lifting equipment of Berth 3 had been modified from two 16 t–30 m gantry cranes to two 40 t–30 m multipurpose gantry cranes. The substitution of the lifting apparatus has the consequence of an augmented load on the lower structure, which may have an effect on the overall stability of the structure. (6) In addition to the influence of external loads, the strength of concrete is of crucial importance. Insufficient strength can readily result in cracking and deformation of concrete components under external loads. (7) A site investigation revealed that the concrete components lacked sufficient thickness of the steel reinforcement protection layer, which directly affected the durability of the concrete components and could lead to corrosion of the steel reinforcement, thereby reducing the load-bearing capacity of the components. (8) The integrity of the pile was found to be defective. The embedded rock pile serves to transfer the load of the upper structure to the stable bedrock. The strength, stiffness, and stability of the pile directly affect the load transfer, while the integrity of the pile also affects the vertical compression, pull-out, and horizontal bearing capacity.

3.2. Cause Elimination

The incident occurred in the early stages of the construction of the wharf, when the wharf was at a high water level and a ship collided with the upper frame structure. Following the collision at the wharf, the concrete components of rows 1 to 5 were significantly damaged, and the external steel reinforcement protection layer was damaged. The steel reinforcement was exposed, but no fracture or damage was found. The damage statistics of all columns, beams, and surface layers at the impact location indicated that there was no structural damage to the main structure. Furthermore, the inspection results of the surface layer, upper structure, track, and ancillary facilities also demonstrated that there was no significant damage. Figure 4e shows the locations of 22 settlement monitoring points that were set at the top of the pile foundation. Figure 5 illustrates the settlement variation curves of four monitoring points (MP1–MP4) situated in rows 1 to 5 following a three-month period after collision. According to the "Engineering survey standards" [24], the settlement displacements of the monitoring points exhibited a very slight increase (0.01 mm/d–0.04 mm/d), and the settlements at each monitoring point after three months were also very small, ranging from 0.5 mm to 0.7 mm. This indicates that the impact of the ship collision on the overall structure of the wharf was minimal.

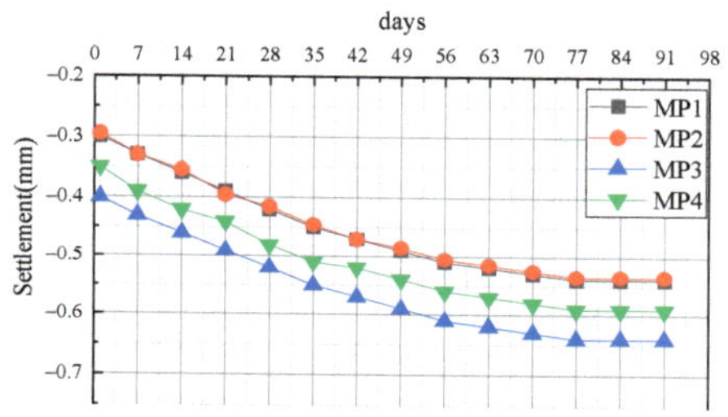

Figure 5. Monitoring displacement of ship collision.

The Xiongjialing site earthwork blasting was conducted using a loosening blasting technology, with the safe distance of blasting effect calculated using Equation (1) [25]:

$$Q = \frac{eqgLW^2 n_c}{(1 + n_c)} \quad (1)$$

where the variables Q, e, g, q, L, and W represent the actual charge of each borehole, the conversion coefficient of explosives, the blockage coefficient of the explosive eye, the amount of explosives required to be installed for media blasting under standard conditions, the length of the borehole, and the distance affected by the explosion, respectively. The action index of blasting, represented by n_c, is also a crucial parameter. In consideration of the site conditions, the values of each parameter were as follows: Q = 15.33 kg, e = 1.5, g = 0.5, q = 0.3 kg, L = 1.5 m, and n_c = 0.6, giving W = 10.5 m.

The closest distance between the site and the wharf was 50 m so there would be no significant impact on the wharf structure.

The following equation for millisecond delayed loosening blasting is used to analyze the effects of underwater blasting [26]:

$$R = \left(\frac{\mu P \sigma_t}{1 - \mu} \right)^{\frac{1}{a}} r_b \quad (2)$$

where R is the radius of the crack zone after blasting, μ is the Poisson's ratio of the rock mass, σ_t is the tensile strength of the rock mass, a is the stress attenuation value and

$a = (2-\mu)/(1-\mu)$, and r_b is the radius of the hole. P is the initial impact value on the hole wall during blast, and it can be calculated as follows:

$$P = \frac{1}{8}\rho_e D^2 \left(\frac{r_c}{r_b}\right)^6 n \tag{3}$$

where ρ_e is the density of explosives, D is the explosive velocity, r_c is the radius of the explosive, and n is the pressure increase coefficient. Based on the actual situation on-site, each parameter was taken as $\mu = 0.25$, $\sigma_t = 5.8$ MPa, $\gamma_b = 23$ mm, $\rho_e = 1.07 \times 10^3$ kg/m^3, $D = 3600$ m/s, $r_c = 19.5$ mm, and $n = 11$, respectively. Thus, $R = 10.1$ m was obtained, which was well below the minimum distance (100 m) from the blast site to the wharf, so it would not have had an impact on the structure.

According to the Seismic Ground Motion Parameter Zoning Map of China [27], the construction of the wharf was designed with seismic intensity of VI, while the 4.6 magnitude earthquake that occurred was calculated to be only II–III. Apparently, this earthquake was not strong enough to cause damage to the wharf structure.

The monitoring of the 22 points at the top of the pile revealed that the final settlement of the MP5 monitoring point was the highest (1.12 mm) due to the upper part bearing the load of the 45 t–26 m-type lifting equipment. In contrast, the settlement of the MP3 monitoring point at the edge of the wharf was the lowest, at −0.79 mm. This resulted in a settlement difference of 1.91 mm between the two points. In accordance with the pertinent Chinese standard, the Code for Design of Building Foundations (GB 50007–2011) [28], the typical standard value for the settlement difference between adjacent pile foundations is 0.002 l_0, where l_0 represents the distance between the adjacent pile foundations. The distance between the adjacent piles in the miscellaneous wharf was 8 m, which permitted a settlement difference of 16 mm. It can be observed that the uneven settlement of the wharf structure was minimal, and thus it can be concluded that it was not the primary factor contributing to the cracking of the concrete components.

In accordance with the Standard for Test Methods of Concrete Physical and Mechanical Properties (GB/T 50081-2019) [29], concrete samples with dimensions of 150 mm × 150 mm × 150 mm were cured under standard conditions for 28 days (Figure 6), during which the temperature was maintained at 20 ± 2 °C and the relative humidity was ≥95%. Strength tests were then conducted, and the average strength of three samples per group was calculated. Individual sample strengths were required to deviate from the group average by no more than 15%. The test results demonstrated an average strength of 44.84 MPa and a minimum strength of 42.20 MPa, which met the design requirements. A random selection of components from the construction site was subjected to the rebound method to test the strength of the sampled concrete components. The test data indicated that the average rebound of the concrete components at 28 days was distributed between 41 and 53. After angle correction and pouring surface correction, it was estimated that the average strength of this batch of concrete components was not less than 40 MPa and not more than 50 MPa, which meets the design requirements. Consequently, the strength of the concrete components was not the primary factor in their cracking.

3.3. Exact Causes

The inspection results of the steel reinforcement protection layer for the column and foundation connecting beam indicated that the qualification rates of the column were 25.6%, 55.6%, and 31.1%, respectively. The qualification rates of foundation connecting beams were 16.7%, 58.9%, and 30.0%, respectively. The maximum negative deviation of the unqualified point in the thickness of the steel reinforcement protective layer was greater than 1.5 times the allowable deviation value. This was due to insufficient thickness, which resulted in multiple connecting beams being exposed and corroded. Therefore, it can be determined that the inadequate thickness of the steel reinforcement protective layer was an important cause of cracking in concrete components.

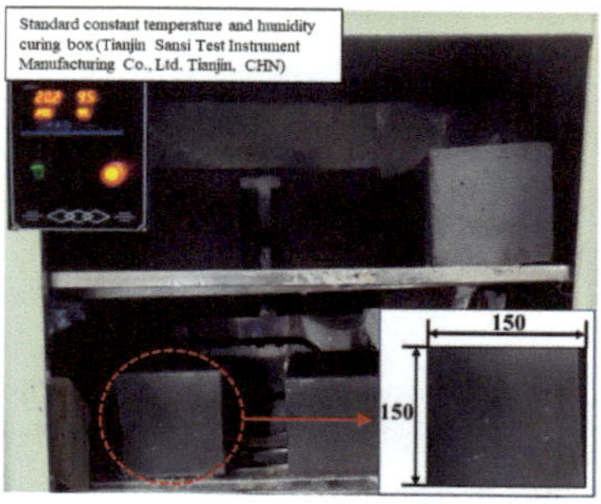

Figure 6. Maintenance of standard concrete specimens.

Two further factors that should be considered to have an impact were changes in the lifting equipment and defects in the integrity of the piles. Figure 4a–c clearly demonstrate the presence of significant cracks in the structure below the location of the lifting equipment. In particular, for rows 30–32 and 36–38, the originally designed 16-ton crane was replaced with a larger 40-ton crane, resulting in the emergence of significant cracks in the three-story structure below. The on-site investigation also revealed that there were integrity defects in the piles at C-3, B-7, C-7, E-7, E-8, E-9, C-10, E-10, and D-11. Furthermore, the C-type and D-type cracks in the foundation connecting beam were primarily concentrated in rows 3, 7, 8, 9, 10, and 11, and their distribution was largely consistent. It is evident that alterations in the lifting apparatus were directly correlated with the occurrence of pile defects, as well as with the emergence of structural cracks.

4. Numerical Simulation

4.1. Numerical Model and Loading Cases

To further investigate the specific impact of changes in lifting equipment and pile defects on the wharf structure, the finite-difference code FLAC^{3D} version 6.0 was employed for numerical simulation. A numerical model of the three-layer frame beam column structure and foundation of the miscellaneous wharf platform was established based on the terrain and geological conditions. The upper structure comprises 40 rows of beams and 204 columns, in addition to a concrete surface layer. The foundation comprises medium to slightly weathered bedrock and a strong weathered zone of upper granite. Each pile foundation is embedded in medium to slightly weathered bedrock to a depth of 5 m. All the upper structures, piles, and foundation were all simulated using solid elements. Given that tetrahedral meshes enable efficient local refinement of fine boundary layers while effectively preventing mesh distortion, they were better suited for the nonlinear large-scale structural model used in this study. The mesh size was determined based on component geometry: cylindrical columns were assigned a 0.5 m mesh, beams and top slabs adopted 1 m elements, and the foundation mesh size was optimized according to its distribution width: larger widths utilized moderately expanded meshes, while narrower regions employed smaller meshes, maintaining an overall range of 0.5–1 m. As illustrated in Figure 7, the model has a length of 332 m, a width of 45.5 m, and a height of 75.47 m. The model comprised a total of 10,409,112 zones and 1,876,477 grid points.

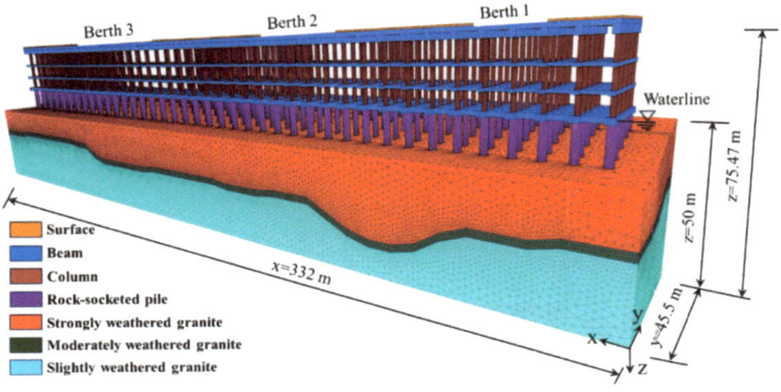

Figure 7. Numerical model.

For the boundary conditions, the four side boundaries and the bottom boundary were fixed with hinges and the top of the foundation was free. The "zone attach" command is used to connect the beams, slabs and columns, so that their translational degrees of freedom are coupled. When subjected to forces, axial forces and shear forces are transmitted, while bending moments are not transmitted. The wharf studied in this paper belongs to the prefabricated assembled structure, which is consistent with its hinged connection mode.

A total of four cases were subjected to simulation. In Case I, the potential impact of pile defects and modified lifting equipment was evaluated. In Case II, the loading of the initial lifting equipment was considered. In Case III, the effects of modified lifting equipment were considered. In Case IV, the impact of pile defects was considered.

All the structures, including the beams, columns and piles, and the rock materials were assumed to obey the Mohr–Coulomb failure criterion. In the calculation of the four cases, a water level of 50 m was considered to apply load to the model, with the foundation and piles below the water level assumed to be saturated. The results of concrete strength testing indicated that the parameters of C40 concrete were employed for concrete beams, piles, and surface layers, while the parameters of C45 concrete were used for concrete columns. The mechanical property parameters of the concrete material were mainly determined by the relevant parameters in the project report during the construction of the wharf and the test results of the concrete curing in Section 3.2. Three types of granite rock layers with varying degrees of weathering were reduced according to the comprehensive granite material parameters outlined in the engineering geological survey report. The three categories of weathering were classified as slight, medium, and strong, with the respective proportions of granite material parameters being 90%, 75%, and 35% [30]. The calculated parameters are presented in Table 1.

Table 1. Numerical model material parameters.

Model Construct	Material Type	Bulk Modulus/GPa	Shear Modulus/GPa	Density (kg/m^3)	Saturation Density (kg/m^3)	Internal Friction Angle (°)	Cohesion (MPa)	Tensile Strength (MPa)
Basis of bedrock layer	Granite	23.81	21.74	2600	2860	51	12	10
beams, piles, and surface layers	C40 concrete	18.00	13.54	2500	2750	45	1.51	1.51
Columns	C45 concrete	18.61	13.96	2500	2750	45	1.6	1.6

To validate the applicability of tetrahedral elements and mesh discretization, numerical simulations under identical loading conditions (considering gravitational load only) were initially conducted using tetrahedral and hexahedral meshes, respectively. Subsequent numerical model computations under Case I were performed with mesh sizes of 0.5 m and 1 m. The comparison of final simulation results presented in Table 2 indicates negligible discrepancies, thereby confirming the rationality of the selected modeling approach.

Table 2. Sensitivity analysis of grid element and mesh size.

Categories		Vertical Displacement/mm	Maximum Tension Stress/MPa	Maximum Compressive Stress/MPa
Mesh Types	Hexahedral mesh	0.029	1.60	1.00
	Tetrahedral mesh	0.022	1.62	1.04
Mesh Sizes	1 m	4.35	1.65	1.47
	0.5 m	4.04	1.65	1.44

The loads of three types of lifting equipment, namely 45 t–26 m, 40 t–30 m, and 16 t–30 m, were calculated based on their maximum operating weights of 595 t, 540 t, and 256 t. The lifting equipment shown in Figure 8 had a length of 9 m and a width of 0.5 m for each track. The contact area between the lifting equipment and the two tracks on the ground was 9 m^2. In the simulation, the position of the lifting equipment load is consistent with the actual situation, with two tracks located between y = 3 and y = 3.5, as well as between y = 13.5 and y = 14 of the surface layer (Figure 9). The estimated stress applied to the surface layer was calculated based on the maximum weight of the lifting equipment and the contact area with the track. The results were 648 kPa, 589 kPa, and 279 kPa, respectively.

Figure 8. Installation of the 40 t–30 m model crane.

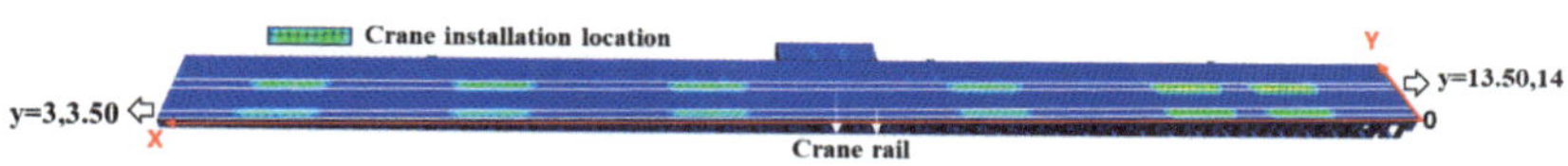

Figure 9. Crane installation location in the model.

A total of nine piles, C-3, B-7, C-7, E-7, E-8, E-9, C-10, E-10, and D-11, were simulated in the model based on the actual situation of piles. This was achieved by removing grid elements at their corresponding positions to simulate actual pile.

Before analyzing the impacts of the crane load and pile defects, the verification analysis was first carried out using working condition 1. Since the actual load conditions endured by the wharf are the same as those of working condition 1, the plastic damage and deformation generated during the operation of working condition 1 can be compared and analyzed with the actual distribution positions of the concrete cracks. If the two are consistent, it indicates that the numerical model is reliable and can be applied to the subsequent load analysis.

4.2. Simulation Results

As illustrated in Figure 10, the wharf structure in Case I exhibited significant plastic damage, with the plastic zone concentrated in the area affected by the crane and defective of piles. The number of columns with plastic damage on the first, second, and third floors was 36, 10, and 3, respectively. A zone of plastic deformation is evident in the vicinity of the wharf, in proximity to the crane installation location (on the river side) on each floor. In contrast, the opposite side exhibits minimal deformation. The number of beams exhibiting plastic failure in the first, second, and third layers is 53, 62, and 12, respectively. Similarly, there are numerous plastic zones on the riverside of each layer, with a high concentration in the area of the first layer comprising defect-embedded rock. This indicates that the plastic zone of columns and beams is primarily located on the first floor of the wharf structure and the riverside. This is consistent with the distribution of damage to columns and beams under actual working conditions. This suggests that the established model can be used for subsequent cases involving crane and pile defects.

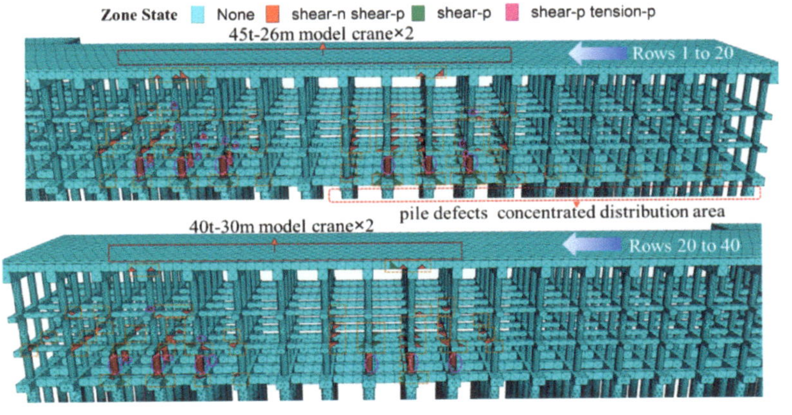

Figure 10. Distribution of plastic zone in Case I.

As illustrated in Figure 11, analysis of displacement contour plots and stress distribution diagrams for Case I reveals significant structural responses at the crane installation location. A maximum displacement amplitude of 4.35 mm was recorded at the installation site, accompanied by pronounced tensile stress concentrations at the pier margin reaching 1.65 MPa. The resultant asymmetric structural deformation of the wharf system demonstrates direct correlation with the geometric configuration of crane placement.

4.2.1. The Impact of Crane Changes

Figure 12 illustrates the vertical displacement of the wharf structure under Cases II and III. As a consequence of the crane's weight, a significant settlement was observed at its location, with the maximum occurring at the surface layer and gradually decreasing towards the next layer. The maximum vertical displacement was observed at the maximum tonnage of the crane, with maximum values of 2.27 mm and 4.35 mm, respectively. Furthermore, due to the crane's non-central position within the panel, asymmetric pressure is generated on the panel as a result of its larger self-weight, resulting in a certain upward bulge deformation on the side away from the crane. The maximum bulge deformation under the three working conditions is 0.70 mm and 0.84 mm, respectively.

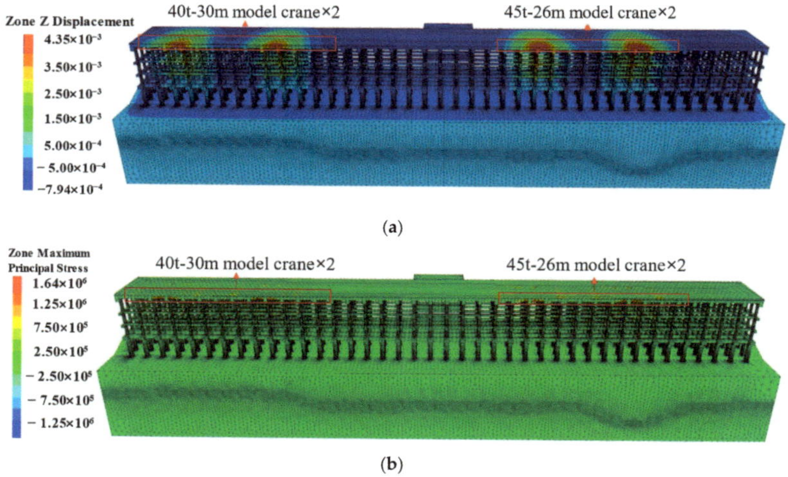

Figure 11. Vertical displacements and maximum principal stress of wharf structure in Case I. (**a**) Vertical displacements; (**b**) maximum principal stress.

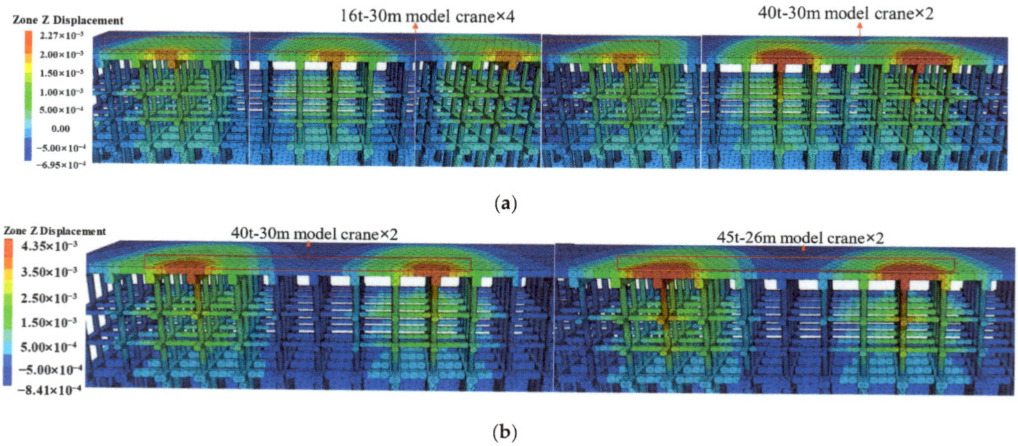

Figure 12. Vertical displacements of wharf structure in different cases (unit: m): (**a**) Case II; (**b**) Case III.

Figure 13 shows the vertical displacement at the same installation position (x = 60 m, 116 m, 188 m, and 240 m) before and after the crane change in Cases II and III. Following the change of the crane, there was a notable increase in vertical displacement, with the greatest vertical displacement increment occurring at the edge of the wharf structure (on the river side). As the value of y increased, the vertical displacement increment exhibited a gradual decrease. A significant turning point was observed at the second crane track, with a rapid decrease in the vertical displacement increment. The maximum vertical displacement increment was observed at x =116 m, with a displacement increment of 2.64 mm and an increase of 161%. The minimum value was generated at x = 60 m, with a displacement increment of 0.58 mm and an increase of 24%. This is due to the inconsistency in the weight increase of the crane following two changes. The weight of the crane at x = 116 m increased by 181.25%, while the weight of the crane at x = 60 m increased by 12.5%. This demonstrates that cranes have a significant impact on the deformation of wharf structures, and that their weight is an important factor affecting structural deformation.

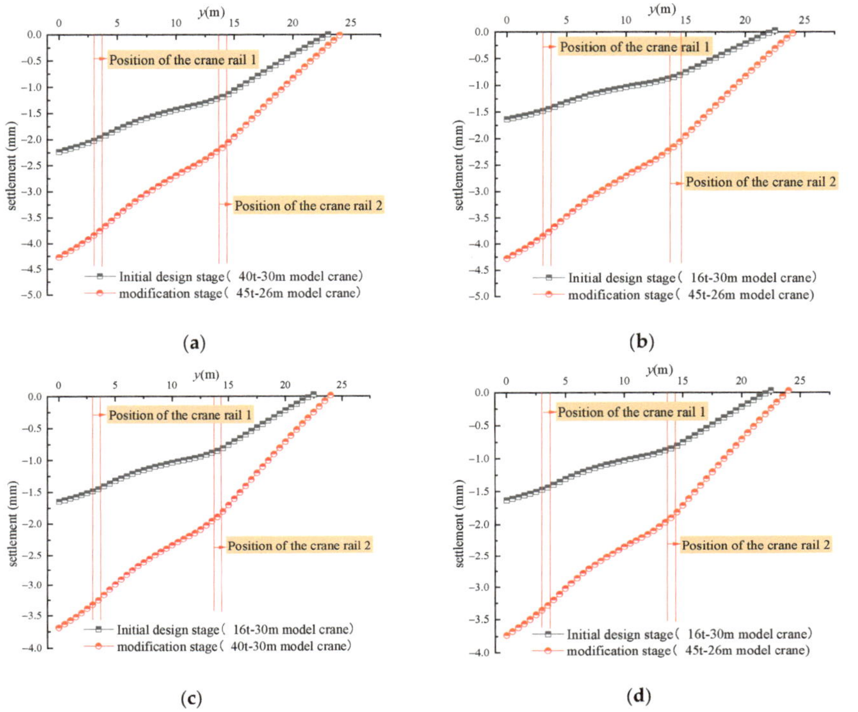

Figure 13. Changes in settlement before and after crane modification: (**a**) $x = 60$ m; (**b**) $x = 116$ m; (**c**) $x = 240$ m; (**d**) $x = 388$ m.

The asymmetric deformation caused by the installation position of the crane results in the generation of significant tensile stress on both sides of the track (orange and yellow distribution in Figure 14), while compressive stress is observed at the track position. The maximum tensile stresses under Cases II and III are 1.49 MPa and 1.65 MPa, respectively. These values are below the designed tensile strength of C40 concrete for the wharf, which is 1.51 MPa. This implies that the introduction of a new crane position may result in the formation of cracks in the concrete.

4.2.2. The Impact of Pile Defects

In Case IV, which considers the influence of pile defects alone, both the defective pile and its corresponding upper components have undergone significant deformation (Figure 15a). This is primarily attributable to the presence of defects in the pile, which has resulted in a reduction in its strength, thereby leading to an uneven distribution of stress on the upper structure and a notable degree of settlement. The greatest vertical displacement is observed at the E-7 rock-socketed pile and connecting beam, where there are significant and dense pile defects, with a maximum vertical displacement of 0.63 mm. It can be observed that the vertical displacement caused by pile defects is comparatively minor in comparison to the displacement caused by changes in the crane, which is 2.08 mm. Local defects in some rock-socketed piles result in significant stress concentration at their corresponding positions, with a maximum tensile stress of 1.1 MPa (Figure 15b). The occurrence of such high tensile stress concentrations in rock-socketed piles renders them susceptible to deformation and failure, which in turn affects the stability of the upper structure.

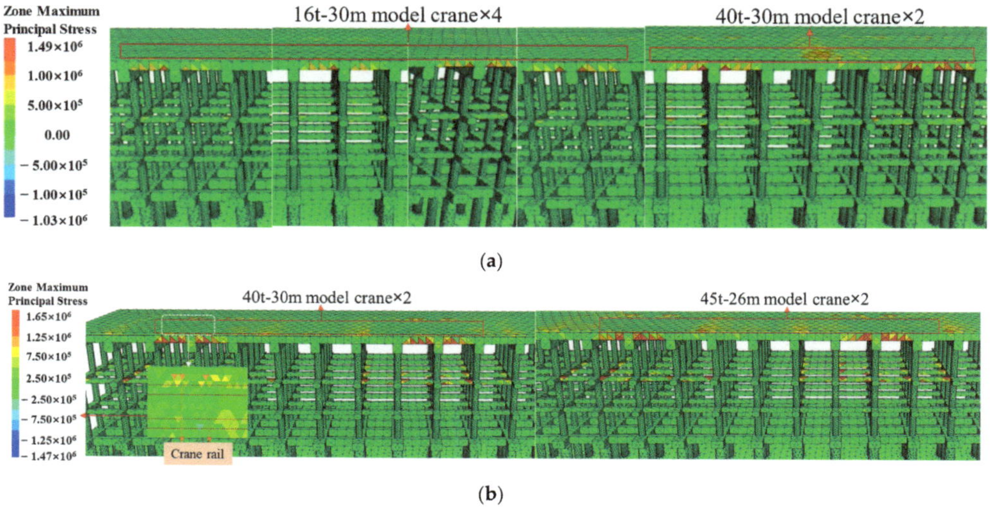

Figure 14. Maximum principal stress of wharf structure in different cases (unit: Pa): (**a**) Case II; (**b**) Case III.

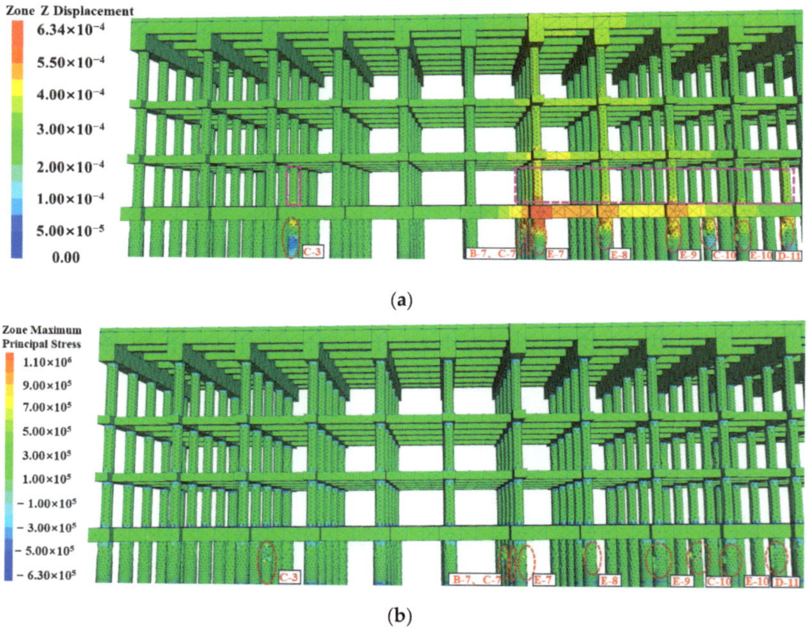

Figure 15. Mechanical behaviors of wharf structure in Case IV: (**a**) vertical displacement (unit: m); (**b**) maximum principal stress (unit: Pa).

Compare the maximum stress and vertical displacement of the wharf structure in four different cases. As illustrated in Figure 16, the maximum tensile stress, compressive stress, and vertical displacement were 1.65 MPa, 1.47 MPa, and 4.35 mm, respectively, in Cases I and III due to the presence of a maximum-tonnage crane. Due to the relatively low tonnage of the crane in the initial design stage, the maximum stress and displacement generated in Case II are smaller than those in Cases I and III, with reductions of 10%, 30%, and 48%, respectively. In the presence of pile defects, Case IV exhibited minimal stress and displacement values of 1.10 MPa, 0.63 MPa, and 0.63 mm, respectively. These values were 33%, 57%, and 85% lower than those observed in Case I.

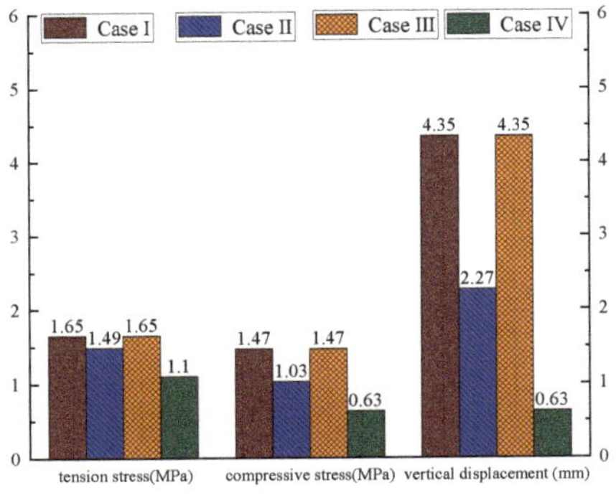

Figure 16. Comparative analysis of four cases.

In comparison, it can be observed that the impact of crane changes is most significant, resulting in notable deformation of the wharf structure. The impact of defective piles is relatively minor, but nevertheless merits consideration. The high stress concentration at the defect site poses considerable harm, and its bearing capacity directly affects the stability of the upper structure. When the two factors are considered together, the deformation and damage at the defect of the pile body are more severe. Consequently, in operational conditions, the combined effect of various loads on the wharf structure results in a more severe situation, necessitating the strengthening and treatment of concrete components to prevent further deterioration.

5. Treatment Measures

In light of the aforementioned analysis, the following treatment measures have been implemented.

1. It is necessary to repair and reinforce the concrete cracks existing on the wharf site. For the repair of shallow cracks or irregular cracking in concrete, it should be noted that these will not directly affect the structural load capacity. Cement, asphalt oil, and other similar materials can be applied directly to the surface of the concrete for repair. In cases where deep or penetrating cracks are more severe, grouting can be employed using pressure equipment to inject bonding materials into the cracks, thereby forming a unified whole and effectively improving the overall stability of the concrete. Distinct repair strategies should be implemented based on crack width. For cracks ≤0.2 mm, no intervention is typically required. Microcracks in concrete ($0.2 < w \leq 0.3$ mm) may be treated using low-viscosity epoxy resin, which penetrates and solidifies within the fissures. Medium cracks (0.3–3 mm) necessitate high-pressure grouting with cementitious or polymeric materials, followed by surface sealing with elastomeric coatings. In cases where grouting is infeasible, structural reinforcement via steel mesh-reinforced shotcrete (C25 fine-aggregate concrete) is recommended. For cracks >3 mm, damaged concrete should be excavated and replaced with non-shrinkage, high-strength concrete to restore load-bearing capacity.
2. Implement construction treatment in instances where the thickness of the steel reinforcement protective layer is not deemed to be adequate. In instances where the protective layer's thickness is inadequate, remeasurement and reconstruction are necessary to augment the layer's thickness and ensure compliance with the specified

requirements. Conversely, in cases where the protective layer's thickness is excessive, mechanical cleaning can be employed to eliminate the surplus layer. In instances where the steel reinforcement protection layer is not deemed to be of an acceptable standard, it is imperative to consider reconstruction in accordance with standard construction methods. Otherwise, it will reduce the bearing capacity of the wharf structure and accelerate the damage of the wharf structure.

3. The lifting equipment is to be replaced in accordance with the actual requirements of the wharf. The installation method combining small and large lifting equipment in Case II is significantly more reasonable than the distribution of overly dense large lifting equipment in Case III. Furthermore, this method will not cause deformation or damage to the local structure of the wharf due to excessive load. Consequently, in order to guarantee the safe operation of the wharf, comprehensive selection, comparison, and optimization analysis of lifting equipment change plans must be conducted, and the most suitable lifting equipment installation plans must be selected for change processing.
4. It is necessary to repair the defects in the piles and to adopt a new method for the daily testing of the pile foundations. This method should be capable of identifying damage to the pile foundation at an early stage and should enable timely measures for repair to be taken, thus avoiding greater losses.
5. It is recommended that information technology be combined with the establishment of a wharf monitoring system to monitor the horizontal displacement and settlement deformation of hydraulic structures in real time. A wharf monitoring and early-warning system should be constructed to provide support and guarantee the construction and operation safety of emergency reinforcement and subsequent system reinforcement.

6. Conclusions

A wharf structure is subject to a number of factors during its operational lifetime. This study employed a potential cause analysis and elimination methodology to elucidate the primary causes of cracking in a high-pile beam plate wharf and proposes corresponding reinforcement measures. Based on the results, the following conclusions can be drawn.

1. The primary cause of the cracking observed in concrete components is the result of alterations made to the crane. Following the modification of the crane, the maximum tensile and compressive stresses of the wharf structure increased by 0.16 MPa and 2.58 MPa, respectively, in comparison to the initial design stage. This represents increases of 11% and 44%, respectively. This resulted in a maximum vertical displacement increase of 2.08 mm, representing a 92% increase. Furthermore, the maximum tensile stress generated by the crane following the modification exceeded the tensile strength of the concrete, leading to the formation of cracks.
2. The impact of pile defects on the stability of wharf structures is relatively minor, yet it is nonetheless a significant factor that cannot be overlooked. The tensile stress at the defect site of the pile is high, with a maximum tensile stress of 1.1 MPa. This makes the defective pile prone to deformation and damage, thereby affecting the stability of the upper structure. The more intact pile resulted in greater deformation of the upper components of the defective pile, with a maximum vertical displacement increase of 0.23 mm, an increase of 58%.
3. A three-dimensional numerical model of a high-pile beam plate wharf was constructed in order to ascertain the complex stress and deformation response of the structure under the combined action of gravity load, water load, crane load, and pile defects. Thus, an effective approach was provided. Given the multitude of factors that influence the safety of wharf structures in practice, it is highly effective and imperative to

establish a comprehensive three-dimensional numerical model that aligns with the actual scale for analysis.

4. In this paper, a large-scale three-dimensional numerical simulation was carried out by modeling according to the actual dimensions. It took into account the deformation and damage of the wharf under the combined action of multiple loads. The simulation results are consistent with the actual situation, which has great reference significance. However, regarding the accurate identification of the existing damage to the wharf structure, since the simulation did not consider the effects of wave loads, wind loads, or water erosion, there still exists a slight difference between the position of the plastic damage obtained from the numerical simulation and the actual damage position. As a result, it is unable to achieve the health monitoring of the wharf structure, and further research on this is still needed.

Author Contributions: Conceptualization, C.Y. and P.H.; methodology, S.W.; validation, C.Y., S.W. and J.W.; investigation, Z.Z.; resources, C.Y.; data curation, S.W.; writing—original draft preparation, P.H.; writing—review and editing, C.Y.; supervision, S.W.; project administration, C.Y.; funding acquisition, C.Y. All authors have read and agreed to the published version of the manuscript.

Funding: This work was supported by the National Natural Science Foundation of China (NSFC) (42177132) and the Open Fund of Hubei Key Laboratory of Disaster Prevention and Mitigation (China Three Gorges University) (2023KJZ17).

Data Availability Statement: The original contributions presented in the study are included in the article. Further inquiries can be directed to the corresponding author.

Conflicts of Interest: The authors declare no conflicts of interest.

Abbreviations

The following abbreviation is used in this manuscript:

MP monitoring point

References

1. Su, J.B.; Yang, G.Q.; Zhang, L.M.; Zhu, R.H.; Qin, W.G. Safety Evaluation Method for High-Piled Wharf Structures Based on the Disturbing-Energy Method. *Struct. Control Health Monit.* **2020**, *146*, 6. [CrossRef]
2. Su, J.B.; Luan, S.L.; Zhang, L.M.; Zhu, R.H.; Qin, W.G. Partitioned genetic algorithm strategy for optimal sensor placement based on structure features of a high-piled wharf. *Struct. Control Health Monit.* **2019**, *26*, e2289. [CrossRef]
3. Chatterjee, K.; Choudhury, D.; Harry, G.; Poulos, H.G. Seismic analysis of laterally loaded pile under influence of vertical loading using finite element method. *Comput. Geotech.* **2015**, *67*, 172–186. [CrossRef]
4. Sun, M.J.; Shan, Z.G.; Wang, W.; Xu, S.M.; Liu, X.L.; Zhang, H.; Guo, X.S. Numerical investigation into the stability of offshore wind power piles subjected to lateral loads in extreme environments. *J. Mar. Sci. Eng.* **2024**, *12*, 915. [CrossRef]
5. Zhao, X.D.; Fan, X.Q.; Tian, M.L. Application of repair and Anti-Corrosion technology on reinforced concrete structures of a domestic wharf. *Adv. Mat. Res.* **2011**, *1165*, 165–168. [CrossRef]
6. Zheng, Y.L.; Zhang, R.X. Experimental study on the damage characteristic and assessment of transverse bent frame of high-piled wharf under impact load. *Dev. Built Environ.* **2023**, *14*, 100124. [CrossRef]
7. Ye, H.L.; Jin, N.G. Degradation mechanisms of concrete subjected to combined environmental and mechanical actions: A review and perspective. *Comput. Concrete* **2019**, *23*, 107–119.
8. Xie, Y.F.; Liu, C.L.; Gao, S.Y.; Tang, J.P.; Chen, Y. Lateral load bearing capacity of offshore high-piled wharf with batter piles. *Ocean. Eng.* **2017**, *1442*, 377–387. [CrossRef]
9. Zhang, X.L.; Duan, B.C.; Wang, C.Z.; Wang, D.Y. Dynamic response analysis of lateral impact force of frame wharf with rock-socketed piles in inland river steel sheath. *Adv. Civ. Eng.* **2019**, *2019*, 6918376. [CrossRef]
10. Yuan, X.J. Analysis of Static and Dynamic Characteristics of High-Piled Wharf. Master's Thesis, Hohai University, Nanjing, China, 2012.
11. Zheng, Y.; Xiao, F.; Zhang, R.; Pan, T.; Lan, X.; Xu, X.; Hou, C. Experimental Investigation and Damage Identification of High-Pile Wharf Framed Bents under Horizontal Impact Loads. *Sensors* **2024**, *24*, 563. [CrossRef]

12. Zhu, R.H.; Wang, Q.M.; Zheng, J.H.; Zeng, H.K.; Zhang, J.B.; Hu, Y.; Li, C.M. Damage detection of foundation pile in high-pile wharf based on statistical high-order moment of dynamic response under regular wave excitation. *Ocean. Eng.* **2023**, *283*, 115180. [CrossRef]
13. Xu, X.; Di, X.; Zheng, Y.; Liu, A.; Hou, C.; Lan, X. Dynamic Response Characteristics and Pile Damage Identification of High-Piled Wharves under Dynamic Loading. *Appl. Sci.* **2024**, *14*, 9250. [CrossRef]
14. You, K.S.; Fang, X.Z.; Wang, F. Stability analysis of high-piled beam-slab wharf based on three-dimensional numerical model. *People 's Pearl River* **2017**, *38*, 54–57+70.
15. Xiao, X.; Li, Q.W. Two-stage deterioration model updating of RC structures in marine environment using long-term field inspection data. *Constr. Build. Mater.* **2023**, *400*, 132817. [CrossRef]
16. Li, Y.S.; Zhu, P.R.; Zhang, G.; Yang, Y. Improving Seaport Wharf Maintenance and Safety with Structural Health Monitoring System in High Salt and Humidity Environment. *Sustainability* **2023**, *15*, 4472. [CrossRef]
17. Li, P.P.; Li, J.P.; Li, L.; Zhou, P.; Liu, J.Y. Overall structural degradation of a high-piled wharf under chloride invasion and the impact force of ship berthing. *Ocean. Eng.* **2024**, *312*, 119320. [CrossRef]
18. Coelho, S.A.; Araújo, D.L. Nonlinear finite element model of the beam-to-column connection for precast concrete frames with high ratio of the continuity tie bars. *Comput. Concrete* **2023**, *31*, 53–69.
19. Mohamed, G.; Moataz, B.; Galal, E.; Mizan, A.; Qing, Q.L.; Mohamed, A.; El, Z. Strengthening of reinforced concrete beams with insufficient lapped splice length of reinforcing bars. *Eng. Struct.* **2024**, *321*, 118922.
20. Magdy, I.S.; Jong, W.H.; Ahmed, A.; Ahmed, H.; Basem, O.R.; Galal, E. Behavior of simple precast high-strength concrete beams connected in the maximum bending moment zone using steel extended endplate connections. *Steel Compos. Struct.* **2024**, *50*, 627–641.
21. Alireza, B.; Mohamed, G.; Galal, E.; Moataz, B.; Ehab, A.M.; Fathi, A.A. arious configurations of externally bonded strain-hardening cementitious composite reducing shear failure risk of defected RC beams. *Sec. Mech. Mater.* **2024**, *11*, 1373292.
22. Moataz, B.; Alireza, B.; Mohamed, G.; Mohamed, E.; Ehab, A.M.; Galal, E. Flexural strengthening of reinforced concrete cantilever beams having insufficient splice length. *Results Eng.* **2024**, *24*, 102869.
23. Li, Y.; Chen, X.H.; Zhang, G.S. A study on effects of water-cement ratio and crack width on chloride ion transmission rate in concrete. *Comput. Concrete* **2017**, *19*, 387–394. [CrossRef]
24. *GB 50026–2020*; Engineering Survey Standards. Ministry of Housing and Urban–Rural Development of the People's Republic of China: Beijing, China, 2020.
25. Wang, P. Application of loose blasting in coal mining. *Energy Sav.* **2021**, *5*, 217–218.
26. Duan, B.F.; Chen, J.H.; Chai, M.X.; Yang, Y.Q.; Liu, K.Q.; Hou, Y.H. Research and Application of Millisecond Delay Loosening Blasting during Fully Mechanized Working Face Passing Collapse Column. *Coal Technol.* **2024**, *43*, 1–5.
27. *GB 18306–2015*; Chinese Standard. Zoning Map of Ground Motion Parameters in China. China Seismological Bureau: Beijing, China, 2015.
28. *GB 50007–2011*; Code for Design of Building Foundations. AQSIQ: Beijing, China, 2011.
29. *GB/T 50081-2019*; Standard for Test Methods of Concrete Physical and Mechanical Properties. Ministry of Housing and Urban–Rural Development of the People's Republic of China: Beijing, China, 2019.
30. Wang, J.T.; Kong, L.W.; Li, C.S.; Liu, B.H.; Yan, J.B. The Small Strain Stiffness Properties of Undisturbed Weathered Granite with Different Weathering Degrees. *Front. Earth Sci.* **2022**, *10*, 946437. [CrossRef]

Disclaimer/Publisher's Note: The statements, opinions and data contained in all publications are solely those of the individual author(s) and contributor(s) and not of MDPI and/or the editor(s). MDPI and/or the editor(s) disclaim responsibility for any injury to people or property resulting from any ideas, methods, instructions or products referred to in the content.

MDPI AG
Grosspeteranlage 5
4052 Basel
Switzerland
Tel.: +41 61 683 77 34

Buildings Editorial Office
E-mail: buildings@mdpi.com
www.mdpi.com/journal/buildings

Disclaimer/Publisher's Note: The title and front matter of this reprint are at the discretion of the Guest Editors. The publisher is not responsible for their content or any associated concerns. The statements, opinions and data contained in all individual articles are solely those of the individual Editors and contributors and not of MDPI. MDPI disclaims responsibility for any injury to people or property resulting from any ideas, methods, instructions or products referred to in the content.

www.ingramcontent.com/pod-product-compliance
Lightning Source LLC
Chambersburg PA
CBHW041414210926
51850CB00013B/3267
9783725844852